Total Synthesis of Natural Products

Jie Jack Li • E.J. Corey
Editors

Total Synthesis of Natural Products

At the Frontiers of Organic Chemistry

Springer

Editors
Jie Jack Li
New Jersey
USA

E.J. Corey
Harvard University
Dept. of Chemistry and Chemical Biology
Cambridge
USA

ISBN 978-3-642-34064-2 ISBN 978-3-642-34065-9 (eBook)
DOI 10.1007/978-3-642-34065-9
Springer Heidelberg New York Dordrecht London

Library of Congress Control Number: 2012955413

Printed on acid-free paper

Springer is part of Springer Science+Business Media (www.springer.com)

Dedicated to
Professor David Y. Gin (1967–2011)

Preface

The last few decades have witnessed some exciting developments of synthetic methodologies in organic chemistry. Chiefly among these developments are ring-closing metathesis (RCM) and transition metal-catalyzed C–H activation, which have emerged as novel and useful tools.

A touchstone for any synthetic methodology is how practical it is in synthesis, especially total synthesis of natural products. Therefore, it is not surprising that books on total synthesis occupy a place on nearly every organic chemist's bookshelf.

This volume is somewhat different from previous books on total synthesis. We have been fortunate enough to enlist eleven current practitioners in the field of total synthesis to describe one of their best total syntheses. These authors leveraged synthetic methodologies developed in their own laboratories as key operations in their construction of natural products. As such, this book reflects a true sense of what is happening at the frontiers of organic chemistry.

Skillman, NJ, USA Jie Jack Li
Cambridge, MA, USA E.J. Corey

Contents

1 **Nominine** 1
Kevin M. Peese and David Y. Gin
1.1 Introduction and Classification 1
1.2 Pharmacology 2
1.3 Biosynthesis 3
1.4 Previous Synthetic Work 3
1.4.1 Total Synthesis of Nominine [19a] 4
1.4.2 Synthetic Studies Toward the Hetisine Alkaloids 6
1.5 Strategy and Retrosynthesis 7
1.6 Synthesis 9
1.7 Complete Synthesis 20
References 21

2 **Nakiterpiosin** 25
Shuanhu Gao and Chuo Chen
2.1 Background 25
2.2 Synthesis of the 6,6,5,6 Steroidal Skeleton 26
2.2.1 The Biomimetic Approaches 27
2.2.2 The Ring-by-Ring Approaches 28
2.2.3 Miscellaneous 28
2.3 Synthesis of Nakiterpiosin 31
2.4 Biology of Nakiterpiosin 34
References 34

3 **The Kinamycins** 39
Seth B. Herzon
3.1 Introduction 39
3.2 Structure Elucidation 41
3.3 Biological Activity and Mechanism of Action Studies 43
3.4 Biosyntheses of the Kinamycins 45

3.5 Syntheses of the Kinamycins . . . 46
3.5.1 Synthesis of (−)-Kinamycin C [24] . . . 46
3.5.2 Synthesis of (±)-*O*-Methyl-Kinamycin C [32] . . . 51
3.5.3 Syntheses of (−)-Kinamycins C, F, and J [39] . . . 54
3.5.4 Synthesis of (−)-Kinamycin F [45] . . . 59
References . . . 64

4 A Short Synthesis of Strychnine from Pyridine . . . 67
David B. C. Martin and Christopher D. Vanderwal
4.1 Introduction . . . 67
4.2 Synthesis of Strychnine: A Historical Perspective . . . 68
4.3 Structural Challenges . . . 71
4.4 Background: Zincke Aldehydes . . . 73
4.5 Background: Intramolecular Cycloadditions of Indoles . . . 75
4.6 Development of the Intramolecular Diels–Alder Cycloaddition of Tryptamine-Derived Zincke Aldehydes . . . 78
4.7 Synthesis of Norfluorocurarine . . . 80
4.8 Protecting Groups Are Not Always Evil . . . 84
4.9 Strategies for D-Ring Formation for Strychnine . . . 87
4.10 Some Unusual Approaches to C15–C20 Bond Formation . . . 92
4.11 A Successful Route to Strychnine . . . 93
4.12 Conclusions . . . 98
References . . . 99

5 Bryostatin 7 . . . 103
Yu Lu and Michael J. Krische
5.1 Introduction . . . 103
5.2 Pharmacology . . . 104
5.3 Biosynthesis . . . 106
5.4 Previous Synthetic Work . . . 108
5.4.1 Total Synthesis of Bryostatin 7 (Masamune 1990) . . . 108
5.4.2 Total Synthesis of Bryostatin 2 (Evans 1998) . . . 110
5.4.3 Total Synthesis of Bryostatin 3 (Nishiyama and Yamamura 2000) . . . 112
5.4.4 Total Synthesis of Bryostatin 16 (Trost 2008) . . . 113
5.4.5 Synthesis of Bryostatin 1 (Keck 2011) . . . 115
5.4.6 Synthesis of Bryostatin 9 (Wender 2011) . . . 117
5.5 Strategy and Retrosynthesis . . . 118
5.6 Synthesis . . . 120
5.6.1 Synthesis of A-Ring Fragment 68 . . . 120
5.6.2 Synthesis of C-Ring Fragment 69 . . . 121
5.6.3 Fragment Union and Total Synthesis of Bryostatin 7 . . . 123
5.7 Conclusion . . . 127
References . . . 127

6 Serratezomine A . . . 131
Julie A. Pigza and Jeffrey N. Johnston
6.1 Introduction and Classification . . . 131
6.2 Pharmacology . . . 132
6.3 Biosynthesis . . . 133
6.4 Previous Synthetic Work . . . 136
6.4.1 Total Synthesis of Serratinine . . . 136
6.4.2 Synthetic Approaches Towards the Framework of Serratinine . . . 138
6.5 Strategy and Retrosynthesis . . . 139
6.6 Synthesis . . . 141
6.7 Complete Synthesis . . . 152
References . . . 152

7 Hypocrellin/Cercosporin . . . 157
Carol A. Mulrooney, Erin M. O'Brien, and Marisa C. Kozlowski
7.1 Introduction . . . 157
7.2 Biological Activity . . . 159
7.3 Previous Synthetic Work . . . 161
7.3.1 Synthesis of (−)-Phleichrome and (−)-Calphostin A,D [30] . . . 161
7.3.2 Synthesis of (−)-Calphostin D [31] . . . 162
7.3.3 Synthesis of (−)-Phleichrome and (−)-Calphostin A [32a] . . . 163
7.3.4 Synthesis of (−)-Calphostin A–D [33a] . . . 164
7.4 Conformational Properties . . . 166
7.5 Strategy and Retrosynthesis . . . 167
7.6 Synthesis . . . 170
7.6.1 Synthesis of (−)-Hypocrellin A . . . 170
7.6.2 Synthesis of (+)-Phleichrome and (+)-Calphostin D . . . 172
7.6.3 Synthesis of (+)-Cercosporin . . . 174
7.7 Synthesis of Perylenequinone Analogs . . . 175
References . . . 179

8 Phomactin A . . . 183
Yu Tang, Kevin P. Cole, and Richard P. Hsung
8.1 Introduction . . . 183
8.1.1 Isolation . . . 183
8.1.2 Biosynthesis . . . 185
8.1.3 Medicinal Chemistry . . . 185
8.1.4 Synthetic Challenges . . . 185
8.2 The Architecturally Distinctive ABD-Tricycle . . . 186
8.2.1 Retrosynthetic Analysis . . . 186
8.2.2 Approaches to the Oxa-Annulation Precursor . . . 188
8.2.3 An Improved Synthesis of Oxa-Annulation Precursor . . . 190

8.2.4 Key Oxa-Annulation and the D-Ring Atropisomerism . . . 191
8.2.5 A Formal Synthesis of (−)-Phomactin A 194
8.3 Lessons Learned from the Challenging Structural Topology 195
8.3.1 Oxidations of C3 and C3a in B-Ring 195
8.3.2 Reduction of C8a and C8b at the AB-Ring Junction 196
8.3.3 Homologation at C5a in the A-Ring 197
8.4 Completion of the Total Synthesis 202
8.4.1 The Diene Route 202
8.4.2 The Allyl Alcohol Route 203
8.4.3 The Vinyl Epoxide Route 203
8.5 Conclusion 207
References 207

9 (+)-11,11′-Dideoxyverticillin A 211
Justin Kim and Mohammad Movassaghi
9.1 Introduction and Classification 211
9.2 Pharmacology 213
9.3 Biosynthesis 214
9.4 Previous Synthetic Work 216
9.4.1 Previous Approaches to the C3–C3′ Dimeric Linkages . . . 217
9.4.2 Previous Approaches to the Epidithiodiketopiperazine Motif 218
9.4.3 Total Synthesis of Epidithiodiketopiperazine Alkaloids . . 220
9.5 Strategy and Retrosynthesis for (+)-11,11′-Dideoxyverticillin A . 222
9.5.1 Synthesis of (+)-11,11′-Dideoxyverticillin A 223
9.5.2 Generalization to the Epipolythiodiketopiperazine Alkaloids 230
9.6 Conclusion 231
References 231

10 Retigeranic acid 235
David R. Adams and Tomáš Hudlický
10.1 Introduction 235
10.2 Isolation and Structure 236
10.3 Biosynthesis 238
10.4 Approaches to Total Synthesis 239
10.4.1 Hudlicky 239
10.4.2 Fallis 241
10.4.3 Fraser-Reid 243
10.4.4 Trauner 244
10.5 Total Syntheses 247
10.5.1 Corey 247
10.5.2 Paquette 249
10.5.3 Wender 252
10.5.4 Hudlicky 255
10.6 Conclusions and Future Perspectives 256
References 256

11 Total Synthesis of the Lycopodium Alkaloid Complanadine A 259
Richmond Sarpong and Daniel F. Fischer
11.1 Introduction 259
11.2 Biosynthesis 259
11.3 Biological Activity 262
11.4 The Siegel Synthesis of Complanadine A 263
11.5 Strategy and Retrosynthesis 264
11.6 Borylative C–H Functionalization 267
11.6.1 Benzene Ring Functionalization: Hartwig Synthesis of Taiwaniaquinol B 267
11.6.2 Pyrrole Ring Functionalization: Gaunt Synthesis of Rhazinicine 268
11.6.3 Indole Ring Functionalization: Movassaghi Synthesis of the Asperazine Core 269
11.7 Completion of the Complanadine A Synthesis 269
11.8 Application of the Strategy to Lycopladines F and G 270
11.9 Conclusion 271
References 271

Index 273

List of Contributors

Chuo Chen Department of Biochemistry, Southwestern Medical Center, University of Texas, Dallas, TX, USA

Kevin P. Cole School of Pharmacy and Department of Chemistry, University of Wisconsin, Madison, WI, USA

Daniel F. Fischer Department of Chemistry, University of California, Berkeley, CA, USA

Shuanhu Gao Department of Biochemistry, Southwestern Medical Center, University of Texas, Dallas, TX, USA

David Y. Gin Molecular Pharmacology and Chemistry Program, Memorial Sloan-Kettering Cancer Center, New York, NY, USA

Seth B. Herzon Yale University, New Haven, CT, USA

Richard P. Hsung School of Pharmacy and Department of Chemistry, University of Wisconsin, Madison, WI, USA

Jeffrey N. Johnston Department of Chemistry, Institute of Chemical Biology, Vanderbilt University, Nashville, TN, USA

Justin Kim Department of Chemistry, Massachusetts Institute of Technology, Cambridge, MA, USA

Marisa C. Kozlowski Department of Chemistry, University of Pennsylvania, Philadelphia, PA, USA

Michael J. Krische Department of Chemistry and Biochemistry, University of Texas, Austin, TX, USA

Yu Lu Department of Chemistry and Biochemistry, University of Texas, Austin, TX, USA

David B. C. Martin Department of Chemistry, University of California, Irvine, CA, USA

Mohammad Movassaghi Department of Chemistry, Massachusetts Institute of Technology, Cambridge, MA, USA

Carol A. Mulrooney Department of Chemistry, University of Pennsylvania, Philadelphia, PA, USA

Erin M. O'Brien Department of Chemistry, University of Pennsylvania, Philadelphia, PA, USA

Kevin M. Peese Discovery Chemistry, Bristol-Myers Squibb Company, Wallingford, CT, USA

Julie A. Pigza Department of Chemistry, Queensborough Community College, Bayside, NY, USA

Richmond Sarpong Department of Chemistry, University of California, Berkeley, CA, USA

Christopher D. Vanderwal Department of Chemistry, University of California, Irvine, CA, USA

Yu Tang School of Pharmacy and Department of Chemistry, University of Wisconsin, Madison, WI, USA

Chapter 1
Nominine

Kevin M. Peese and David Y. Gin

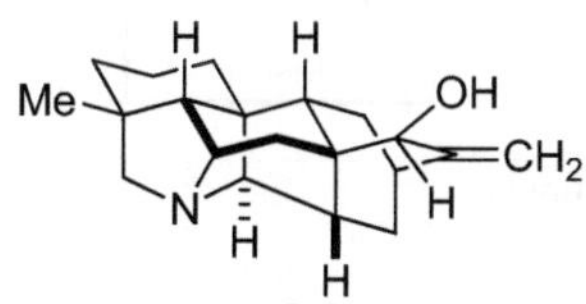

nominine

1.1 Introduction and Classification

The genera of *Aconitum* (commonly known as Monkshood) and *Delphinium*, and to a lesser extent *Rumex*, *Consolida*, and *Spiraea*, have long been recognized as a rich source of alkaloid natural products [1]. The diterpenoid alkaloids are generally classified into two major groups: the C_{19}-diterpenoid alkaloids (sometimes referred to as the C_{19}-norditerpenoid alkaloids) and the C_{20}-diterpenoid alkaloids. Within the C_{20}-diterpenoid alkaloids, at least 11 separate classes have been isolated, including the hetisine alkaloids (Chart 1.1).

Among the first hetisine alkaloids isolated were nominine (**1**) [2], kobusine (**2**) [3], pseudokobusine (**3**) [4], hetisine (**4**) [5], and ignavine (**5**) [6] in the 1940s and 1950s (Chart 1.2). Since these early isolations, over 100 distinct hetisine alkaloids have

K.M. Peese (✉)
Discovery Chemistry, Bristol-Myers Squibb Company, 5 Research Parkway, Wallingford, CT 06492, USA
e-mail: Kevin.peese@bms.com

D.Y. Gin
Molecular Pharmacology and Chemistry Program, Memorial Sloan-Kettering Cancer Center, 1275 York Avenue, New York, NY 10065, USA

J.J. Li and E.J. Corey (eds.), *Total Synthesis of Natural Products*,
DOI 10.1007/978-3-642-34065-9_1,

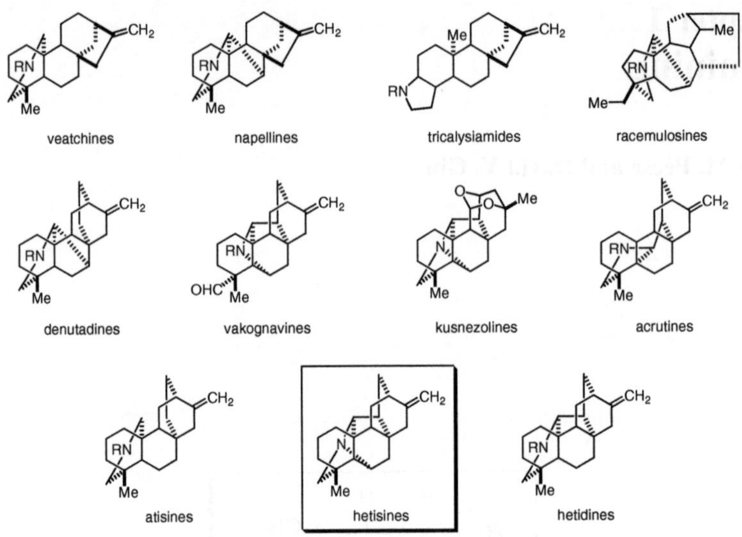

Chart 1.1 C_{20}-diterpenoid alkaloids

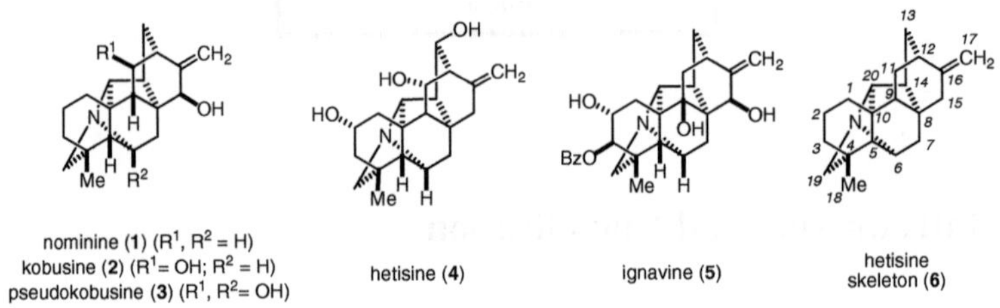

Chart 1.2 Hetisine alkaloids

been isolated and characterized with new alkaloids continuing to be discovered. Structurally, the hetisines are characterized by a highly fused heptacyclic ring system with an embedded tertiary nitrogen core. The separate members of the hetisine alkaloids are distinguished by the number, location, and stereochemical placement of oxygen functionality, primarily alcohols and simple esters.

1.2 Pharmacology

The hetisine alkaloids have long been recognized to be active constituents of traditional eastern herbal medicines [7]. Pharmacological investigations of the hetisine alkaloids have shown a diverse range of bioactivities [1b, 7]. Significantly, guan-fu base A (**7**) is reported to be under clinical development in China for arrhythmia [1a]. In addition, kobusine (**2**), pseudokobusine (**3**), as well as multiple

Chart 1.3 Pharmacologically significant hetisine alkaloids

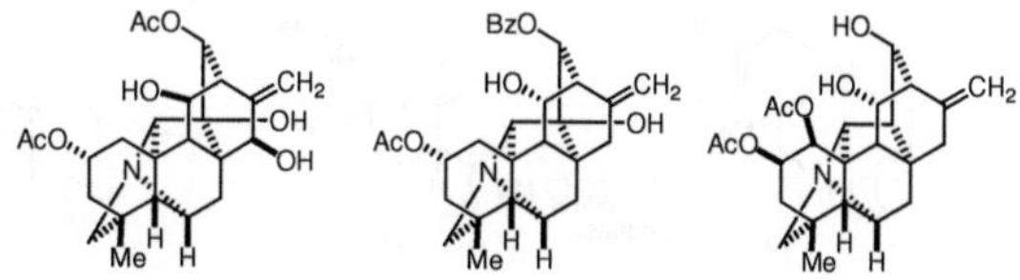

guan–fu base A (**7**) zeravshanisine (**8**) tadzhaconine (**9**)

O-acyl derivatives thereof have shown potent vasodilatory activity [8]. A number of other hetisine alkaloids have shown diverse biological activities. These include nominine [9] (**1**) (local anesthetic, anti-inflammatory, and antiarrhythmic), hetisine [9a] (**4**) (hypotensive), ignavine [10] (**5**) (analgesic, anti-inflammatory, antipyretic, sedative, antidiuretic), zeravshanisine [9a] (**8**) (antiarrhythmic and local anesthetic), and tadzhaconine [9a, 11], (**9**) (antiarrhythmic) (Chart 1.3).

1.3 Biosynthesis

The biosynthesis of the atisane class of C_{20}-diterpene alkaloids, including the hetisine family, has been proposed to take place in two principal phases [1a]. The first phase encompasses biogenesis of most of the diterpene framework via a standard, diterpene biosynthesis (Scheme 1.1) [12]. Beginning with geranylgeranyl diphosphate (**10**), cyclization with *ent*-copalyl diphosphate synthase produces *ent*-copalyl diphosphate (**11**). The exocyclic alkene of **11** then undergoes annulation with the allylic diphosphate to afford, after a series of carbocation rearrangements, *ent*-atisir-16-ene (**19**). Noteworthy in this cascade is the nonclassical carbocations **15** and **16**. The second phase of the biosynthesis of the atisane class has been hypothesized, but is not well understood [13]. It has been proposed that an oxidation event occurs on *ent*-atisir-16-ene (**19**) to give a dialdehyde or its synthetic equivalent (**20**). Following a condensation event with a nitrogen source and reduction, the atisine skeleton (**21**) of C_{20}-diterpene alkaloids is accessed. Carbon–carbon bond formation between C-14 and C-20, possibly through a Prins-type intermediate, produces the hetidine skeleton (**22**) of C_{20}-diterpene alkaloids. Finally, bond formation between the nitrogen and C-6 generates the hetisine-type skeleton (**6**).

1.4 Previous Synthetic Work

The C_{20}-diterpene alkaloids have long served as classic targets within the field of natural product synthesis [14]. Total syntheses of four C_{20}-diterpene alkaloids have thus far been reported: atisine [15], veatchine [16], garryine [17], and napelline [18]. In spite of this progress, synthetic efforts toward the hetisine alkaloids have been relatively sparse. Prior to our work in the area, these efforts include one total synthesis and five synthetic studies.

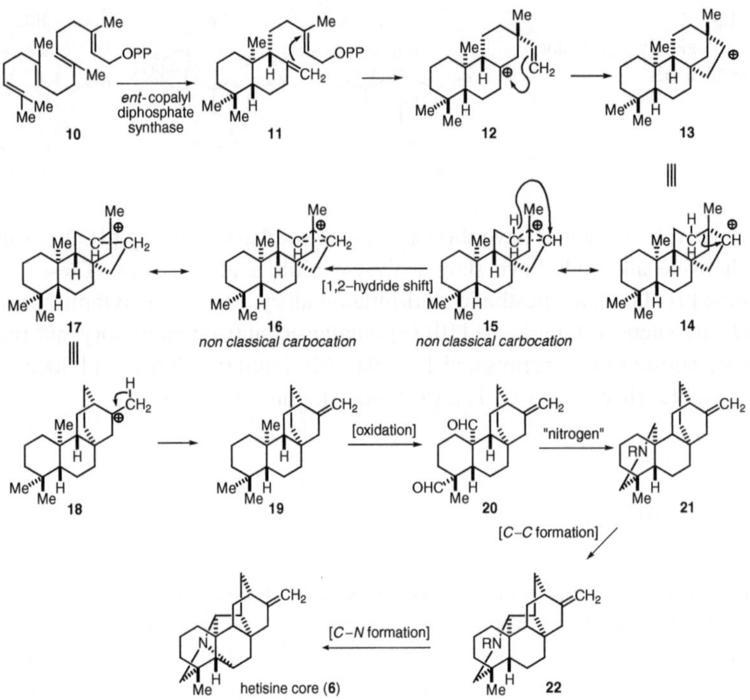

Scheme 1.1 Postulated biosynthesis of hetisine alkaloids

1.4.1 Total Synthesis of Nominine [19a]

In 2004, Muratake and Natsume reported the landmark total synthesis of (±)-nominine (**1**), the first total synthesis of a hetisine alkaloid (Scheme 1.2) [19]. Their approach was based upon two key reactions: α-arylation of an aldehyde [20] for formation of the C-9 and C-10 carbon–carbon bond (i.e., **24**→**25**) and Lewis acid-catalyzed acetal-ene reaction [21] for formation of the key C-14 and C-20 carbon–carbon bond (i.e., **26**→**27**). Beginning with 3-methoxyactetic acid (**23**), aryl bromide aldehyde **24** was prepared in a straightforward nine-step sequence. In the first key reaction of the synthesis, treatment of aryl bromide **24** with $PdCl_2(PPh_3)_2$ and Cs_2CO_3 in refluxing THF-delivered tricyclic **25** in 65 % yield, 4.2:1 *dr*. Next, elaboration through a six-step sequence produced intermediate alkene **26**. Acetal-ene reaction of **26** using $BF_3{\cdot}OEt_2$ in toluene at −18 °C afforded ether **27** in 66 % after subsequent deketalization with *p*-TsOH in acetone. Installation of the nitrogen atom began six steps later with the conjugate cyanation of enone **28** with Et_2AlCN (Nagata reagent) [22] in toluene resulting in β-cyanoketone **29**. Then, following protection of the ketone as the TMS enol ether, the cyano group was reduced to the primary amine

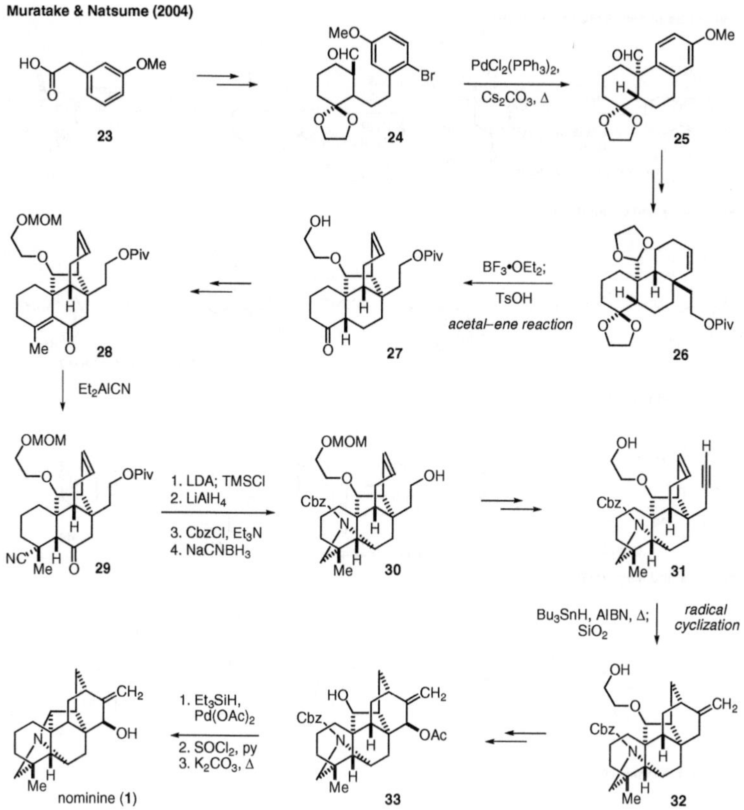

Scheme 1.2 Total synthesis of nominine, first total synthesis of a hetisine alkaloid

with $LiAlH_4$. The amine was condensed with the proximal ketone functionality to furnish the enamine which was immediately protected with a Cbz group leading to the protected enamine. Reduction of the enamine with $NaCNBH_3$ produced Cbz-protected pyrrolidine **30**. Following a ten-step interlude to complete construction of the [2.2.2] bicyclo-octane system, completion of the aza-ring system was addressed. Deprotection of the Cbz group of **33** was accomplished with Et_3SiH, $Pd(OAc)_2$, and NEt_3. The last critical *C–N* bond of the pyrrolidine was then formed via alkylation of the amine with the adjacent alcohol by first activation of the alcohol with $SOCl_2$ and then annulation. The synthesis was then completed with the deprotection of the allylic acetate to the allylic alcohol with K_2CO_3 in refluxing methanol-giving nominine. Overall, Muratake and Natsume were able to accomplish a 40-step synthesis of (±)-nominine in 0.15 % yield.

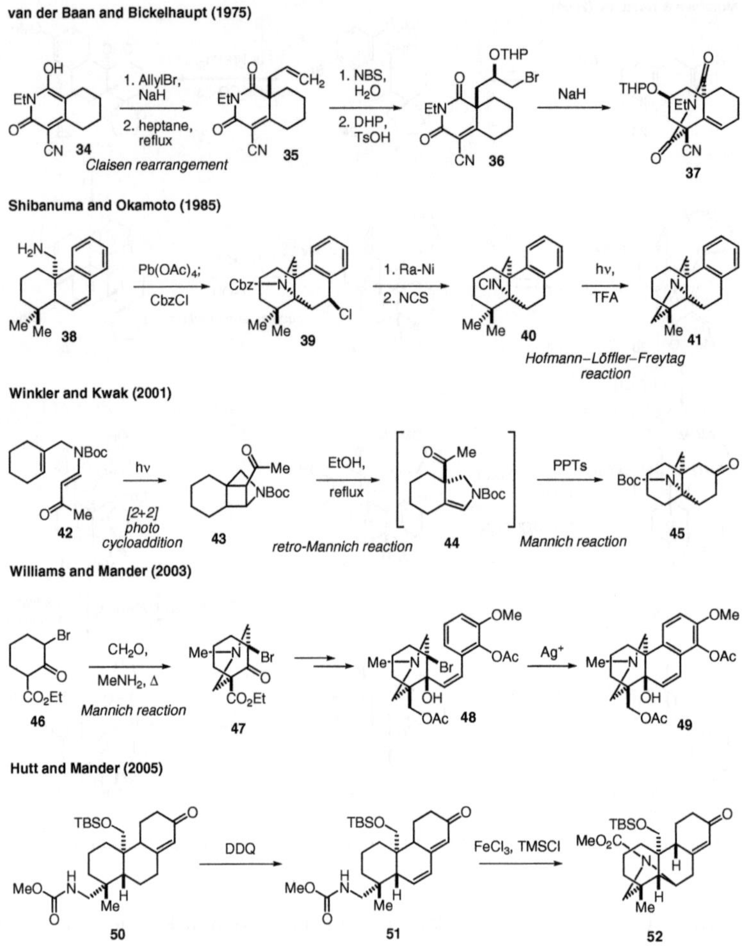

Scheme 1.3 Previous synthetic studies

1.4.2 *Synthetic Studies Toward the Hetisine Alkaloids*

In 1975, van der Baan and Bickelhaupt reported the synthesis of imide **37** from pyridone **34** as an approach to the hetisine alkaloids, using an intramolecular alkylation as the key step (Scheme 1.3) [23]. Beginning with pyridone **34**, alkylation with sodium hydride/allyl bromide followed by a thermal [3,3] Claisen rearrangement gave alkene **35**. Next, formation of the bromohydrin with *N*-bromosuccinimide and subsequent protection of the resulting alcohol as the tetrahydropyranyl (THP) ether produced bromide **36**, which was then cyclized in an intramolecular fashion to give tricylic **37**.

Ten years later in 1985, Shibanuma and Okamoto reported the synthesis of pentacyclic intermediate **41** (Scheme 1.3) [24]. This approach to the hetisine alkaloids, a refinement of work previously reported by Okamoto [25], utilized a Hofmann–Löffler–Freytag reaction to form the polycyclic substructure surrounding the tertiary amine of the hetisine alkaloids. In the key sequence of the synthesis, styrene **38** was subjected to lead tetraacetate [$Pb(OAc)_4$] oxidation to give an aziridine which was immediately fragmented by treatment with benzyl chloroformate (CbzCl) to give benzylic chloride **39**. Reductive cleavage of the benzylic chloride and concomitant cleavage of the carbamate with Raney nickel followed by *N*-chlorination of the amine provided the key Hofmann–Löffler–Freytag precursor *N*-chloramine **40**. Photolysis in acidic media provided the Hofmann–Löffler–Freytag reaction product **41**.

More recently in 2001, Winkler and Kwak reported methodology designed to access the pyrrolidine core of the hetisine alkaloids via a photochemical [2+2], retro-Mannich, Mannich sequence (Scheme 1.3) [26]. In a representative example of the methodology, vinylogous amide **42** was photo-irradiated to give the [2+2] cycloaddition product **43**. Heating cyclobutane **43** in ethanol provided enamine **44** via a retro-Mannich reaction. Exposure of enamine **44** to acidic conditions then effected a Mannich reaction, resulting in pyrrolidine **45**.

In 2003, Williams and Mander reported a method designed to access the hetisine alkaloids (Scheme 1.3) [27]. This approach, based upon a previously disclosed strategy by Shimizu et al. [28], relied on arylation of a bridgehead carbon via a carbocation intermediate in the key step. Beginning with β-keto ester **46**, double Mannich reaction provided piperidine **47**. Following a straightforward sequence, piperidine **47** was transformed to the pivotal bromide intermediate **48**. In the key step, bromide **48** was treated with silver (I) 2,4,6-trinitrobenzenesulfonate in nitromethane (optimized conditions) to provide **49** as the most advanced intermediate of the study, in 54 % yield.

Finally in 2005, Hutt and Mander reported their strategy for the synthesis of nominine (Scheme 1.3) [29]. The approach relies upon construction of the steroidal ABC carbocyclic ring structure followed by stepwise preparation of the fused aza-ring system. In the key sequence of the synthetic study, enone **50** was oxidized to dienone **51** with DDQ followed by Lewis acid-catalyzed intramolecular conjugate addition of the methylcarbamate to the newly formed dienone to deliver pyrrolidine **52**.

1.5 Strategy and Retrosynthesis

The highly fused and bridged architecture of the carbon–nitrogen skeleton within the hetisine alkaloids presents a formidable challenge for the synthetic chemist. While the placement and orientation of the oxygen functionalities of the various hetisine alkaloids presents its own hurdles, the key synthetic challenge of the hetisine family, exemplified by nominine as the simplest member, is construction of the polycyclic ring system, especially the scaffold surrounding the nitrogen.

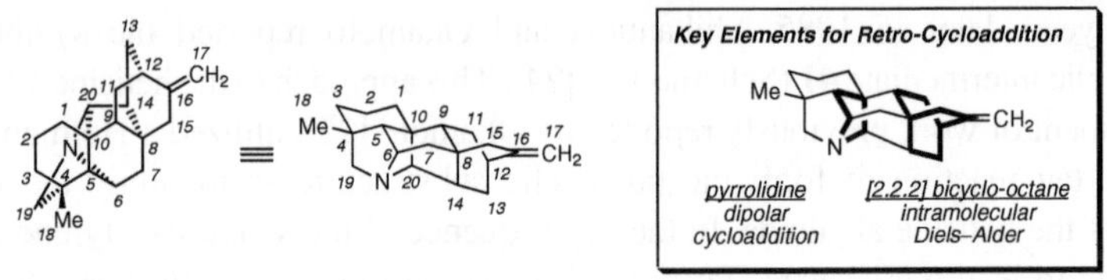

Chart 1.4 Key strategic retrosynthetic elements

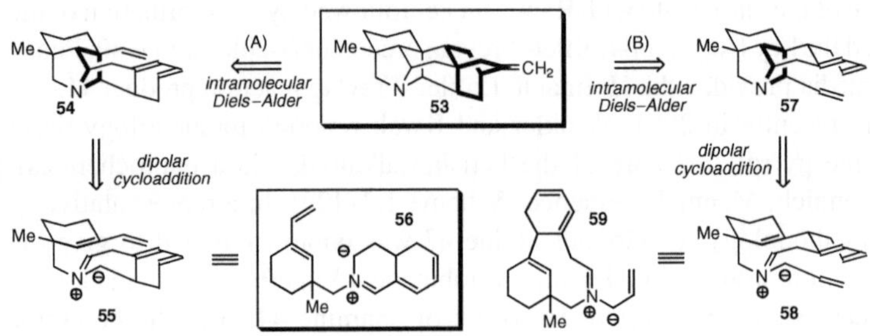

Scheme 1.4 Retrosynthetic analysis

Intramolecular cycloadditions are among the most efficient methods for the synthesis of fused bicyclic ring systems [30]. From this perspective, the hetisine skeleton encompasses two key retro-cycloaddition key elements: (1) a bridging pyrrolidine ring accessible via a [3+2] azomethine dipolar cycloaddition and (2) a [2.2.2] bicyclo-octane accessible via a [4+2] Diels–Alder carbocyclic cycloaddition (Chart 1.4). While intramolecular [4+2] Diels–Alder cycloadditions to form [2.2.2] bicycle-octane systems have extensive precedence [3+2], azomethine dipolar cycloadditions to form highly fused aza systems are rare [31–33]. The staging of these two operations in sequence is critical to a unified synthetic plan. As the proposed [3+2] dipolar cycloaddition is expected to be the more challenging of the two transformations, it should be conducted in an early phase in the forward synthetic direction. As a result, a retrosynthetic analysis would entail initial consideration of the [4+2] cycloaddition to arrive at the optimal retrosynthetic C–C bond disconnections for this transformation.

Two possible intramolecular disconnections are available for the [2.2.2] bicyclo-octane ring system (path A and path B, Scheme 1.4). The choice between the initial [4+2] disconnections A and B at first appears inconsequential leading to idealized intermediates of comparable complexity (**54** and **57**). However, when the [4+2] and [3+2] disconnections are considered in sequence, the difference becomes clear. For path A, retrosynthetic [3+2] disconnection of intermediate **54** leads to the conceptual precursor **56**, which embodies a considerable simplification. In contrast, path B reveals a retrosynthetic [3+2] disconnection of intermediate **57** to provide the precursor **59**, a considerably less simplified medium-ring bridged macrocycle. Thus, unification of the [3+2]/[4+2] dual cycloaddition strategy, using the staging

Scheme 1.5 Simplified retrosynthesis

and disconnection approach of path A, leads to a highly streamlined retrosynthetic strategy (Scheme 1.5).

The use of biomimetic strategies in natural product synthesis has been an effective guide in development of total synthesis in recent years [34]. Biomimetic strategies are often the most elegant when the biosynthesis of a natural product imparts most of a molecule's complexity in one reaction or a tandem sequence that has a potential parallel in chemical synthesis. Examples of this can be found in Shair's synthesis of longithorone A [35] and Sorensen's elegant synthesis of (+)-FR182877 [36]. In the case of the proposed biosynthesis of the hetisine alkaloids, however, complexity is generated stepwise over an extended sequence. In fact, for the biosynthesis of the heptacyclic hetisine skeleton, no more than two rings are proposed to be generated in any particular step. Accordingly, following a biomimetic strategy is not likely to hold a privileged position relative to other synthetic strategies.

1.6 Synthesis

Synthetic work commenced with evaluation of an azomethine ylide dipole for the proposed intramolecular dipolar cycloaddition. A number of methods exist for the preparation of azomethine ylides, including, *inter alia*, transformations based on fluoride-mediated desilylation of α-silyliminium species, electrocyclic ring opening of aziridines, and tautomerization of α-amino acid ester imines [37]. In particular, the fluoride-mediated desilylation of α-silyliminium species, first reported by Vedejs in 1979 [38], is among the most widely used methods for the generation of non-stabilized azomethine ylides (Scheme 1.6).

The key cycloaddition reaction to form the pyrrolidine ring within the hetisine alkaloids involves the generation and reaction of an extended *endocyclic* azomethine ylide, a class of reactive intermediate with relatively little precedent [32, 33]. As a consequence, a suitable model system was explored to assess the feasibility of this cycloaddition approach (Scheme 1.7). To this end, 3-methylcylcohexen-2-one **65** underwent conjugate addition with cyanide using $Al(CN)Et_2$ in benzene affording an aluminum enolate [22]. Treatment of the enolate with TBAT generated an activated aluminum enolate, which upon treatment with Tf_2O, furnished vinyl triflate **66** in 81 % yield. Reduction of the nitrile with diisobutylaluminum hydride (DIBAL-H) followed by $NaBH_4$/MeOH produced amine **67** in 88 % yield. To construct the endocyclic

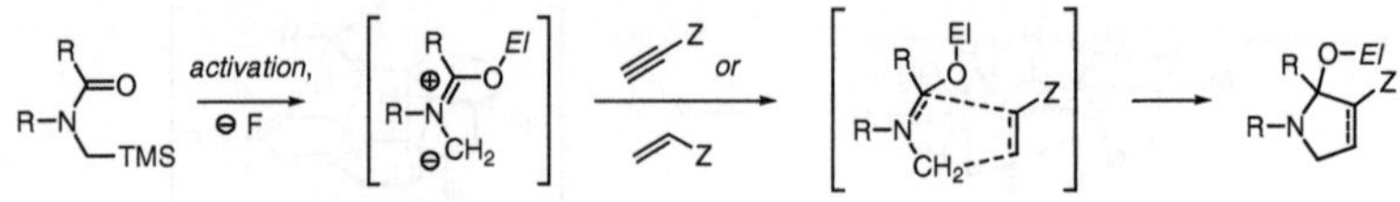

Scheme 1.6 Azomethine dipolar cycloaddition utilizing desilylation

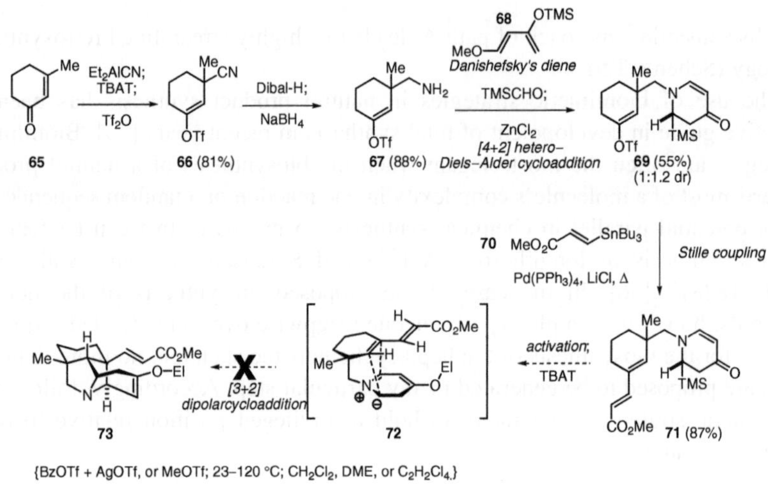

Scheme 1.7 Intramolecular azomethine dipolar cycloaddition

dipole portion of the cycloaddition model substrate, a novel hetero Diels–Alder approach was developed to prepare a suitable 2,3-dihydro-2-silylpyridin-4-one. Condensation of amine **67** with TMSCHO generated the corresponding *C*-silyl aldimine, which was trapped in situ with Danishefsky's diene (**68**) in a hetero Diels–Alder cycloaddition under $ZnCl_2$ catalysis to provide the 2,3-dihydro-2-silylpyridin-4-one **69** [39]. The *C*-silyl vinylogous amide **69** was isolated in 55 % yield as an inseparable yet inconsequential 1.2:1 mixture of diastereomers. Subsequent introduction of the dipolarophile functionality in this model substrate was performed by Stille coupling [40] of vinyl triflate **69** with stannane **70** to afford the conjugated dienoate **71** (87 %), the requisite precursor to the proposed endocyclic azomethine ylide formation and intramolecular dipolar cycloaddition.

The feasibility of azomethine ylide generation from **7** and intramolecular dipolar cycloaddition was examined under a variety of conditions. For example, activation of vinylogous amide **71** with BzOTf [41] followed by desilylation with TBAT led to complex mixtures of products. Likewise, using MeOTf as the activating agent yielded similar results. Significantly, none of these protocols furnished the desired pyrrolidine **73**. Only decomposition of the silylpyridinone to form unidentified products was observed, despite the fact that quantitative *O*-methylation of the

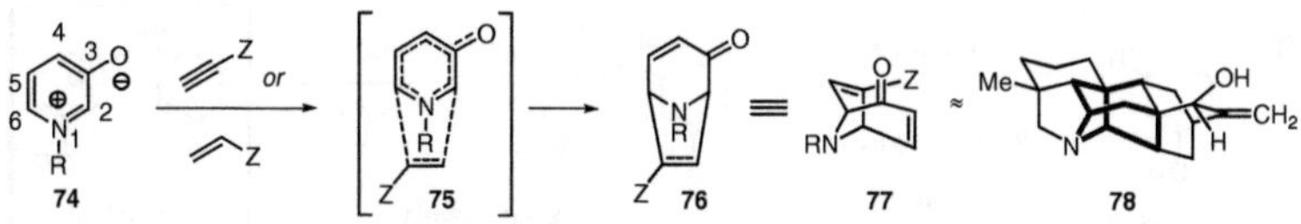

Scheme 1.8 Dipolar cycloaddition of 3-oxidopyridinium betaine

vinylogous amide was verified to occur by NMR analysis in the first stage of the process.

With the failure of **72** to undergo the desired intramolecular dipolar cycloaddition, the strategy of employing an endocyclic non-stabilized azomethine ylide in the key cycloaddition step was reevaluated. Since the difficulties in this approach were likely associated with lack of stability of the azomethine ylide, a new route involving the generation and cycloaddition of a more stabilized substrate was pursued. In this context, oxidopyridinium ylides display reactivity patterns similar to azomethine ylides with the exception that oxidopyridinium ylides tend to be less reactive due to their enhanced stability. Dipolar cycloadditions of 3-oxidopyridinium betaines (**74**) were introduced by Katritzky in 1970 [42] and have since been shown to be useful in numerous synthetic transformations (Scheme 1.8) [43]. Oxidopyridinium betaines **74** are moderately reactive aza–1,3–dipoles that undergo dipolar cycloaddition reactions at the 2,6-positions of the pyridinium ring with electron-deficient alkene and alkyne dipolarophiles to afford tropane cycloadducts **76** = **77**. These cycloadditions are generally under *HOMO*–dipole *LUMO*–dipolarophile control and preferentially provide *endo* products.

To investigate the feasibility of employing 3-oxidopyridinium betaines as stabilized 1,3-dipoles in an intramolecular dipolar cycloaddition to construct the hetisine alkaloid core (Scheme 1.8, **77** ≈ **78**), a series of model cycloaddition substrates were prepared. In the first (Scheme 1.9a), an ene-nitrile substrate (i.e., **83**) was selected as an activated dipolarophile functionality. Nitrile **66** was subjected to reduction with DIBAL-H, affording aldehyde **79** in 79 % yield. This was followed by reductive amination of aldehyde **x** with furfurylamine (**80**) to afford the furan amine **81** in 80 % yield. The ene-nitrile was then readily accessed via palladium-catalyzed cyanation of the enol triflate with KCN, 18–crown–6, and $Pd(PPh_3)_4$ in refluxing benzene to provide ene-nitrile **82** in 75 % yield. Finally, bromine-mediated aza-Achmatowicz reaction [44] of **82** then delivered oxidopyridinium betaine **83** in 65 % yield.

The second cycloaddition substrate took to form of **91** (Scheme 1.9b), incorporating a vinyl sulfone dipolarophile. Beginning with cyano ketone **84**, which was readily prepared from 1,5-dicyanopentane via a previously reported three-step sequence [45], condensation with thiophenol produced vinyl sulfide **85** in 84 % yield. Vinyl sulfide **85** underwent bromination in acetonitrile to afford bromo-vinyl sulfide **86** (86 %), which was then treated with isopropylmagnesium chloride [46] to effect metal-halogen exchange affording an intermediate vinyl magnesium bromide species. Subsequent alkylation with MeI in the presence of catalytic CuCN provided the alkylated vinyl sulfide **87** in 93 % yield. The nitrile within vinyl

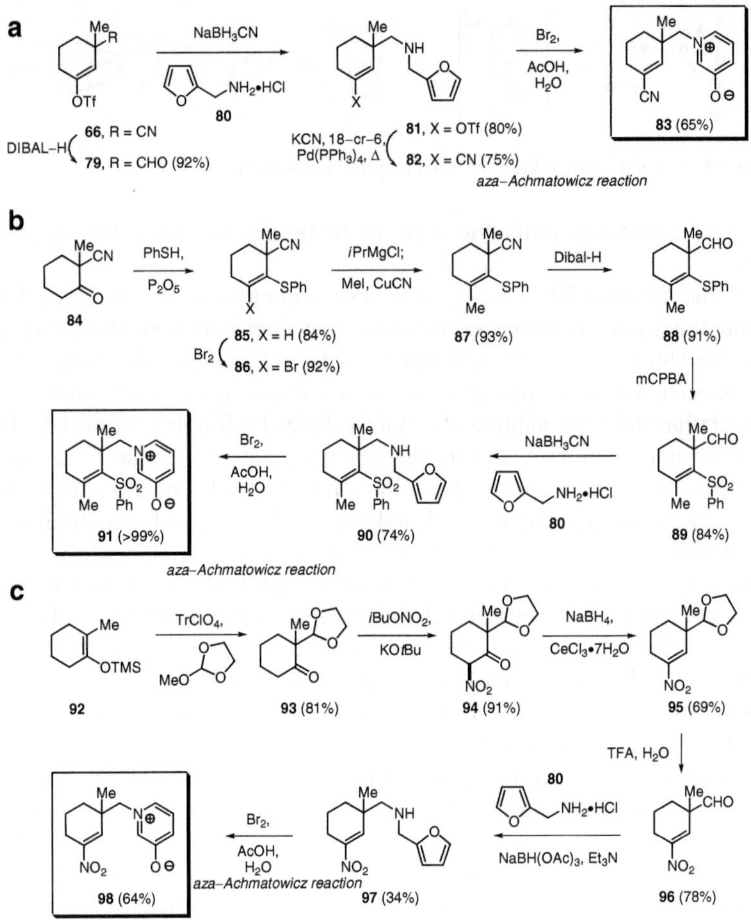

Scheme 1.9 Preparation of model cycloaddition substrates

sulfide **87** was reduced with DIBAL-H to furnish aldehyde **88** (91 %), which was followed by oxidation of the sulfide with *m*-CPBA-generated sulfone **89** (84 %), and subsequent reductive amination with furfuryl amine hydrochloride (**80**) to afford substituted furfuryl amine **90** (74 %). Finally, bromine-mediated aza-Achmatowicz [44] reaction of furfuryl amine **90** produced oxidopyridinium betaine **91** in quantitative yield.

The third cycloaddition substrate explored the feasibility of a vinyl nitro functionality as an activated dipolarophile (**98**, Scheme 1.9c). Preparation of nitroalkene oxidopyridinium betaine **98** began with silylenol ether **92**, which was treated with methoxydioxolane in the presence of Lewis acid catalyst, $TrClO_4$, to afford keto dioxolane **93** in 58 % yield [47]. Ketone **93** then underwent α-nitration by treatment with *i*-$BuONO_2$ and KO*t*-Bu to provide nitro ketone **84** (91 %), which was then converted to the nitroalkene functionality via reduction under Luche conditions to

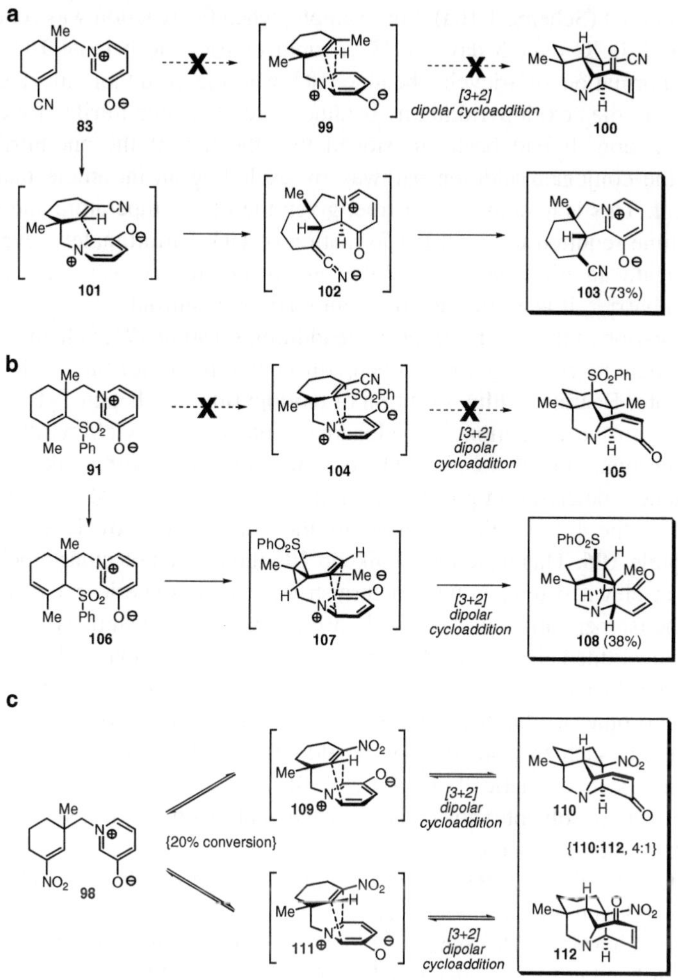

Scheme 1.10 Evaluation of intramolecular dipolar cycloadditions of model substrates

furnish nitroalkene **95** (69 %) [48]. Deprotection of the dioxolane group with aqueous trifluoroacetic acid produced aldehyde **96** (78 %), which underwent reductive amination of with furfuryl amine hydrochloride (**80**) to furnish the substituted furfuryl amine **97** in 34 % yield. Bromine-mediated aza-Achmatowicz reaction [44] of furfuryl amine **97** provided oxidopyridinium betaine **98** (64 %) to serve as the 1,3-dipole.

Each of the 3-oxidopyridinium betaine substrates **83**, **91**, and **98** were extensively investigated for their potential to engage in intramolecular dipolar cycloaddition (Scheme 1.10). Heating a solution of ene-nitrile **83** in variety of solvents failed to effect the desired intramolecular [3+2] dipolar cycloaddition to form the bridged pyrrolidine **100**, as tricyclic oxidopyridinium betaine **103** was the only isolated product (Scheme 1.10a). For example, when the reaction was conducted in toluene at 170 °C over 5 days, a 73 % yield of tricyclic betaine **103** could be isolated. Formation of tricyclic betaine **103** was the result of direct conjugate addition of the oxidopyridinium betaine into the ene-nitrile followed by re-aromatization. It had been envisioned that the use of the ene-nitrile would disfavor the conjugate addition pathway by excluding an intramolecular protonation event. This is to some extent true, given the high temperature and extended reaction time required to effect the formation of **103**. Unfortunately, even though the conjugate addition pathway of the ene-nitrile moiety was entailed a high activation barrier, it was still the dominant reaction manifold.

Investigation of the vinyl sulfone cycloaddition substrate (**91**, Scheme 1.10b) led to an alternate reaction manifold, albeit still inappropriate for the synthesis of the hetisine core. Heating a dilute solution of oxidopyridinium betaine **91** in refluxing toluene led to the formation of cycloadduct **108** in 38 % yield as the principal product. Formation of this undesired isomeric cycloadduct **108** is likely the result of an alkene isomerization process wherein the α,β-π-system of dipolarophile **91** migrates to the β,γ-position relative to the sulfone to provide tri-substituted dipolarophile **106**. This unactivated alkene **106** then intramolecular dipolar cycloaddition at elevated temperature to furnish undesired isomeric cycloadduct **108**. The alkene isomerization event is likely driven by a relief of steric strain present in the tetra-substituted alkene **91**. Moreover, it has been shown that sulfones generally have no significant conjugative stabilization with adjacent alkenes due to lack of significant orbital overlap between the sulfone and the alkene π-system, so it is not surprising that such an isomerization would occur [49]. Indeed, this reactivity has on occasion been synthetically exploited [50]. Unfortunately, an extensive survey of a wide array of solvents showed no improvement; in no experiment could desired cycloadduct **105** be observed.

Despite the lack of success in the attempts at intramolecular cycloaddition with substrates **83** and **91**, a moderately promising outcome was observed for the nitroalkene substrate (**98**, Scheme 1.10c). Heating a dilute solution of oxidopyridinium betaine **98** in toluene to 120 °C produced a 20 % conversion to a 4:1 mixture of two cycloadducts (**110** and **112**), in which the major cycloadduct was identified as **110**. While initially very encouraging, it became apparent that the dipolar cycloaddition reaction proceeded to no greater than 20 % conversion, an outcome independent of choice of reaction solvent. Further investigation, however, revealed that the reaction had reached thermodynamic equilibrium at 20 % conversion, a fact verified by resubmission of the purified major cycloadduct **110** to the reaction conditions to reestablish the same equilibrium mixture at 20 % conversion.

In an effort to shift the cycloaddition equilibrium toward the cycloaddition products, a strategy was formulated to lower the energetic cost of breaking aromaticity of the betaine moiety. It was surmised that if an aromatic ring were fused to the oxidopyridinium betaine, the energetic cost of de-aromatization of the betaine would be lowered. In this context, use of a 4-oxidoisoquinolinium betaine in an aza–1,3–dipolar cycloaddition was first reported by Katritzky in 1972 [51] during his seminal studies on oxidopyridinium betaines. Dipolar cycloadditions of

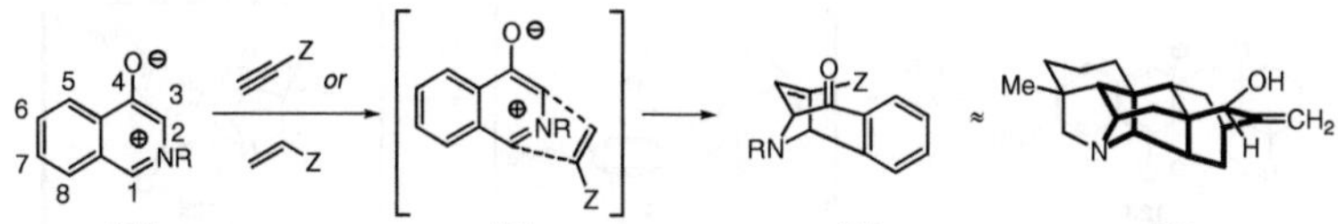

Scheme 1.11 Dipolar cycloaddition of 4-oxidoisoquinolinium betaine

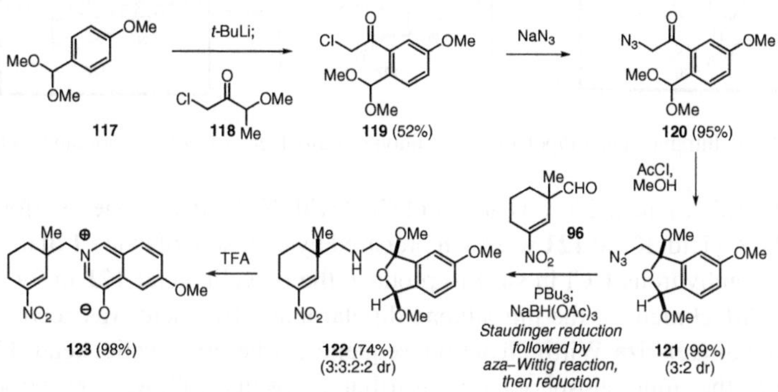

Scheme 1.12 Preparation of nitroalkene oxidoisoquinolinium betaine **123**

4-oxidoisoquinolinium betaines comprise a synthetically valuable subset of dipolar cycloadditions of oxidopyridinium betaines in which the oxidopyridinium betaine ring is benzo fused across the 4,5-positions of the oxidopyridinium betaine ring (Scheme 1.11). In an manner analogous to oxidopyridinium betaines, 4-oxidoisoquinolinium betaines **113** undergo dipolar cycloadditions with electron-deficient alkene and alkyne dipolarophiles at the 1,3-positions of the oxidoisoquinolinium betaine ring and generally afford *endo* products **115**, a substructure that maps very well onto the hetisine skeleton **116**. Despite the synthetic potential of 4-oxidoisoquinolinium betaines, however, only a small handful of studies of their use in dipolar cycloadditions have been reported [52].

To investigate whether the use of a 4-oxidoisoquinolinium betaine would be able to favorably shift the equilibrium of the intramolecular dipolar cycloaddition, the preparation of nitroalkene oxidoisoquinolinium betaine **123** (Scheme 1.12) was targeted. Evaluation of nitroalkene oxidoisoquinolinium betaine **123** would allow a direct comparison with nitroalkene oxidopyridinium betaine **98** (cf. Scheme 1.10c). Considering possible methods for the preparation of oxidoisoquinolinium betaines, an approach analogous to the aza-Achmatowicz [44] strategy for the generation of oxidopyridinium betaines was attractive. If a similar method could be implemented for the preparation of oxidoisoquinolinium betaines, it might be synthetically valuable.

These efforts began with directed lithiation [53] of commercially available 4-methoxybenzaldehyde dimethyl acetal (**117**, Scheme 1.12), followed by quenching with amide **118** to produce chloro acetophenone **119** (52 %). Conversion

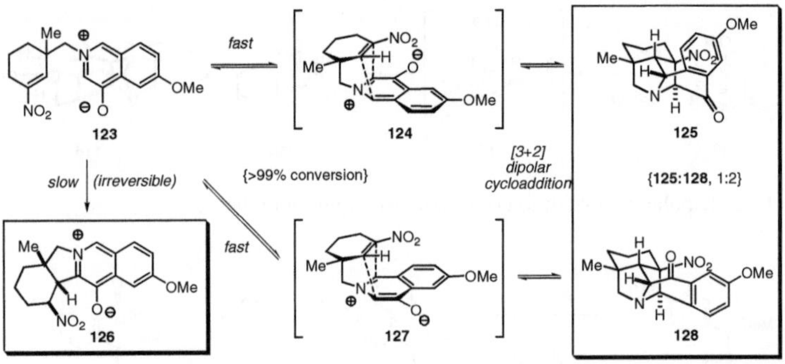

Scheme 1.13 Intramolecular dipolar cycloaddition of nitroalkene oxidoisoquinolinium betaine **123**

of the chloride to the azide was accomplished with NaN_3 in acetone to afford azide **120** in 95 % yield. Ketal **121** was then prepared by exposure of azide **120** to AcCl in methanol (anhydrous HCl in situ) to produce the bicyclic ketal **121** in 99 % yield (3:2 dr). Attachment of the nitroalkene dipolarophile fragment was accomplished via a Staudinger–aza-Wittig reaction sequence, whereby cyclic ketal **121** was reduced to the iminophosphorane with tributylphoshine followed by introduction of the aldehyde **96** to access the imine. This intermediate imine was then reduced with $NaBH(OAc)_3$ resulting in amine **122** in 74 % yield as a 3:3:2:2 mixture of diastereomers. Methanol extrusion and rearrangement of cyclic ketal amine **122** to access oxidoisoquinolinium betaine **123** was accomplished by treatment of the amine with a 9:1 mixture of CH_2Cl_2/TFA.

With oxidoisoquinolinium betaine **123** in hand, the intramolecular dipolar cycloaddition was examined (Scheme 1.13). Heating a solution of oxidoisoquinolinium betaine **123** in toluene led to complete conversion to cycloadducts **125** and **128** in a 2:1 kinetic ratio, which was found to reverse (1:2, **125**:**128**) over time. In addition to the two cycloadducts **125** and **128**, a third product, tetracyclic oxidoisoquinolinium betaine **126**, was also observed to slowly accumulate, presumably resulting from the conjugate addition pathway. Through further investigation, the process of conjugate addition was found to be irreversible such that all material eventually funneled to the production of tetracyclic betaine **126**. Unfortunately, due to competing conjugate addition as well as the low regioselectivity of the cycloaddition itself, the desired cycloadduct **128** was isolable in no greater than ~35 % yield.

Clearly, the nitroalkene dipolarophile oxidoisoquinolinium betaine **123** is nonideal for the synthesis of the hetisine alkaloids, as mass throughput for the needed cycloadduct would be low, and conversion of the tertiary nitro group to carbon-based functionality, as would be required in the latter stages of the synthesis, could be problematic. On the other hand, an ene-nitrile dipolarophile has several potential advantages over nitroalkene dipolarophile. Most importantly, the ene-nitrile cycloadduct has carbon functionality installed at the C-10 position. Second, the conjugate addition byproduct pathway that occurs so readily for the nitroalkene oxidoisoquinolinium betaine **123** system (see Scheme 1.13) should be much slower

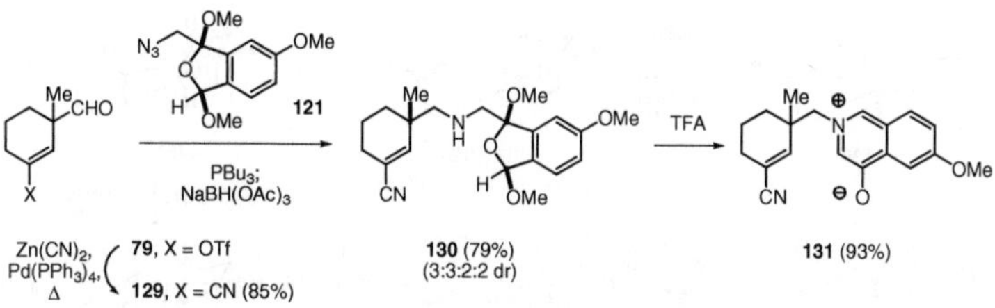

Scheme 1.14 Preparation of ene-nitrile oxidoisoquinolinium betaine **131**

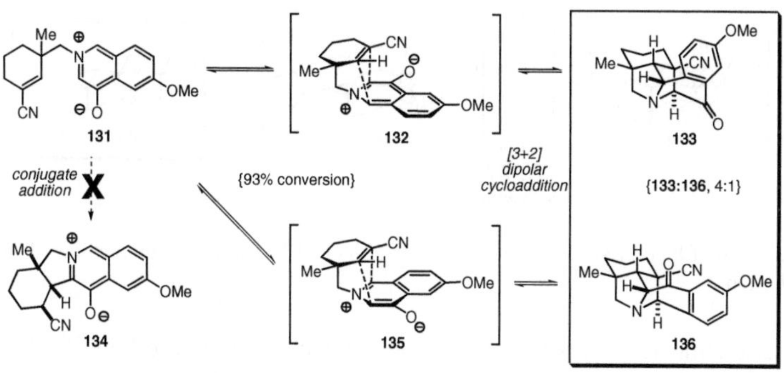

Scheme 1.15 Intramolecular dipolar cycloaddition of ene-nitrile oxidoisoquinolinium betaine **131**

for the ene-nitrile substrate. Indeed, in the case of the ene-nitrile oxidopyridinium betaine **83** (see Scheme 1.10a), the conjugate addition process is slow at 170 °C and likely occurs only in the absence of the desired cycloaddition pathway. Based on this hypothesis, efforts focused on preparing ene-nitrile oxidoisoquinolinium betaine **131**.

Ene-nitrile oxidoisoquinolinium betaine **131** was readily prepared from vinyl triflate aldehyde **79** (Scheme 1.14). Palladium-catalyzed cyanation of vinyl triflate **79** with $Zn(CN)_2$ in DMF at 60 °C produced ene-nitrile aldehyde **129** in 85 % yield [54]. Using the previously developed Staudinger–aza-Wittig reduction sequence, aldehyde **129** was coupled with cyclic ketal azide **121** to afford a 79 % yield of amine **130**. The cyclic ketal amine **130** was then treated with 9:1 mixture of CH_2Cl_2/TFA to provide ene-nitrile oxidoisoquinolinium betaine **131** in 93 % yield.

When ene-nitrile oxidoisoquinolium betaine **131** was heated as a dilute solution in toluene to 120 °C (Scheme 1.15), near quantitative conversion to the cycloadduct **133**, resulting from the undesired regioselectivity, was observed. While the near complete conversion to cycloadduct **133** of oxidoisoquinolinium betaine **131** indeed demonstrated complete avoidance of the conjugate addition pathway in favor of cycloaddition, initial production of undesired isomeric cycloadduct **133** (instead of **136**) was disappointing. Notably, cycloadduct **133** is expected to be less kinetically favored based on frontier molecular orbital (FMO) analysis (assuming dipole HOMO-controlled cycloaddition) of the putative transition state. This result stands in contrast to the cycloaddition of nitroalkene oxidoisoquinolinium betaine

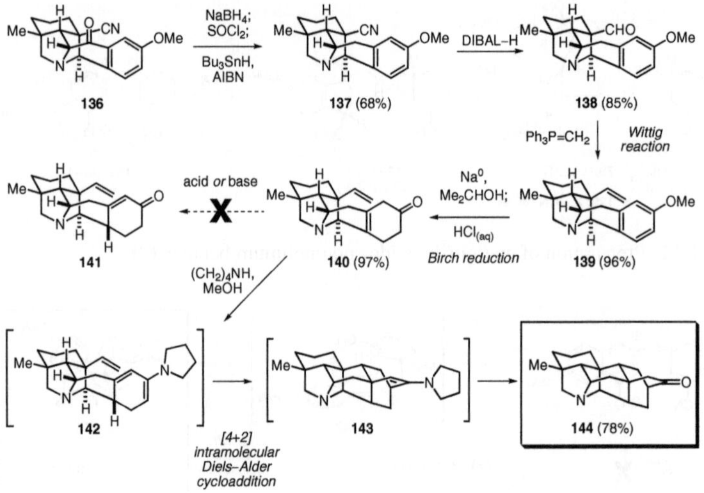

Scheme 1.16 Elaboration of dipolar cycloadduct **136** and Intramolecular Diels-Alder cycloaddition

123 (cf. Scheme 1.13) in which the predicted cycloadduct based on FMO analysis, cycloadduct **128**, was observed to be the major product. Importantly, when a dilute solution of ene-nitrile oxidoisoquinolium betaine **131** was heated for an extended period of time (15 h) at 180 °C, a ratio of 1:3.6 **136** (desired): **133** (undesired) was obtained, which represented the system's thermodynamic selectivity. This result demonstrated that the dipolar cycloaddition reaction was indeed reversible at high temperature. Under this protocol, the desired cycloadduct could now be isolated in 20 % yield. The undesired isomer could then be resubjected to the equilibration conditions, thereby regenerating the equilibrium mixture of 1:3.6 **136** (desired):**133** (undesired). Through this iterative process, a recycling procedure was established that allowed for material throughput, yielding ~20 % of the desired cycloadduct **136** per equilibration.

With the establishment of the intramolecular dipolar cycloaddition of oxidoisoquinolinium betaine **131** to provide dipolar cycloadduct **136**, the critical challenge of indentifying a viable [3+2] dipolar cycloaddition for the synthesis of the hetisine pyrrolidine core was solved. Attention now turned toward the second cycloaddition of the dual cycloaddition strategy, namely, the proposed intramolecular [4+2] Diels–Alder reaction. Advancement of dipolar cycloadduct **136** toward nominine began with the ketone-to-methylene reduction (Scheme 1.16). Initial efforts for this transformation involved attempted reduction of ketone **136** under Lewis acidic-ionizing conditions in the presence of a hydride source (TFA/Et_3SiH, $BF_3{\cdot}OEt_2$/Et_3SiH, $BF_3{\cdot}OEt_2$/$NaBH(OAc)_3$, $BF_3{\cdot}OEt_2$/$NaBH_3CN$), all of which failed to effect full reduction to afford **137**. Success was finally achieved when the ketone was reduced to the corresponding alcohol, which was then chlorinated with thionyl chloride to allow for radical dehalogenation to provide **137** in 68 % yield over three steps. Reduction of nitrile **137** proceeded smoothly with DIBAL-H to give aldehyde

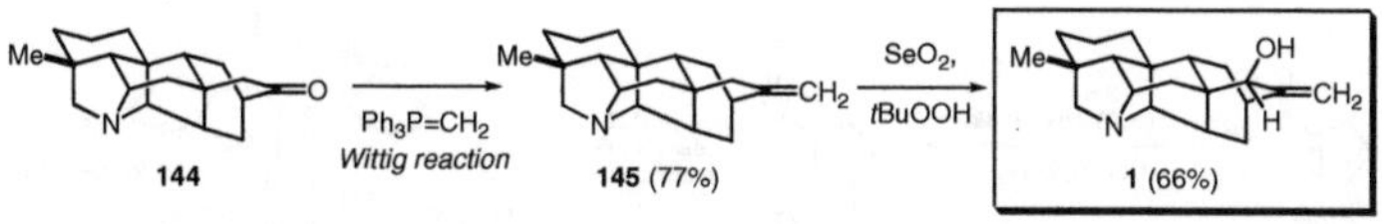

Scheme 1.17 Completion of nominine synthesis

138 (85 %), allowing for Wittig olefination with $Ph_3P{=}CH_2$ to provide alkene **139** in 96 % yield. With the alkene dienophile functionality in place, the arene was reduced via a Birch protocol (Na^0, *i*-PrOH, THF, NH_3; HCl (aq)) to afford cyclohexenone **140** in 97 % yield [55].

Isomerization of the alkene within the nonconjugated enone **140** to the corresponding α,β-unsaturated enone **141** proved to be challenging; evaluation of a range of acidic and basic conditions all failed to deliver the conjugated product. During the course of these isomerization studies, however, an attempt was made to prepare and isomerize an enimine intermediate [56]. When a solution of nonconjugated enone **141** in 9:1 methanol/pyrrolidine was heated to 60 °C for 4 h, the nonconjugated enone underwent complete conversion to heptacyclic ketone **144** (78 %), possessing the full carbon–nitrogen ring system of the hetisine alkaloids. It is likely that treatment of the nonconjugated enone **140** with pyrrolidine produced a rapidly equilibrating mixture of dienamine isomers. Of these, dienamine isomer **142** is the only intermediate that was competent to undergo an intramolecular [4+2] Diels–Alder cycloaddition with the proximal alkene to produce enamine cycloadduct **143**. Hydrolysis of enamine cycloadduct **143** was subsequently effected upon exposure of the cycloadduct to silica gel to deliver ketone **144**.

The late stages of the synthesis (Scheme 1.17) proceeded with Wittig methylenation of ketone **144** with $Ph_3P{=}CH_2$ at 70 °C to furnish exocyclic alkene **145** in 77 % yield. Finally, the alcohol was installed via a SeO_2-mediated allylic hydroxylation [57] of the exocyclic alkene **145** to afford (±)-nominine (**1**) in 66 % and 7:1 dr. The structure of nominine (**1**) was verified via an X-ray crystal structure determination, thereby completing the racemic total synthesis of (±)-nominine (**1**).

The synthesis of nominine was readily rendered asymmetric via early establishment of absolute chirality. In this capacity, methodology developed by Hoveyda for the formation all carbon quaternary stereogenic centers was shown to be applicable [58]. In this method (Scheme 1.18), treatment of cyclic keto enoates **146** with dialkyl or diarylzinc reagents in the presence of CuOTf and the chiral *N*-heterocyclic carbene ligand complex **148** [(*S*,*S*)-NHC] provided the conjugate addition products **147** with good to excellent levels of asymmetric induction. In the application of this methodology, methyl 1-cyclohexene-1-carboxylate (**149**) was oxidized with CrO_3 in AcOH, Ac_2O, and dichloromethane to provide γ-keto methyl enoate **150** in 56 % yield (Scheme 1.19). Utilizing the Hoveyda method, γ-keto methyl enoate **150** underwent conjugate addition of $ZnMe_2$ mediated by the Cu-(*S*,*S*)-NHC **148** complex. The intermediate zinc enolate was then trapped with triflic anhydride to provide enol triflate (+)-**151** in 91 % yield and 92:8 *er* [8]. Reduction of the ester with DIBAL-H provided the aldehyde (+)-**79** (92 %). Following Pd-catalyzed

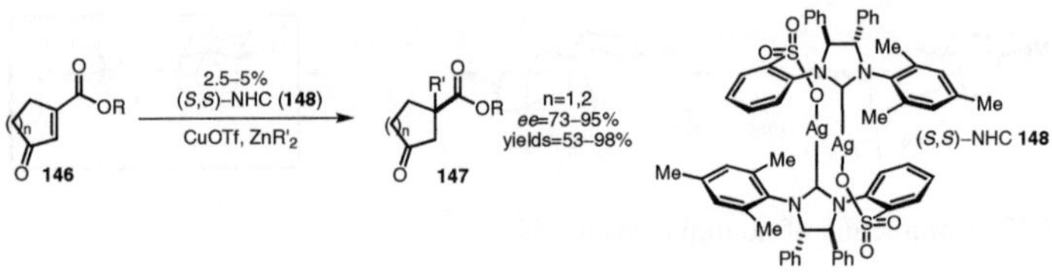

Scheme 1.18 Preparation of Hoveyda catalyst (*S*,*S*)-NHC **148**

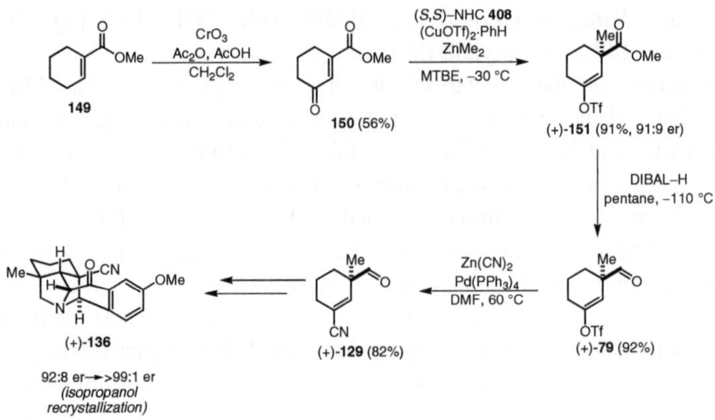

Scheme 1.19 Asymmetric synthesis of nominine

cyanation of enol triflate (+)-**79** (82 %), ene-nitrile (+)-**129**, an intermediate in the racemic total synthesis, was accessed. This intermediate was then carried on to enantio-enriched cycloadduct (+)-**136**. Recrystallization then provided enantio-pure material which was advanced on to (+)-nominine.

1.7 Complete Synthesis

Overall, the racemic synthesis was accomplished in a 15-step sequence with only a single protecting group manipulation and marked the second total synthesis of (±)-nominine (**1**) (Scheme 1.20) [59], the first having been completed by Muratake and Natsume in 2004 [19]. The asymmetric synthesis of nominine was accomplished in a 16-step sequence in a similar fashion [60].

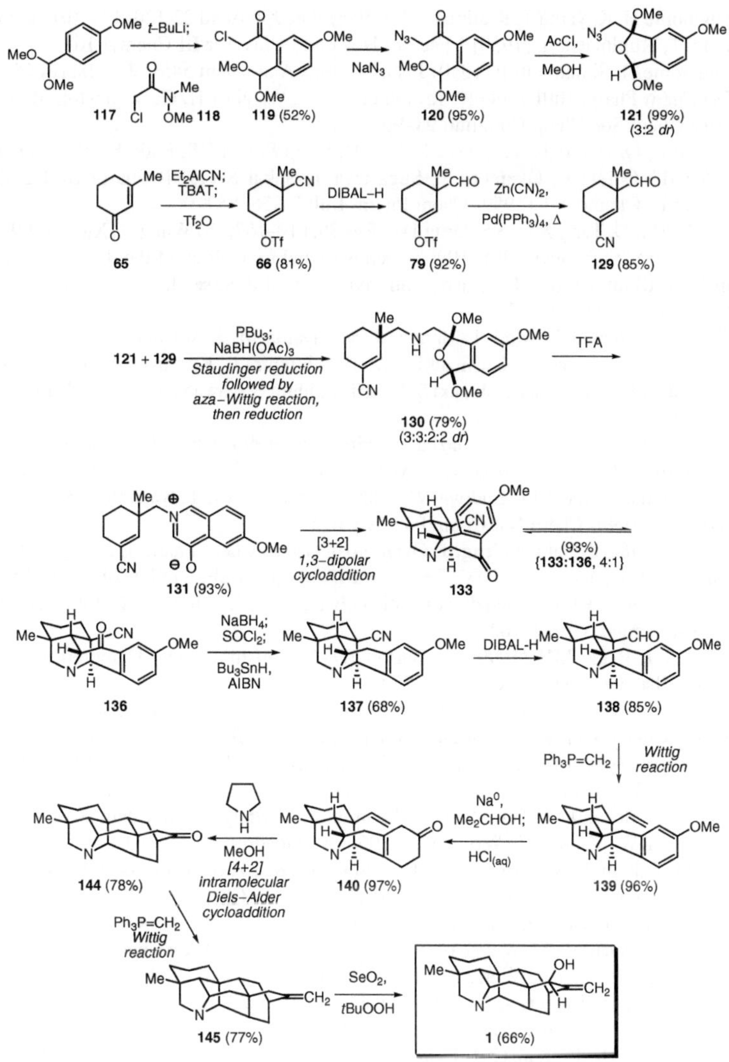

Scheme 1.20 Total synthesis of nominine

References

1. (a) Wang F-P, Chen Q-H, Liu X-Y (2010) Nat Prod Rep 27:529–570; (b) Wang F-P, Liang X (2002) C_{20}-diterpenoid alkaloids. The alkaloids, vol 59. Academic, New York, pp 1–264
2. (a) Ochiai E, Okamoto T, Sakai S, Saito A (1956) Yakugaku Zasshi 76:1414–1418; (b) Sakai S, Yamamoto I, Yamaguchi K, Takayama H, Ito M, Okamoto T (1982) Chem Pharm Bull 30:4579–4582
3. (a) Jacobs WA, Craig LC (1942) J Biol Chem 143:605–609; (b) Przybylska M (1962) Can J Chem 40:566–568

4. (a) Suginome H, Koyama T, Kunimatsu Y (1946) Proc Jpn Acad 22:120–121; (b) Suginome H, Koyama T, Kunimatsu Y (1950) J Fac Sci Hokkaido Univ Ser III Chem 4:16
5. (a) Suginome H, Simamouti F (1940) Justus Liebigs Ann Chem 545:220–228; (b) Okamoto T (1959) Chem Pharm Bull 7:44–49; (c) Pelletier SW, Wright LH, Gary Newton M, Wright H (1970) J Chem Soc Chem Commun 98–99
6. (a) Ochiai E, Okamoto T, Sugasawa T, Tani H, Sakai S, Hai HS, Endo H (1953) Pharm Bull 1:60–65; (b) Ochiai E, Okamoto T, Sugasawa T, Sakai S (1954) Pharm Bull 2:388–397; (c) Ochiai E, Okamoto T (1959) Chem Pharm Bull 7:556–559
7. (a) Zhu D, Bai D, Tang X (1996) Drug Dev Res 39:147–157; (b) Wang X, Xie H (1999) Drugs Fut 24:877–882; (c) Jacyno JM (1996) Chemistry and toxicology of the diterpenoid alkaloid, Chap 5. In: Blum MS (ed) Chemistry and toxicology of diverse classes of alkaloids. Alaken, Fort Collins, CO, pp 301–336
8. (a) Wada K, Ishizuki S, Mori T, Bando H, Murayama M, Kawahara N (1997) Biol Pharm Bull 20:978–982; (b) Wada K, Ishizuki S, Mori T, Fujihira E, Kawahara N (1998) Biol Pharm Bull 21:140–146; (c) Wada K, Ishizuki S, Mori T, Fujihira E, Kawahara N (2000) Biol Pharm Bull 23:607–615
9. (a) Salimov BT, Kuzibaeva ZK, Dzhakhangirov FN (1996) Chem Nat Compd 32:366–368; (b) Shakirov R, Telezhenetskaya MV, Bessonova IA, Aripova SF, Israilov IA, Sultankhodzhaev MN, Vinogradova VI, Akhmedzhanova VI, Tulyaganov TS, Salimov BT, Tel'nov VA (1996) Chem Nat Compd 32:596–675
10. Saito H, Ueyama T, Naka N, Yagi J, Okamoto T (1982) Chem Pharm Bull 30:1844–1850
11. Shakirov R, Telezhenetskaya MV, Bessonova IA, Aripova SF, Israilov IA, Sultankhodzhaev MN, Vinogradova VI, Akhmedzhanova VI, Tulyaganov TS, Salimov BT (1996) Tel'nov VA. Chem Nat Compd 32:932–1028
12. MacMillan J, Beale MH (1999) In: Barton JD, Nakanishi K (eds) Comprehensive natural products. Elsevier, New York, pp 230–233
13. Sultankhodzhaev MN, Nishanov AA (1995) Chem Nat Compd 31:283–298
14. Corey EJ, Cheng X (1989) The logic of total synthesis. Wiley, New York, p 404
15. (a) Nagata W, Sugasawa T, Narisada M, Wakabayashi T, Hayase Y (1963) J Am Chem Soc 85:2342–2343; (b) Nagata W, Sugasawa T, Narisada M, Wakabayashi T, Hayase Y (1967) J Am Chem Soc 89:1483–1499; (c) Masamune S (1964) J Am Chem Soc 86:291–292; (d) Guthrie RW, Valenta Z, Wiesner K (1966) Tetrahedron Lett 7:4645–4654; (e) Ihara M, Suzuki M, Fukumoto K, Kametani T, Kabuto C (1988) J Am Chem Soc 110:1963–1964; (f) Ihara M, Suzuki M, Fukumoto K, Kabuto C (1990) J Am Chem Soc 112:1164–1171
16. (a) Nagata W, Narisada M, Wakabayasi T, Sugawasa T (1964) J Am Chem Soc 86:929–930; (b) Nagata W, Narisada M, Wakabayasi T, Sugawasa T (1967) J Am Chem Soc 89:1499–1504; (c) Valenta Z, Wiesner K, Wong CM (1964) Tetrahedron Lett 5:2437–2442; (d) Wiesner K, Uyeo S, Philipp A, Valenta Z (1968) Tetrahedron Lett 9:6279–6282
17. Masamune S (1964) J Am Chem Soc 86:290–291
18. (a) Wiesner K, Ho P, Tsai CSJ, Lam Y (1974) Can J Chem 52:2355–2357; (b) Sethi SP, Atwal KS, Marini-Bettolo RM, Tsai TYR, Wiesner K (1980) Can J Chem 58:1889–1891
19. (a) Muratake H, Natsume M (2004) Angew Chem Int Ed 43:4646–4649; (b) Muratake H, Natsume M (2002) Tetrahedron Lett 43:2913–2917; (c) Muratake H, Natsume M (2006) Tetrahedron 62:7056–7070; (d) Muratake H, Natsume M (2006) Tetrahedron 62:7071–7092; (e) Muratake H, Natsume M, Nakai H (2006) Tetrahedron 62:7093–7112
20. Muratake H, Nakai H (1999) Tetrahedron Lett 40:2355–2358
21. For examples, see: (a) Johnson WS, Kinnel RB (1966) J Am Chem Soc 88:3861–3862; (b) Ladépêche A, Tam E, Ancel J, Ghosez L (2004) Synthesis 9:1375–1380
22. Nagata W, Yoshiokas M, Murakami M (1972) J Am Chem Soc 94:4635–4643
23. van der Baan JL, Bickelhaupt F (1975) Recl Trav Chim Pays-Bas 94:109–112
24. Shibanuma Y, Okamoto T (1985) Chem Pharm Bull 33:3187–3194
25. Somei M, Okamoto T (1970) Chem Pharm Bull 18:2135–2138
26. Kwak Y, Winkler JD (2001) J Am Chem Soc 123:7429–7430

27. (a) Williams CM, Mander LN (2003) Org Lett 5:3499–3502; (b) Williams CM, Mander LN, Bernhardt PV, Willis AC (2005) Tetrahedron 61:3759–3769
28. (a) Shimizu B, Ogiso A, Iwai I (1963) Chem Pharm Bull 11:766–769; (b) Shimizu B, Ogiso A, Iwai I (1963) Chem Pharm Bull 11:333–336
29. (a) Hutt OE, Mander LN, Willis AC (2005) Tetrahedron Lett 45:4569–4572; (b) Hutt OE, Mander LN (2007) J Org Chem 72:10130–10140
30. Nicolaou KC, Snyder SA, Montagnon T, Vassilikogiannakis G (2002) Angew Chem Int Ed 41:1668–1698
31. Coldham I, Hufton R (2005) Chem Rev 105:2765–2810
32. (a) Fowler FW (1971) Angew Chem Int Ed Engl 10:135; (b) Fowler FW (1971) Angew Chem Int Ed Engl 10:135–136
33. (a) Pandey G, Lakshmaiah G, Ghatak A (1993) Tetrahedron Lett 34:7301–7304; (b) Pandey G, Bagul TD, Sahoo AK (1998) J Org Chem 63:760–768; (c) Pandey G, Laha JK, Mohanakrishnan AK (1999) Tetrahedron Lett 40:6065–6068; (d) Pandey G, Sahoo AK, Bagul TD (2000) Org Lett 2:2299–2301; (e) Pandey G, Laha JK, Lakshmaiah G (2002) Tetrahedron 58:3525–3534
34. (a) de la Torre MC, Sierra MA (2004) Angew Chem Int Ed 43:160–181; (b) Nicolaou KC, Montagnon T, Synder SA (2003) Chem Commun 5:551–564
35. Layton ME, Morales CA, Shair MD (2002) J Am Chem Soc 124:773–774
36. Vosburg DA, Vanderwal CD, Sorensen EJ (2002) J Am Chem Soc 124:4552–4553
37. Najera C, Sansano JM (2003) Curr Org Chem 7:1105–1150
38. Vedejs E, Martinez GR (1979) J Am Chem Soc 101:6452–6454
39. Kerwin JF Jr, Danishefsky S (1982) Tetrahedron Lett 23:3739–3742
40. Peña MR, Stille JK (1989) J Am Chem Soc 111:5417–5424
41. (a) Eckelbarger JD, Wilmot JT, Gin DY (2006) J Am Chem Soc 128:10370–10371; (b) Eckelbarger JD, Wilmot JT, Epperson MT, Thakur CS, Shum D, Antczak C, Tarassishin L, Djaballah H, Gin DY (2008) Chem Eur J 14:4293–4306
42. Katritzky AR, Takeuchi Y (1970) J Am Chem Soc 92:4134–4136
43. (a) Dennis N, Katritzky AR, Takeuchi Y (1976) Angew Chem Int Ed Engl 15:1–9; (b) Lown JW (1984) In: Padwa A (ed) 1,3-Dipolar cycloaddition chemistry, vol 1. Wiley, New York, pp 653–732; (c) Crabb JN, Storr RC (1984) In: Padwa A (ed) 1,3-Dipolar cycloaddition chemistry, vol 2. Wiley, New York, pp 543–595; (d) Katritzky AR, Dennis N (1989) Chem Rev 89:827–861; (e) Śliwa W (1996) Heterocycles 43:2005–2029
44. (a) Petersen JB, Norris K, Clauson-Kaas N, Svanholt K (1969) Acta Chem Scand 23:1785–1790; (b) Shono T, Matsumura Y, Tsubata K, Inoue K, Nishida R (1983) Chem Lett 74:21–24; (c) Ciufolini MA, Hermann CYW, Dong Q, Shimizu T, Swaminathan S, Xi N (1998) Synlett 11:105–114; (d) Müller C, Diehl V, Lichtenthaler FW (1998) Tetrahedron 54:10703–10712; (e) Zhou W, Lu Z, Xu Y, Liao L, Wang Z (1999) Tetrahedron 55:11959–11983
45. Dehli JR, Gotor V (2002) J Org Chem 67:1716–1718
46. Abarbri M, Dehmel F, Knochel P (1999) Tetrahedron Lett 40:7449–7453
47. Meulemans TM, Stork GA, Hansen BJM, de Groot A (1998) Tetrahedron Lett 39:6565–6568
48. Stork G, Clark G, Weller T (1984) Tetrahedron Lett 25:5367–5370
49. O'Connor DE, Lyness WI (1964) J Am Chem Soc 86:3840–3846
50. (a) Hamann PR, Toth JE, Fuchs PL (1984) J Org Chem 49:3865–3867; (b) Lee SW, Fuchs PL (1991) Tetrahedron Lett 32:2861–2864 l; (c) Nickel A, Maruyama T, Tang H, Murphy PD, Greene B, Yusuff N, Wood JL (2004) J Am Chem Soc 126:16300–16301
51. Dennis N, Katritzky AR, Takeuchi YJ (1972) J Chem Soc Perkin Trans 1:2054–2057
52. (a) Garling DL, Cromwell NH (1973) J Org Chem 38:654–658; (b) Dennis N, Katritzky AR, Parton SK (1975) Chem Pharm Bull 23:2899–2903; (c) Dennis N, Katritzky AR, Parton SK (1976) J Chem Soc Perkin Trans 1:2285–2288; (d) Hanaoka M, Wada A, Yasuda S, Mukai C, Imanishi T (1979) Heterocycles 12:511–514; (e) Cushman M, Patrick DA, Toma PH, Byrn SR (1989) Tetrahedron Lett 30:7161–7164; (f) Padwa A, Dean DC, Osterhout MH, Precedo L, Semones MA (1994) J Org Chem 59:5347–5357; (g) Dicesare JC, Burgess JP, Mascarella SW,

Carroll FI (1994) J Heterocyl Chem 37:187–192; (h) Edmunds JJ, Cheng X, Tobias B (1996) J Chem Soc Perkin Trans 1:2005–2008; (i) Constable KP, Blough BE, Carroll FI (1996) Chem Commun 6:717–718

53. (a) Plaumann HP, Keay BA, Rodrigo R (1979) Tetrahedron Lett 20:4921–4924; (b) Karl J, Gust R, Spruss T, Schneider MR, Shönenberger H, Engel J, Wrobel K, Lux F, Haeberlin ST (1988) J Med Chem 31:72–83
54. Tschaen DM, Desmond R, King AO, Fortin MC, Pipik B, King S, Verhoeven TR (1994) Synth Commun 24:887–890
55. (a) Rabideau PW, Marcinow Z (1992) Org React 42:1–334; (b) Wentland MP, Albertson NF, Pierson AK (1980) J Med Chem 23:71
56. Gao H, Su X, Li Z (1997) Steroids 62:398–402
57. (a) Umbreit MA, Sharpless KB (1977) J Am Chem Soc 99:5526–5528; (b) Furber M, Mander LN (1987) J Am Chem Soc 109:6389–6396
58. (a) Brown MK, May TL, Baxter CA, Hoveyda AH (2007) Angew Chem Int Ed 46:1097–1100; (b) Lee K, Brown MK, Hird AW, Hoveyda AH (2006) J Am Chem Soc 128:7182–7184
59. (a) Peese KM, Gin DY (2006) J Am Chem Soc 128:8734–8735; (b) Peese KM, Gin DY (2005) Org Lett 7:3323–3325
60. Peese KM, Gin DY (2008) Chem Eur J 14:1654–1665

Chapter 2
Nakiterpiosin

Shuanhu Gao and Chuo Chen

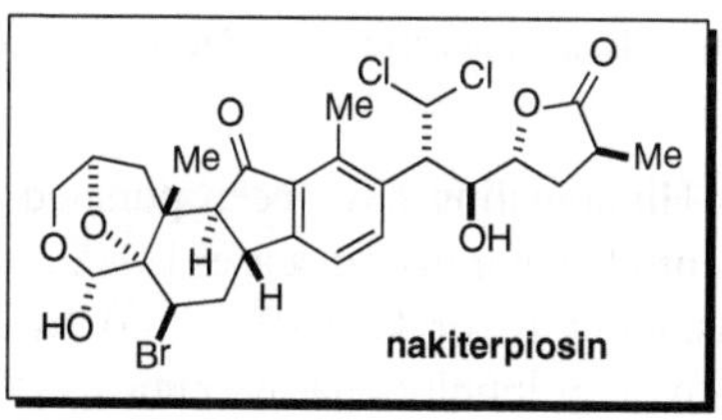

2.1 Background

For decades, ranchers in central Idaho were puzzled by a mysterious birth defect in their flocks of sheep. A percentage of their lambs, ranging from 1 % to 20 %, were born with only one eye. The Poisonous Plant Research Laboratory of the US Department of Agriculture started to investigate this "malformed lamb disease" in 1954. During the 11 years of work, they found that ewes grazing on corn lily (*Veratrum californicum*) on the *14th* day of gestation gave birth to cyclopic lambs, while the ewes were left unaffected [1]. They further found that cyclopamine (**3**) was responsible for the one-eyed face malformation and veratramine (**4**) led to leg deformity (Chart 2.1). The molecular target of **3** was identified 30 years later to be smoothened (Smo) [2]. Smo is a seven-pass transmembrane protein that regulates the activity of the Hedgehog (Hh) signal transduction pathway. Since Hh signaling is central to stem cell differentiation and tissue homeostasis, and 10 % of basal cell carcinoma and medulloblastoma patients carry hyperactive mutant Smo, small

S. Gao • C. Chen (✉)
Department of Biochemistry, University of Texas, Southwestern Medical Center, 5323 Harry Hines Boulevard, Dallas, TX 75390, USA
e-mail: Chuo.Chen@UTSouthwestern.edu

J.J. Li and E.J. Corey (eds.), *Total Synthesis of Natural Products*,
DOI 10.1007/978-3-642-34065-9_2, © Springer-Verlag Berlin Heidelberg 2012

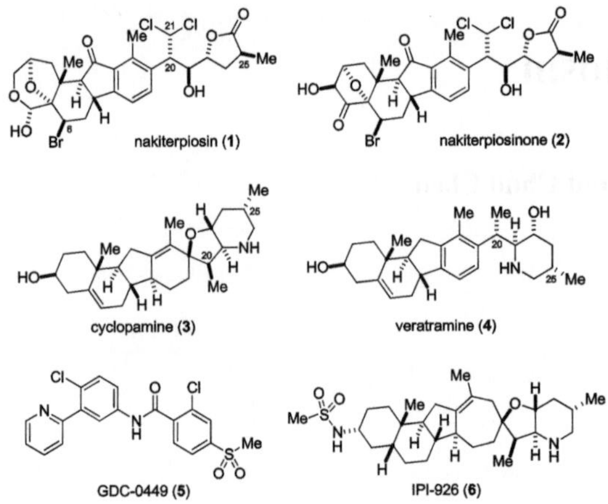

Chart 2.1 C-nor-D-homosteroids and the first two Hedgehog antagonists in clinical trials

molecules that suppress Hh signaling have been pursued as a new class of therapeutics for cancer and neurodegenerative diseases [3]. Several small-molecule Smo inhibitors, including vismodegib (or GDC-0449, **5**) by Genentech [4], IPI-926 (**6**, a cyclopamine derivative) by Infinity Pharmaceuticals [5], BMS-833923 (XL139) by Bristol-Myers Squibb [6], LDE225 and LEQ506 by Novartis [7], PF-04449913 by Pfizer, and TAK-441 by Millennium Pharmaceuticals, are now under clinical evaluation [8].

Structurally, **3** and **4** belong to a special class of steroids in which the C-ring is contracted and D-ring expanded by one carbon via a C-13 → C-12 migration. For decades, the C-nor-D-homosteroids were found only in plants. It was not until 2003 that the first marine-originated members, nakiterpiosin (**1**) and nakiterpiosinone (**2**), were reported by Uemura and coworkers [9] as part of a study of coral black disease. From 1981 to 1985, large patches (up to 1,000 m in length) of cyanobacteriosponges *Terpios hoshinota* were observed in Okinawa [10]. These thin, encrusting sponges aggressively compete with corals for space by epizoism. Uemura and coworkers hypothesized that *T. hoshinota* killed the covered corals by secreting toxic compounds. Searching for these toxins, they isolated 0.4 mg of **1** and 0.1 mg of **2** from 30 kg of the sponges. Both compounds inhibited the growth of P388 mouse leukemia cells with an IC_{50} of 10 ng/mL.

2.2 Synthesis of the 6,6,5,6 Steroidal Skeleton

The unique molecular skeleton of the C-nor-D-homosteroids represents significant challenges for organic chemists. The structure elucidation and the total synthesis of cyclopamine (**3**, also known as 11-deoxojervine), jervine (11-oxo-**3**), and veratramine (**4**) are important milestones in steroid chemistry. Many synthetic strategies were developed in the 1960s–1970s for these targets. Notably, Masamune

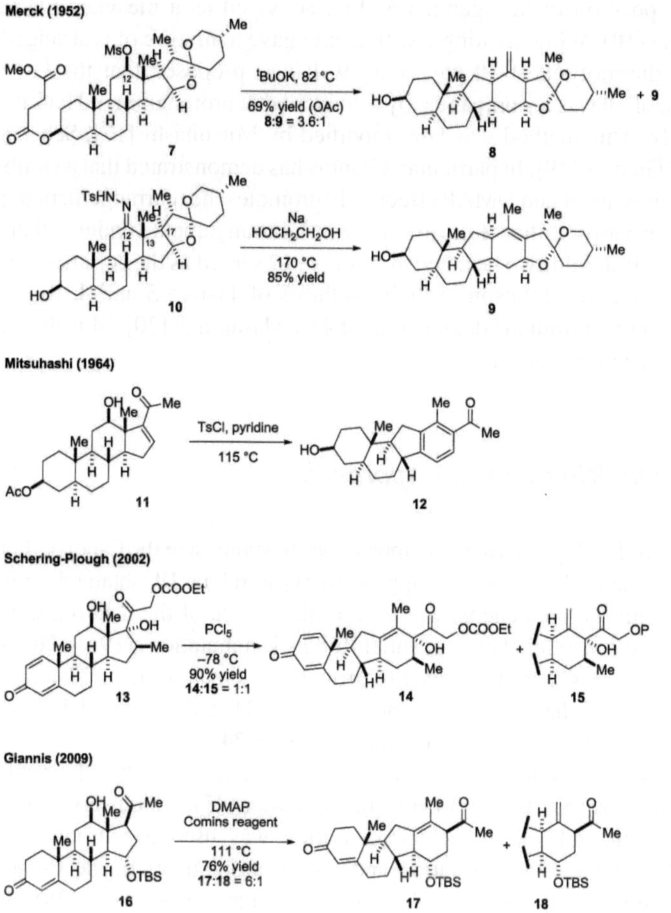

Scheme 2.1 The biomimetic approaches to C-nor-D-homosteroids

and Johnson documented the synthesis of jervine and veratramine, respectively, in 1967 [11, 12]. Together with Masamune's previous report of the conversion of jervine to cyclopamine by a Wolff reduction, these reports are the first syntheses of these three steroidal alkaloids [13]. In addition, a formal synthesis was reported by Kutney in 1975 [14] and an efficient approach by Giannis in 2009 [15]. The development of the synthetic approaches to this unique 6,6,5,6 steroidal skeleton is summarized below.

2.2.1 The Biomimetic Approaches

The biomimetic approach to the core skeleton of C-nor-D-homosteroid was first developed by the Merck research group (Scheme 2.1) [16]. In the Merck procedure, the C-12 position of hecogenin was first activated as a mesylate (**7**) or a tosylhydrozone (**10**). While treating **7** with a base gave a mixture of rearranged products **8** and **9**, thermolysis of **10** gave only **9**. It was proposed that the C-13 → C-12 migration of **10** was accompanied by a concerted deprotonation of H-17 to provide **9** selectively. This method was later modified by Mitsuhashi [17], Schering-Plough [18], and Giannis [19]. In particular, Giannis has demonstrated that a combination of the Comins reagent and DMAP effectively promotes the rearrangement of a series of steroid derivatives that fail to undergo rearrangement under other reported conditions. It should also be noted that ketone **12** served as the common intermediate for Masamune and Johnson in their synthesis of 11-oxo-**3** and **4**. Ketone **12** was initially obtained from the degradation of **4** by Masamune [20]. Mitsuhashi prepared **11** by degrading hecogenin.

2.2.2 *The Ring-by-Ring Approaches*

Johnson has developed two linear approaches to synthesize the C-nor-D-homosteroid skeleton (Scheme 2.2). In his first approach [21], tetralone **19**, obtained from reduction of 2,5-dimethoxynaphthalene, was used as the source of the C,D-rings. The B- and A-rings were constructed by sequential Robinson annulations (**19** → **20** → **21**). The C11,12 olefin was then introduced to provide **22**. Ozonolysis of **22** followed by an aldol reaction of the resulting dialdehyde gave **23**. Subsequent deformylation and deoxygenation afforded the cyclopamine skeleton **24**.

In Johnson's second approach [12b], the C-ring was introduced directly with the desired ring size. Starting from Hagemann's ester (**25**), which served as the source of the D-ring, a Knoevenagel condensation was used to introduce the C-ring (**25** → **26**). After decarboxylation and D-ring aromatization, the B- and A-rings were introduced stepwise by Robinson annulations (**26** → **27** → **28**). A series of reduction and aromatization reactions were then performed to deliver racemic **12**. Johnson's asymmetric synthesis of veratramine (**4**) was accomplished by adopting Mitsuhashi's procedure [17a]. Finally, the side chain of **12** was functionalized by an epoxide–aldehyde rearrangement.

In contrast to the Johnson's D → A-ring construction approach, Brown devised an A → D-ring construction approach [22]. Starting from Wieland–Miescher ketone (**30**), a common source of the A, B-rings in the de novo synthesis of steroids, the C-ring was introduced via hydrazone allylation, ozonolysis, aldol condensation, and olefin isomerization (**31** → **32**). The D-ring was assembled by a reductive alkylation of enone **32** followed by an aldol condensation to give **33** after deprotection.

2.2.3 *Miscellaneous*

During the synthesis of an indenone derivative, Hoornaert found that $AlCl_3$ catalyzed the dimerization of indenone **34** to form truxone **35** (Scheme 2.3) [23]. Attempts to

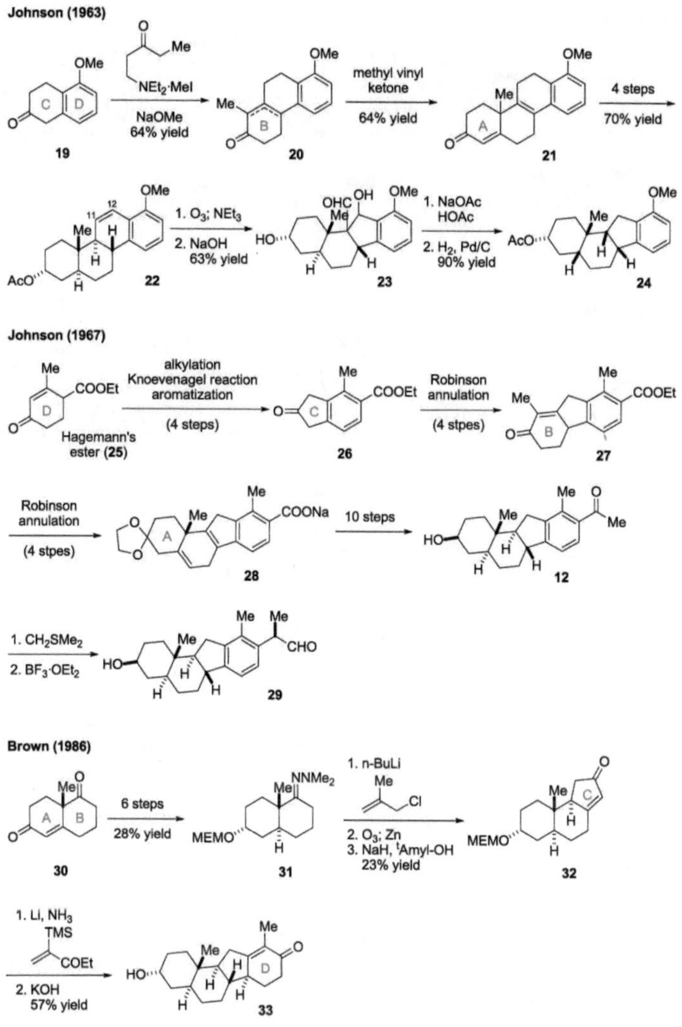

Scheme 2.2 The ring by-ring approaches to C-nor-D-homosteroids

induce the retrodimerization of **35** by photolysis resulted in a decarbonylation through a Norrish type I cleavage to give **36**. The subsequent photolytic, disrotatory retro-electrocyclization reaction and a thermal, suprafacial 1,5-sigmatropic benzoyl shift afforded **38** that bears a C-nor-D-homosteroid skeleton.

The thermally and Lewis acid-promoted transannular Diels–Alder reactions have proven to be a powerful tool for the synthesis of steroids and other natural products [24]. A research team led by Takamura, Arimoto, and Uemura utilized this reaction to assemble the polycyclic skeleton of nakiterpiosin (**1**) [25]. Heating macrolide **39** at 160 °C gave **40** and **41** as a mixture of diastereomers in good yields.

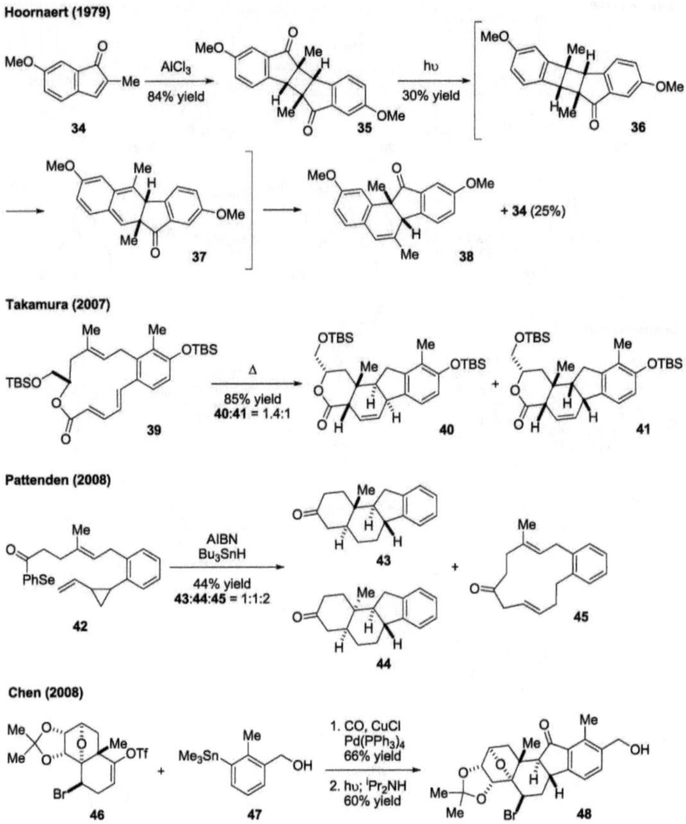

Scheme 2.3 Miscellaneous approaches to C-nor-D-homosteroids

Pattenden reported a tandem cyclization approach for the synthesis of estrone in 2004. Later, they further demonstrated that this strategy could be used to generate the veratramine skeleton [26].

We recently developed a convergent approach that comprises a carbonylative Stille coupling [27, 28] and a photo-Nazarov cyclization reaction [29–31] for the synthesis of nakiterpiosin [32]. Several highly acid- and base-sensitive functional groups were tolerated under these nearly neutral reaction conditions. We found that using a stoichiometric amount of $Pd(PPh_3)_4$ and 1 atmosphere of CO, triflate **46**, and stannane **47** could be coupled to give the corresponding enone in 66 % yield. The steric hindrance of both coupling components rendered the carbonylative coupling significantly challenging. The employment of CuCl as an additive and DMSO as the solvent accelerated the reaction considerably, thus making the desired reaction outcompete the decomposition pathways. Attempts to add LiCl to facilitate the reaction led to the elimination of the bromide. The beneficial role of CuCl in Stille reactions was first discovered by Liebeskind and later studied by Corey [33].

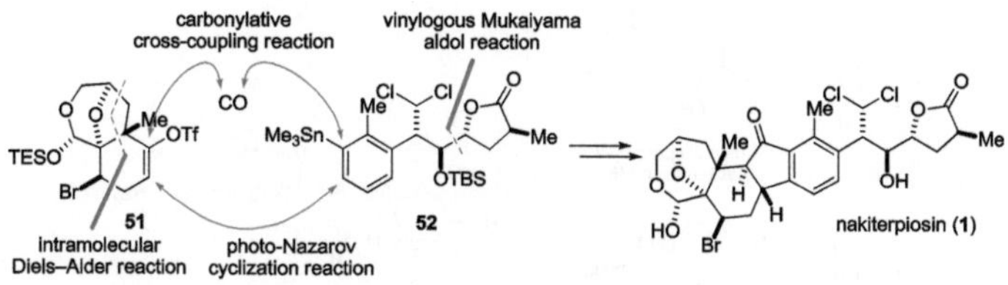

Chart 2.2 The synthetic strategy for nakiterpiosin

It is believed that the copper salts facilitate transmetalation by generating a highly reactive organocuprate intermediate.

The Nazarov cyclization of vinyl aryl ketones involves a disruption of the aromaticity, and therefore, the activation barrier is significantly higher than that of the divinyl ketones. Not surprisingly, the Lewis acid-catalyzed protocols [30] resulted only in decomposition to the enone derived from **46**, **47**, and CO. Pleasingly, however, photolysis [31] readily delivered the desired annulation product **48** in 60 % yield. The photo-Nazarov cyclization reaction of aryl vinyl ketones was first reported by Smith and Agosta. Subsequent mechanistic studies by Leitich and Schaffner revealed the reaction mechanism to be a thermal electrocyclization induced by photolytic enone isomerization. The mildness of these reaction conditions and the selective activation of the enone functional group were key to the success of this reaction.

2.3 Synthesis of Nakiterpiosin

As described above, our synthetic strategy involves the convergent construction of the central cyclopentanone ring with a carbonylative cross-coupling reaction and a photo-Nazarov cyclization reaction (Chart 2.2). The electrophilic coupling component **51** was synthesized by an intramolecular Diels–Alder reaction [34] and the nucleophilic coupling component **52** by a vinylogous Mukaiyama aldol reaction [35].

The structure of nakiterpiosin was originally assigned as **49** by Uemura based on NMR experiments [9]. Puzzled by the inconsistency of the C-20 stereochemistry of **49** with that of cyclopamine (**3**) and veratramine (**4**), we first set out to probe the relative stereochemistry of nakiterpiosin. Our model studies indicated the potential misassignment of the C-6, C-20, and C-25 stereogenic centers [32]. We next considered the biogenesis of the halogen atoms of nakiterpiosin to rationalize the C-6 and C-20 stereochemistry (Chart 2.3) [36]. We envisioned that the C-21 chlorine atoms of nakiterpiosin might be introduced by radical chlorination, and the C-6 bromine atom by bromoetherification (as shown in **50**) to result in retention of the C-20 configuration and the anti C-5,6 bromohydrin stereochemistry. Taken together, these considerations led us to propose **1** as the correct structure of nakiterpiosin, which was later confirmed via the total synthesis of **49** and **1**.

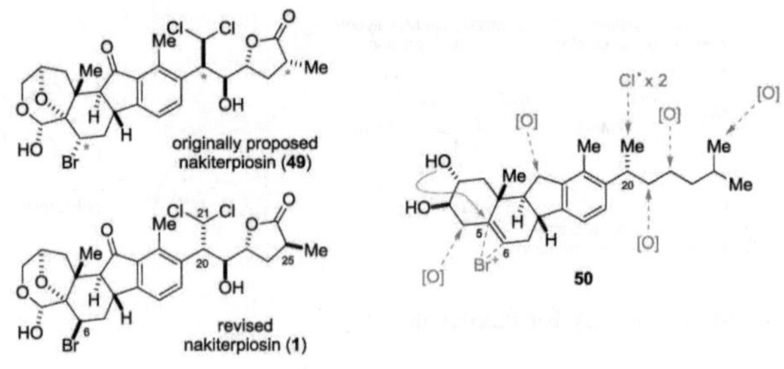

Chart 2.3 Structural revision and biosynthesis analysis of nakiterpiosin

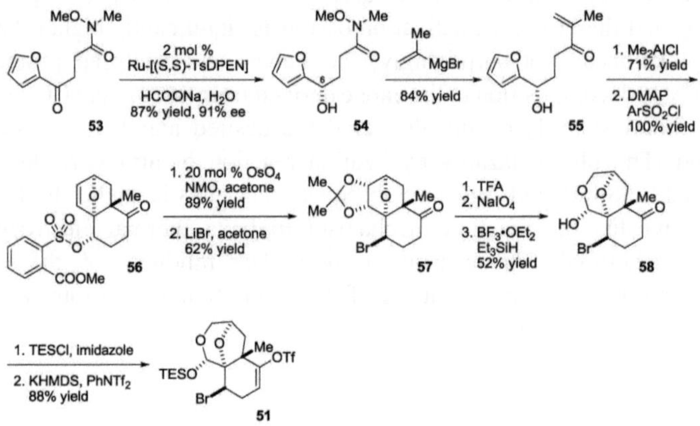

Scheme 2.4 Synthesis of the electrophilic coupling component **51**

The synthesis of the electrophilic coupling component **51** commenced with a Friedel–Craft acylation of furan with succinic anhydride (Scheme 2.4) [37]. The resulting acid was converted to a Weinreb amide (**53**). The Noyori reduction [38] with the Xiao modification [39] was then used to set the C-6 stereochemistry, affording **54**. A Grignard reaction then gave the enone (**55**). The subsequent intramolecular Diels–Alder reaction proceeded with good stereochemical control [40] to give the exo product exclusively. The sterically congested C-6 hydroxyl group was then activated with an unusual, electron-deficient aryl sulfonate group to afford **56**.

To avoid the retro-Diels–Alder reaction, **56** was dihydroxylated prior to the introduction of the bromine atom (**57**). Removal of the acetonide group followed by cleavage of the diol afforded a bis-hemiacetal. Selective reduction of the less-hindered hemiacetal group gave **58**. The remaining hemiacetal was protected, and the ketone was converted to an enol triflate, thus concluding the synthesis of the electrophilic coupling component **51**.

The synthesis of the nucleophilic coupling component **52** started with the reduction of 3-bromo-2-methylbenzenecarboxylic acid, and followed with a Horner–Wadsworth–Emmons reaction of the corresponding aldehyde, and a 1,2-reduction of

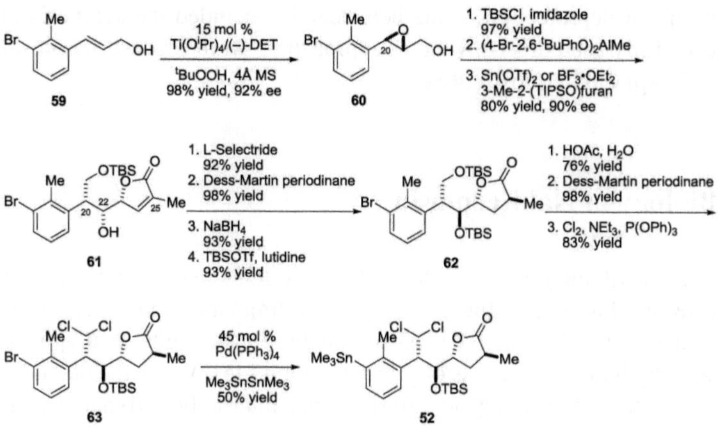

Scheme 2.5 Synthesis of the nucleophilic coupling component **52**

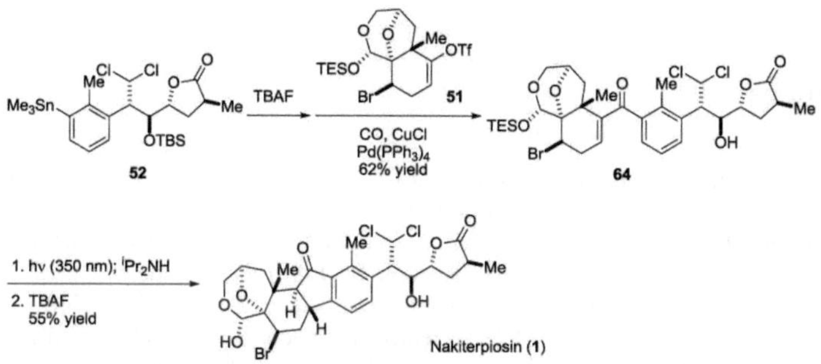

Scheme 2.6 Completion of the synthesis of nakiterpiosin (**1**)

the resulting enoate to afford **59** (Scheme 2.5). A Sharpless epoxidation [41] was then used to set the C-20 stereochemistry, giving epoxide **60** with 92 % ee. After the protection of the hydroxyl group, a pinacol-type rearrangement using Yamamoto's catalyst [42] followed by a vinylogous Mukaiyama aldol reaction afforded **61** without significant erosion of the enantiomeric purity.

With the complete carbon framework of the side chain in place, we next sought to set its anti–anti–trans configuration. The C-25 stereochemistry could be established by either a directed hydrogenation [43] or a conjugate reduction. The C-22 stereochemistry was inverted by reduction of the C-22 ketone to afford the requisite anti–anti–trans configuration. Subsequent protection of the hydroxyl group gave **62**. To introduce the *gem*-dichloromethyl group, we selectively deprotected the primary alcohol, oxidized it to an aldehyde, and chlorinated it with $Cl_2/P(OPh)_3$ [44]. Bromide **63** was then stannylated to provide the nucleophilic coupling component **52**.

To complete the synthesis of nakiterpiosin (**1**), we first deprotected **52** and then coupled it to **51** under the previously described carbonylative conditions (Scheme 2.6). Photolysis of **64** readily provided the desired annulation product.

The subsequent deprotection of the hemiacetal concluded the synthesis of **1**. We also successfully used this convergent approach to synthesize nakiterpiosinone (**2**) and 6,20,25-*epi*-nakiterpiosin (**49**).

2.4 Biology of Nakiterpiosin

The strong growth inhibitory activity of **1** toward P388 cells prompted us to further investigate its biological functions. Our preliminary studies showed that **1** suppressed Hh signaling in NIH3T3 mouse fibroblasts with an IC_{50} of 0.6 μM, presumably by inducing the loss of primary cilium [45]. While the detailed mechanism is not clear, **1** is likely to influence the microtubule dynamics through a different mode of action from common antimitotic agents such as taxol and nocodazole. Further work is needed to elucidate its molecular target.

References

1. (a) Binns W, James LF, Keeler RF, Balls LD (1968) Cancer Res 28:2323–2326; (b) Keeler RF (1978) Lipids 13:708–715; (c) James LF (1999) J Nat Toxins 8:63–80
2. (a) Cooper MK, Porter JA, Young KE, Beachy PA (1998) Science 280:1603–1607; (b) Incardona JP, Gaffield W, Kapur RP, Roelink H (1998) Development 125:3553–3562; (c) Chen JK, Taipale J, Cooper MK, Beachy PA (2002) Genes Dev 16:2743–2748
3. (a) Scalesa SJ, de Sauvage FJ (2009) Trends Pharmacol Sci 30:303–312; (b) Epstein EH (2008) Nat Rev Cancer 8:743–754; (c) van den Brink GR (2007) Physiol Rev 87:1343–1375; (d) Romer J, Curran T (2005) Cancer Res 65:4975–4978; (e) Hooper JE, Scott MP (2005) Nat Rev Mol Cell Biol 6:306–317; (f) Borzillo GV, Lippa B (2005) Curr Top Med Chem 5:147–157; (g) Dellovade T, Romer JT, Curran T, Rubin LL (2006) Annu Rev Neurosci 29:539–563; (h) Yauch RL, Gould SE, Scales SJ, Tang T, Tian H, Ahn CP, Marshall D, Fu L, Januario T, Kallop D, Nannini-Pepe M, Kotkow K, Marsters JC, Rubin LL, de Sauvage FJ (2008) Nature 455:406–410
4. (a) Smaele ED, Ferretti E, Gulino A (2010) Curr Opin Investig Drugs 11:707–718; (b) Yauch RL, Dijkgraaf GJP, Alicke B, Januario T, Ahn CP, Holcomb T, Pujara K, Stinson J, Callahan CA, Tang T, Bazan JF, Kan Z, Seshagiri S, Hann CL, Gould SE, Low JA, Rudin CM, de Sauvage FJ (2009) Science 326:572–574
5. (a) Olive KP, Jacobetz MA, Davidson CJ, Gopinathan A, McIntyre D, Honess D, Madhu B, Goldgraben MA, Caldwell ME, Allard D, Frese KK, DeNicola G, Feig C, Combs C, Winter SP, Ireland-Zecchini H, Reichelt S, Howat WJ, Chang A, Dhara M, Wang L, Rückert F, Grützmann R, Pilarsky C, Izeradjene K, Hingorani SR, Huang P, Davies SE, Plunkett W, Egorin M, Hruban RH, Whitebread N, McGovern K, Adams J, Iacobuzio-Donahue C, Griffiths J, Tuveson DA (2009) Science 324:1457–1461; (b) Tremblay MR, Lescarbeau A, Grogan MJ, Tan E, Lin G, Austad BC, Yu L-C, Behnke ML, Nair SJ, Hagel M, White K, Conley J, Manna JD, Alvarez-Diez TM, Hoyt J, Woodward CN, Sydor JR, Pink M, MacDougall J, Campbell MJ, Cushing J, Ferguson J, Curtis MS, McGovern K, Read MA, Palombella VJ, Adams J, Castro AC (2009) J Med Chem 52:4400–4418
6. Siu LL, Papadopoulos K, Alberts SR, Kirchoff-Ross R, Vakkalagadda B, Lang L, Ahlers CM, Bennett KL, Tornout JMV (2010) J Clin Oncol (Meeting Abstracts) 28(Suppl):2501

7. (a) Buonamici S, Williams J, Morrissey M, Wang A, Guo R, Vattay A, Hsiao K, Yuan J, Green J, Ospina B, Yu Q, Ostrom L, Fordjour P, Anderson DL, Monahan JE, Kelleher JF, Peukert S, Pan S, Wu X, Maira S-M, García-Echeverría C, Briggs KJ, Watkins DN, Yao Y-M, Lengauer C, Warmuth M, Sellers WR, Dorsch M (2010) Sci Transl Med 2:51ra70; (b) Pan S, Wu X, Jiang J, Gao W, Wan Y, Cheng D, Han D, Liu J, Englund NP, Wang Y, Peukert S, Miller-Moslin K, Yuan J, Guo R, Matsumoto M, Vattay A, Jiang Y, Tsao J, Sun F, Pferdekamper AC, Dodd S, Tuntland T, Maniara W, Kelleher JF III, Yao Y-M, Warmuth M, Williams J, Dorsch M (2010) ACS Med Chem Lett 1:130–134
8. (a) http://clinicaltrials.gov/; (b) Low JA, de Sauvage FJ (2010) J Clin Oncol. doi:10.1200/JCO.2010.1227.9943; (c) Mahindroo N, Punchihewa C, Fujii N (2009) J Med Chem 52:3829–3845
9. (a) Teruya T, Nakagawa S, Koyama T, Suenaga K, Kita M, Uemura D (2003) Tetrahedron Lett 44:5171–5173; (b) Teruya T, Nakagawa S, Koyama T, Arimoto H, Kita M, Uemura D (2004) Tetrahedron 60:6989–6993
10. (a) Plucer-Rosario G (1987) Coral Reefs 5:197–200; (b) Rützler K, Smith KP (1993) Sci Mar 57:381–393; (c) Rützler K, Smith KP (1993) Sci Mar 57:395–403; (d) Liao M-H, Tang S-L, Hsu C-M, Wen K-C, Wu H, Chen W-M, Wang J-T, Meng P-J, Twan W-H, Lu C-K, Dai C-F, Soong K, Chen C-A (2007) Zool Stud 46:520; (e) Lin W-j (2009) M.S. Thesis, National Sun Yat-sen University, Kaohsiung, Taiwan, ROC
11. (a) Masamune T, Takasugi M, Murai A, Kobayashi K (1967) J Am Chem Soc 89:4521–4523; (b) Masamune T, Takasugi M, Murai A (1971) Tetrahedron 27:3369–3386
12. (a) Johnson WS, de Jongh HAP, Coverdale CE, Scott JW, Burckhardt U (1967) J Am Chem Soc 89:4523–4524; (b) Johnson WS, Cox JM, Graham DW, Whitlock HW Jr (1967) J Am Chem Soc 89:4524–4526
13. Masamune T, Mori Y, Takasugi M, Murai A, Ohuchi S, Sato N, Katsui N (1965) Bull Chem Soc Jpn 38:1374–1378
14. (a) Kutney JP, By A, Cable J, Gladstone WAF, Inaba T, Leong SY, Roller P, Torupka EJ, Warnock WDC (1975) Can J Chem 53:1775–1795; (b) Kutney JP, Cable J, Gladstone WAF, Hanssen HW, Nair GV, Torupka EJ, Warnock WDC (1975) Can J Chem 53:1796–1817
15. Giannis A, Heretsch P, Sarli V, Stößel A (2009) Angew Chem Int Ed 48:7911–7914
16. (a) Hirschmann R, Snoddy CS Jr, Wendler NL (1952) J Am Chem Soc 74:2693–2694; (b) Hiskey CF, Hirschmann R, Wendler NL (1953) J Am Chem Soc 75:5135–5136; (c) Hirschmann R, Snoddy CS Jr, Hiskey CF, Wendler NL (1954) J Am Chem Soc 76:4013–4025
17. (a) Mitsuhashi H, Shibata K (1964) Tetrahedron Lett 5:2281–2283; (b) Mitsuhashi H, Shimizu Y, Moriyama T, Masuda M, Kawahara N (1974) Chem Pharm Bull 22:1046–1052; (c) Mitsuhashi H, Shimizu Y (1961) Tetrahedron Lett 2:777–780
18. Fu X, Chan T-M, Tann C-H, Thiruvengadam TK (2002) Steroid 67:549–554
19. Heretsch P, Rabe S, Giannis A (2010) J Am Chem Soc 132:9968–9969
20. Franck RW, Johnson WS (1963) Tetrahedron Lett 4:545–547
21. (a) Johnson WS, Szmuszkovicz J, Rogier ER, Hadler HI, Wynberg H (1956) J Am Chem Soc 78:6285–6289; (b) Johnson WS, Rogier ER, Szmuszkovicz J, Hadler HI, Ackerman J, Bhattacharyya BK, Bloom BM, Stalmann L, Clement RA, Bannister B, Wynberg H (1956) J Am Chem Soc 78:6289–6302; (c) Franck RW, Johnson WS (1963) Tetrahedron Lett 4:545–547; (d) Johnson WS, Cohen N, Habicht ER Jr, Hamon DPG, Rizzi GP, Faulkner DJ (1968) Tetrahedron Lett 9:2829–2833
22. (a) Brown E, Lebreton J (1986) Tetrahedron Lett 27:2595–2598; (b) Brown E, Lebreton J (1987) Tetrahedron 43:5827–5840
23. Ceustermans RAE, Martens HJ, Hoornaert GJJ (1979) J Org Chem 44:1388–1391
24. (a) Marsault E, Toró A, Nowak P, Deslongchamps P (2001) Tetrahedron 57:4243–4260; (b) Porco JA Jr, Schoenen FJ, Stout TJ, Clardy J, Schreiber SL (1990) J Am Chem Soc 112:7410–7411; (c) Roush WR, Koyama K, Curtin ML, Moriarty KJ (1996) J Am Chem Soc 118:7502–7512; (d) Vosburg DA, Vanderwal CD, Sorensen EJ (2002) J Am Chem Soc 124:4552–4553; (e) Evans DA, Starr J (2002) Angew Chem Int Ed 41:1787–1790

25. Ito T, Ito M, Arimoto H, Takamura H, Uemura D (2007) Tetrahedron 48:5465–5469
26. (a) Stoker DA (2008) Ph.D. Thesis, University of Nottingham, Nottingham, England; (b) Pattenden G, Gonzalez MA, McCulloch S, Walter A, Woodhead SJ (2004) Proc Natl Acad Sci USA 101:12024–12029
27. (a) Brennführer A, Neumann H, Beller M (2009) Angew Chem Int Ed 48:4114–4133; (b) Barnard CFJ (2008) Organometallics 27:5402–5422; (c) Skoda-Földes R, Kollár L (2002) Curr Org Chem 6:1097–1119
28. (a) Tanaka M (1979) Tetrahedron Lett 20:2601–2602; (b) Beletskaya IP (1983) J Organomet Chem 250:551–564; (c) Sheffy FK, Godschalx JP, Stille JK (1984) J Am Chem Soc 106:4833–4840; (d) Goure WF, Wright ME, Davis PD, Labadie SS, Stille JK (1984) J Am Chem Soc 106:6417–6422
29. (a) Nakanishi W, West FG (2009) Curr Opin Drug Discov Dev 12:732–751; (b) Frontier AJ, Collison C (2005) Tetrahedron 61:7577–7606; (c) Pellissier H (2005) Tetrahedron 61:6479–6517; (d) Tius MA (2005) Eur J Org Chem 2193–2206
30. (a) Liang G, Xu Y, Seiple IB, Trauner D (2006) J Am Chem Soc 128:11022–11023; (b) He W, Herrick IR, Atesin TA, Caruana PA, Kellenberger CA, Frontier AJ (2008) J Am Chem Soc 130:1003–1011; (c) Marcus AP, Lee AS, Davis RL, Tantillo DJ, Sarpong R (2008) Angew Chem Int Ed 47:6379–6383
31. (a) Crandall JK, Haseltine RP (1968) J Am Chem Soc 90:6251–6253; (b) Noyori R, Katô M (1968) Tetrahedron Lett 9:5075–5077; (c) Smith AB III, Agosta WC (1973) J Am Chem Soc 95:1961–1968; (d) Leitich J, Heise I, Werner S, Krueger C, Schaffner K (1991) J Photochem Photobiol A Chem 57:127–151; (e) Leitich J, Heise I, Rust J, Schaffner K (2001) Eur J Org Chem 2001:2719–2726
32. (a) Gao S, Wang Q, Chen C (2009) J Am Chem Soc 131:1410–1412; (b) Gao S, Huang QWLJ-S, Lum L, Chen C (2010) J Am Chem Soc 132:371–383
33. (a) Liebeskind LS, Fengl RWJ (1990) J Org Chem 55:5359–5364; (b) Farina V, Kapadia S, Krishnan B, Wang C, Liebeskind LSJ (1994) J Org Chem 59:5905–5911; (c) Han X, Stoltz BM, Corey EJ (1999) J Am Chem Soc 121:7600–7605
34. (a) Takao K-I, Munakata R, Tadano K-I (2005) Chem Rev 105:4779–4807; (b) Keay BA, Hunt IR (1999) Adv Cycloaddit 6:173–210; (c) Roush WR (1990) Adv Cycloaddit 2:91–146; (d) Craig D (1987) Chem Soc Rev 16:187–238
35. (a) Denmark SE, Heemstra J, John R, Beutner GL (2005) Angew Chem Int Ed 44:4682–4698; (b) Casiraghi G, Zanardi F, Appendino G, Rassu G (2000) Chem Rev 100:1929–1972; (c) Rassu G, Zanardi F, Battistinib L, Casiraghi G (2000) Chem Soc Rev 29:109–118
36. (a) Neumann CS, Fujimori DG, Walsh CT (2008) Chem Biol 15:99–109; (b) Vaillancourt FH, Yeh E, Vosburg DA, Garneau-Tsodikova S, Walsh CT (2006) Chem Rev 106:3364–3378; (c) Butler A, Walker JV (1993) Chem Rev 93:1937–1944
37. Chen L (2004) Yin B-l, Xu H-h, Chiu M-h, Wu Y-l. Chin J Chem 22:92–99
38. (a) Hashiguchi S, Fujii A, Takehara J, Ikariya T, Noyori R (1995) J Am Chem Soc 117:7562–7563; (b) Fujii A, Hashiguchi S, Uematsu N, Ikariya T, Noyori R (1996) J Am Chem Soc 118:2521–2522
39. (a) Wu X, Li X, Hems W, King F, Xiao J (2004) Org Biomol Chem 2:1818–1821; (b) Wu X, Li X, King F, Xiao J (2005) Angew Chem Int Ed 44:3407–3411; (c) Ohkuma T, Utsumi N, Tsutsumi K, Murata K, Sandoval C, Noyori R (2006) J Am Chem Soc 128:8724–8725
40. (a) Roush WRJ (1979) J Org Chem 44:4008–4010; (b) Roush WR, Hall SE (1981) J Am Chem Soc 103:5200–5211; (c) Roush WR, Kageyama M, Riva R, Brown BB, Warmus JS, Moriarty KJJ (1991) J Org Chem 56:1192–1210; (d) Taber DF, Gunn BP (1979) J Am Chem Soc 101:3992–3993; (e) Taber DF, Saleh SA (1980) J Am Chem Soc 102:5085–5088; (f) Boeckman RK Jr, Napier JJ, Thomas EW, Sato RIJ (1983) J Org Chem 48:4152–4154; (g) Boeckman RK Jr, Barta TEJ (1985) J Org Chem 50:3421–3423
41. (a) Katsuki T, Martin VS (1996) Org React 48:1–300; (b) Katsuki T, Sharpless KB (1980) J Am Chem Soc 102:5974–5976; (c) Gao Y, Klunder JM, Hanson RM, Masamune H, Ko SY, Sharpless KB (1987) J Am Chem Soc 109:5765–5780

42. Maruoka K, Ooi T, Yamamoto H (1989) J Am Chem Soc 111:6431–6432
43. Hoveyda AH, Evans DA, Fu GC (1993) Chem Rev 93:1307–1370
44. (a) Spaggiari A, Vaccari D, Davoli P, Torre G, Prati FJ (2007) J Org Chem 72:2216–2219; (b) Hoffmann RW, Bovicelli P (1990) Synthesis 657–659
45. (a) Eggenschwiler JT, Anderson KV (2007) Annu Rev Cell Dev Biol 23:345–373; (b) Quarmby LM, Parker JDK (2005) J Cell Biol 169:707–710

Chapter 3
The Kinamycins

Seth B. Herzon

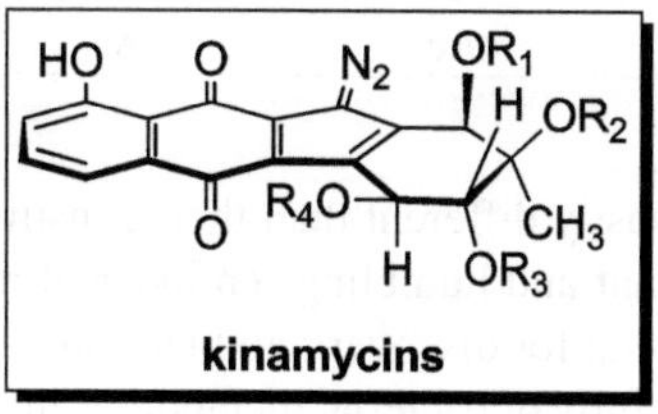

3.1 Introduction

Chemists are continuously confronted with the challenge of classifying and accurately assessing the complexity elements presented by natural products, and these elements are often assigned familiar categories such as topological, stereochemical, and size complexities. An interesting additional element of complexity involves the fusion of well-known functional groups into larger entities, with the individual constituents in communication by a combination of resonance and inductive effects. α,β-Unsaturated carbonyls (carbonyl + alkene) and enamines (amine + alkene) may be the simplest examples of such functionality. However, progressively more complex "macrofunctional groups" can be constructed, and at the far end of this spectrum lie arrangements of atoms within molecules that are simply inconceivable to the practicing chemist and which are only found within highly complex and heteroatom-rich natural products. Such macrofunctional groups often possess reactivity and

S.B. Herzon (✉)
Yale University, 225 Prospect Street, New Haven, CT 06520-8107, USA

J.J. Li and E.J. Corey (eds.), *Total Synthesis of Natural Products*,
DOI 10.1007/978-3-642-34065-9_3,

Table 3.1 Structures of selected kinamycins

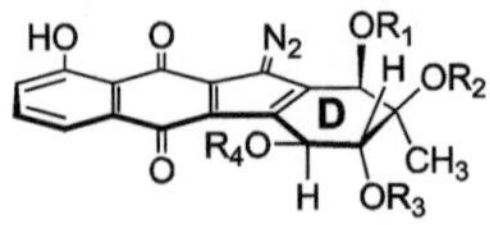

Kinamycin	R_1	R_2	R_3	R_4	Refs.
A (**1**)	H	Ac	Ac	Ac	[1–3]
B (**2**)	H	Ac	H	H	[1–3]
C (**3**)	Ac	H	Ac	Ac	[1–3]
D (**4**)	Ac	H	Ac	H	[1–3]
E (**5**)	Ac	H	H	H	[4a,c]
F (**6**)	H	H	H	H	[4a,c]
G (**7**)	Ac	C(O)*i*-Pr	Ac	Ac	[4b]
H (**8**)	C(O)*i*-Pr	H	Ac	Ac	[4b]
I (**9**)	C(O)*i*-Pr	H	C(O)*i*-Pr	Ac	[4d]
J (**10**)	Ac	Ac	Ac	Ac	[1–3]

stability profiles that are vastly different than their constituent components. Targets of this type are an important and humbling reminder of nature's synthetic prowess and present an exciting arena for discovery and learning.

The kinamycins are a family of bacterial metabolites that exemplify the concept of a macrofunctional group. Kinamycins A–D (**1–4**) were first isolated in 1970 by Ōmura and coworkers as orange, crystalline solids from the fermentation broth of *Streptomyces murayamaensis* (Table 3.1) [1–3, 5]. Many members of the kinamycin family are now known, and these are distinguished primarily by the substitution pattern about the D-ring (Table 3.1) [4, 6]. The macrofunctional group within the kinamycins comprises a diazocyclopentadiene fused to an oxygenated naphthoquinone and two pseudo-benzylic oxygen substituents. The kinamycins are often referred to as diazofluorenes (diazotetrahydrobenzo[*b*]fluorenes), a descriptor that is inclusive of the hemisaturated six-membered ring. The presence of an isolable diazo functional group within a natural product is unusual and may reflect an increase in stability that occurs on conjugation with the cyclopentadiene and naphthoquinone rings. This macrofunctional group presented significant challenges to the structural elucidation of the kinamycins (vide infra).

The related dimeric diazofluorenes lomaiviticins A (**11**) and B (**12**) were isolated by He, Ireland, and coworkers in 2001 (Chart 3.1) [7]. Although the lomaiviticins bear some structural homology to the kinamycins, significant differences exist between the two classes of natural products. For example, the lomaiviticins contain a dihydroxynaphthoquinone residue, rather than a juglone residue, as found in the kinamycins. The lomaiviticins also contain 2–4 2,6-dideoxyglycoside residues, *N*,*N*-dimethylpyrrolosamine and oleandrose, rather than acyl substituents, as found in the kinamycins. Finally, the lomaiviticins contain an ethyl substituent, rather than a methyl group, within the cyclohexenone ring. Given these differences, it seems

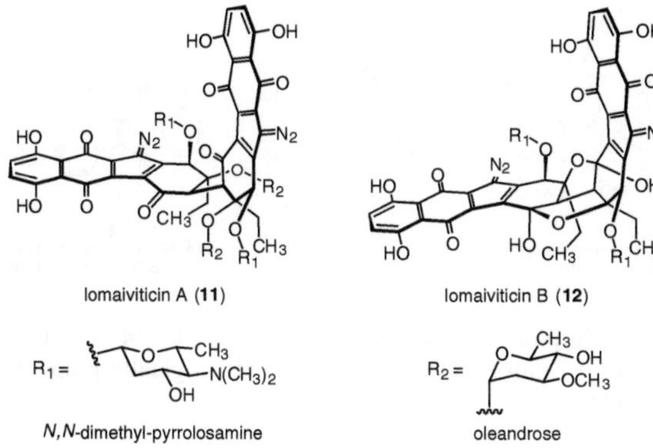

Chart 3.1 Structures of lomaiviticins A (**11**) and B (**12**)

unlikely that the kinamycins serve as biosynthetic precursors to the lomaiviticins, although the two families may share similar biosynthetic pathways.

3.2 Structure Elucidation

The structure elucidation of the kinamycins was a formidable challenge, and the information presented below draws from the work of several research groups over a period of more than 20 years. As will be shown, the originally proposed structure of the kinamycins contained a cyanamide rather than a diazo function. Subsequent synthetic and biosynthetic studies led to replacement of the cyanamide with a diazo function. The structural elucidation was challenging, in part, because of the high degree of unsaturation of the kinamycins, which limits the utility of ^{1}H and 2D NMR analysis. In addition, because these structures were unprecedented, there were no clear benchmarks for comparison at the time. The pathway from isolation to determination of the correct structure is described below.

Initial spectroscopic and degradation studies of the kinamycins were conducted by Ōmura and coworkers [3]. These investigations established the constitution and relative stereochemistry of the D-ring and the juglone fragments of the kinamycins. Based on MS, IR (~2,150 cm^{-1}), and the results of alkaline degradation (liberation of ammonia), a cyanamide function was suggested to bridge the juglone and D-ring substructures, leading to formulation of an *N*-cyanocarbazole function (**13**, Chart 3.2). Shortly thereafter, an X-ray analysis of a *para*-bromobenzoate derivative of (−)-kinamycin C (**3**) was obtained, which established the absolute stereochemistry [8]. The quality of the X-ray data did not permit rigorous assignment of the putative cyanamide function. Instead, the presence of this function was inferred based on Ōmura's earlier studies.

Subsequent biosynthetic and synthetic studies, however, provided data that were inconsistent with the *N*-cyanocarbazole function. Gould synthesized $^{15}N_2$-kinamycin

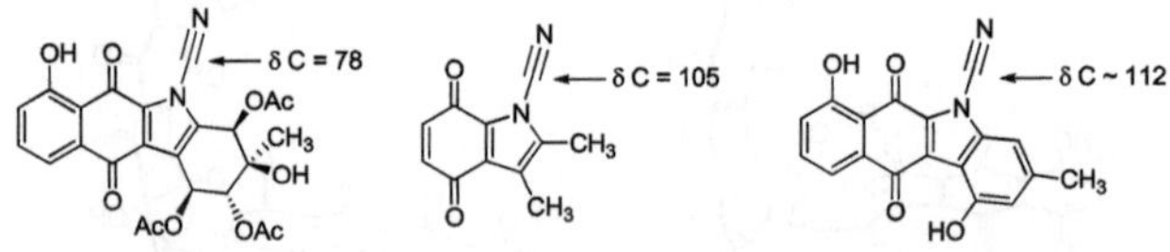

13
originally proposed structure
for kinamycin C

14

15
originally proposed structure
for the metabolite prekinamycin

Chart 3.2 **13**: Originally proposed structure for kinamycin C; **14**: structure an *N*-cyanoindoloquinone synthesized by Dmitrienko and coworkers; **15**: originally proposed structure for the metabolite prekinamycin

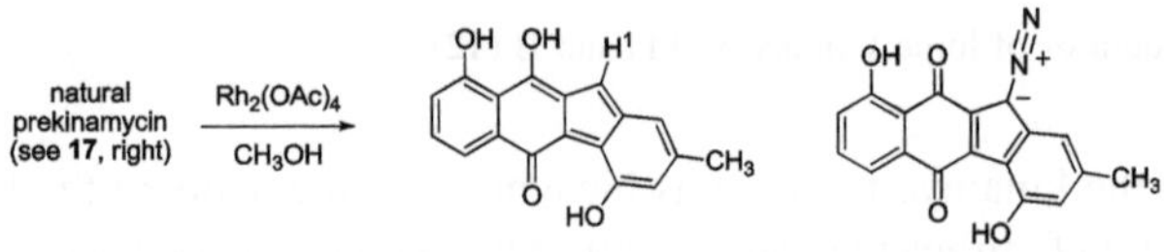

16

prekinamycin (**17**)
[revised (correct) structure]

Scheme 3.1

D [9], which allowed for unequivocal location of the putative cyanamide carbon by NMR spectroscopy. This carbon was observed to resonate at 78.5 ppm ($J_{C,N} = 21.2$, 5.4 Hz). Close inspection of ^{13}C NMR data of natural kinamycin D (**4**) revealed a small singlet at the same resonance, nearly obscured by the solvent signal. Gould noted that this was nearly 30 ppm upfield of what would have been expected for a typical cyanamide carbon. In order to obtain spectra on cyanamides more closely resembling the kinamycins, Dmitrienko and coworkers synthesized a series of *N*-cyanoindoloquinones, as exemplified by structure **14** [10]. These workers observed that the resonance of the cyanamide carbon was in the ~105 ppm range, as expected. Additionally, the cyanamide stretch of **14** in the infrared was in the 2,237–2,259 cm^{-1} range vs. 2,119–2,170 cm^{-1} for the kinamycins. Finally, Echavarren and coworkers prepared the structure **15**, which was originally assigned as the related metabolite prekinamycin [11]. In this instance, ^{1}H and ^{13}C NMR data of synthetic material did not match those reported for natural prekinamycin. As before, the putative cyanamide carbon of synthetic **15** resonated downfield of that in the natural material. Collectively, these studies called into question the original structural assignment of the kinamycins.

The matter was settled in 1994 in back-to-back communications by Gould [12] and Dmitrienko [13]. Gould showed that treatment of natural prekinamycin with dirhodium tetraacetate in methanol yielded the fluorene **16** (Scheme 3.1). The vinyl proton formed in this reaction (H-1) provided a critical spectroscopic handle and allowed unambiguous determination of the carbocyclic structure, excluding the presence of an indole heterocycle. In parallel, his research group obtained a high-quality crystal structure of a kinamycin derivative. The refined data set was shown to best accommodate a diazo rather than cyanamide (or isonitrile) function.

Contemporaneously, Dmitrienko prepared a synthetic derivative of the proposed structure of prekinamycin (**15**) and provided evidence that the NMR and IR spectroscopic data for the kinamycins and prekinamycin agree best with those of known diazofluorenes. In particular, his research group noted that the nitrogen chemical shifts and coupling constants of *N*-15-enriched kinamycin D [9] were in good agreement with those of doubly *N*-15-labeled ethyldiazoacetate but agreed less well with doubly *N*-15-labeled cyanoindole **14**.

As a result of these careful investigations, the structures of the kinamycins and prekinamycin were reassigned as diazotetrahydro[*b*]benzofluorenes (diazofluorenes, see Table 3.1 and structure **17**, Scheme 3.1). Like the *N*-cyanoindole originally proposed, this substructure was without close precedent in natural products chemistry. In contemplating syntheses of the kinamycins, the reactivity of the diazo group must be considered in the context of the cyclopentadiene and naphthoquinone functional groups to which it is directly bonded (and undoubtedly which contribute to the enhanced stability of these metabolites, as compared to simple diazoalkanes, such as diazomethane).

3.3 Biological Activity and Mechanism of Action Studies

The kinamycins and lomaiviticins are potent cytotoxic agents. The kinamycins are high nanomolar inhibitors of cultured human cancer cells and have also demonstrated low-micromolar activity against Gram-positive and Gram-negative bacteria [2]. Kinamycin C has been evaluated against a 60 cell line panel at the National Cancer Institute, and it exhibited a mean GI_{50} of 340 nM. The lomaiviticins, however, are several orders of magnitude more cytotoxic than the kinamycins. Against a panel of 25 human cancer cell lines, lomaiviticin A (**11**) exhibited 50 % inhibitory potencies in the single-digit nanomolar range, with activities extending down to the picomolar range against ovarian cell lines [7]. In addition, both lomaiviticins A and B have demonstrated powerful antimicrobial activity, with efficacy at low ng/mL concentrations (plate assay).

An understanding of the molecular mechanisms underlying the observed cytotoxic effects of the kinamycins and lomaiviticins is not fully complete, although several different proposals have been advanced. Many of these studies have focused on the diazofluorene functional group. In 1995, Arya and Jebaratnam reported that DNA was efficiently cleaved when incubated in the presence of the diazofluorene **18** and cupric acetate (Scheme 3.2a) [14]. Control experiments indicated that both reagents were required for cleavage. The authors proposed that the diazofluorene **18** may undergo oxidation to the free radical **19** and that the latter might be the active DNA-cleaving agent. Arya and Jebaratnam suggested that the quinone function of the kinamycins may act as an internal oxidant toward the diazo group, triggering the formation of a radical similar to **19**. However, a large body of evidence, discussed below, suggests the biological effects of the kinamycins and lomaiviticins may be

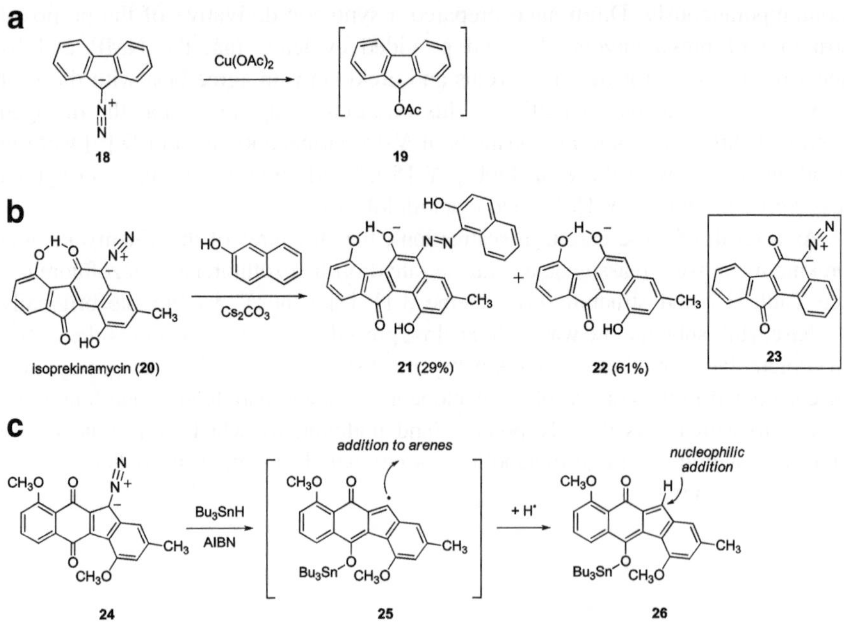

Scheme 3.2 (**a–c**) Mechanism of action studies

mediated by a reducing cofactor, which argues against an oxidatively triggered pathway.

Dmitrienko and coworkers suggested an alternative mode of reactivity for the diazofluorene functional group (Scheme 3.2b) [15]. As a model for the kinamycins, these researchers studied isoprekinamycin (**20**), a metabolite obtained from the bacterial strain that produces the kinamycins. It was shown that isoprekinamycin (**20**) underwent addition of β-napthol to form the alkylation product **21**, as well as a substantial amount of the reduced product **22**. The authors proposed that a hydrogen-bonding interaction between the ketone and the adjacent phenol (see structure **20**) may increase the electrophilicity of the α-diazoketone function of isoprekinamycin (**20**). The related synthetic intermediate **23**, which lacks this interaction, was unreactive under otherwise identical conditions. The authors suggested that the kinamycins and lomaiviticins may exert their cytotoxic effects by undergoing addition of a biological nucleophile to the diazo function or, alternatively, by forming vinyl radicals via loss of dinitrogen from diazenes resembling **21**.

Feldman and Eastman have suggested that the kinamycins may by reductively activated to form reactive vinyl radical (**25**) and *ortho*-quinone methide (**26**) intermediates (Scheme 3.2c) [16]. The authors provided convincing evidence that the alkenyl radical **25** is generated when the model substrate dimethyl prekinamycin (**24**) is exposed to reducing conditions (tri-*n*-butyltin hydride, AIBN). Products that may arise from addition of this radical (**25**) to aromatic solvents (benzene, anisole, and benzonitrile) were isolated. The *ortho*-quinone methide **26** was also formed,

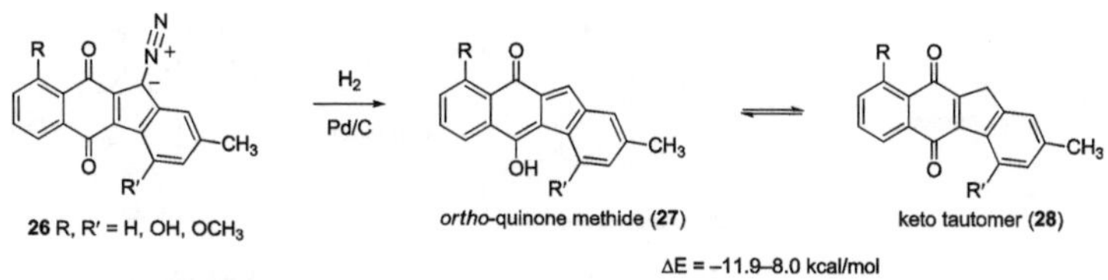

Scheme 3.3 Hydrogenation of prekinamycin analogs by Skibo and Khdour

presumably by hydrogen atom abstraction by **25**. In addition, alkylation products derived from the nucleophilic addition to the indicated position of **26** were isolated. Accordingly, it was proposed that the kinamycins and lomaiviticins may exert their cytotoxic effects by generation of free radical or electrophilic *ortho*-quinone methide intermediates.

In a separate study, Skibo and Khdour examined the reduction of prekinamycin as well as a series of related analogs (Scheme 3.3) [17]. It was established that catalytic hydrogenation of prekinamycin derivatives (**26**) forms *ortho*-quinone methide intermediates such as **27**. Interestingly, these *ortho*-quinone methides were found to be in equilibrium with their keto tautomers (**28**), and the position of equilibrium was shown to be substituent dependant. Prekinamycin analogs that exhibited the most stable *ortho*-quinone methide tautomers were found to be the most cytotoxic, leading the authors to suggest these intermediates may be involved in the biological effects of the kinamycins and lomaiviticins.

Several additional studies have shed light on the potential mechanism of action of the kinamycins and lomaiviticins. Melander and coworkers have shown that the kinamycins cleave DNA in the presence of a reducing cofactor [18] and that this cleavage appears to be non-sequence specific, suggesting a free radical mechanism may be operative. In addition, this group has shown that simple diazofluorenes are able to cleave dsDNA [19], which suggests the diazofluorene may be a general DNA-cleaving pharmacophore. Hasinoff and Dmitrienko have suggested that the cytotoxicity of the kinamycins may be due to generation of ROS in vivo [20]. In addition, they have shown that the production of cyclin D3 is diminished in cells treated with kinamycin F (**6**) [20], leading them to suggest the kinamycins may target a protein involved in the cyclin D3 production pathway.

3.4 Biosyntheses of the Kinamycins

Gould and coworkers have extensively studied the biosyntheses of the kinamycins, and this work was recently reviewed [5a]. Feeding studies established that the carbocyclic skeletons of the kinamycins are constructed from 10 equivalents of *S*-acetyl coenzyme A, and the pathway shown in Scheme 3.4 was proposed. The pathway begins with formation of the natural product dehydrorabelomycin (**29**). A novel ring contraction then occurs to form the cyclopentadienone **30**. Feeding studies with *N*-15-ammonium sulfate established that the diazo functional group is then installed

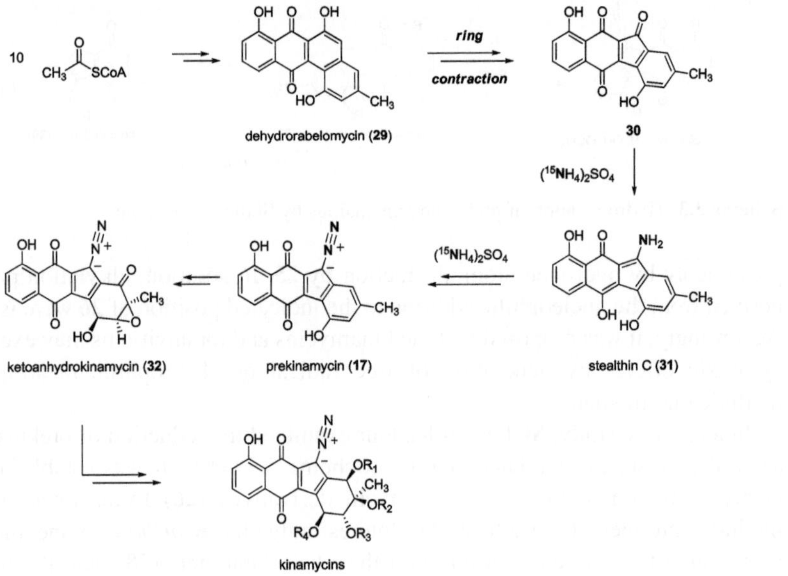

Scheme 3.4 Biosyntheses of the kinamycins

by a two-step process, proceeding via the metabolite stealthin C (**31**) [21]. Although it is clear that the diazo forms from two equiv of ammonia, the mechanism of the oxidation of **31** to **17** is not known. Dearomatization of the D-ring then generates ketoanhydrokinamycin (**32**) [6]. Finally, a sequence of oxidation state changes, rearrangements, and acyl transfer reactions generates the kinamycins. A biomimetic approach to the kinamycins, which hinges on the manipulation of synthetic constructs resembling ketoanhydrokinamycin (**32**), has been described by Dmitrienko and coworkers [22].

3.5 Syntheses of the Kinamycins

Prior to our work in this area, three groups reported syntheses of various kinamycins. These are presented in chronological order below. A large number of synthetic studies toward the lomaiviticins have been reported, although these are not discussed here [23].

3.5.1 Synthesis of (−)-Kinamycin C [24]

Porco and Lei reported the first synthesis of (−)-kinamycin C (**3**) [24]. Their route constitutes the first completed pathway to any of the kinamycins and provides several powerful insights into the strategies that are viable for construction of

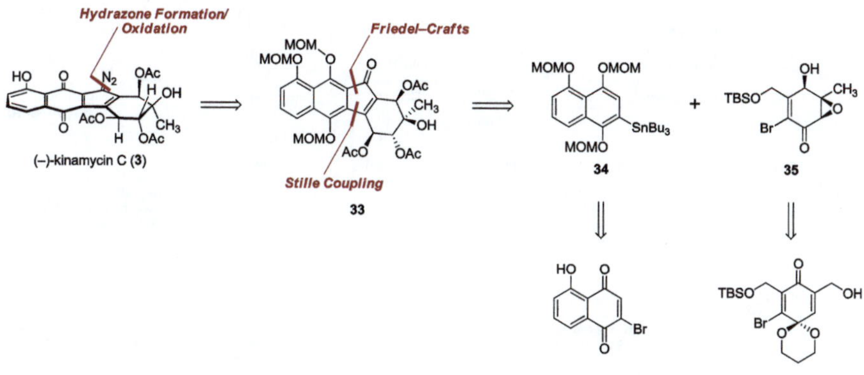

Scheme 3.5 Retrosynthesis of (−)-kinamycin C (**3**) by Porco and Lei

these molecules. Porco's retrosynthetic analysis is shown in Scheme 3.5. The target **3** was deconstructed to the tetracyclic ketone **33** by manipulation of the oxidation state of the naphthoquinone and cleavage of the diazo function. In the forward sense, it was envisioned that the diazo group may arise by formation of the hydrazone of the ketone **33**, followed by oxidation. The tetracyclic skeleton of the ketone **33** was deconstructed by Friedel–Crafts acylation and Stille coupling to form the aryl stannane **34** and the vinyl bromide **35** as synthetic precursors. The aryl stannane **34** was envisioned to arise from 2-bromojuglone (**36**). The vinyl bromide was derived from the acetal **37**, itself ultimately prepared from an aromatic precursor.

Porco's route to (−)-kinamycin C (**3**) began with 2,5-dihydroxybenzaldehyde (**38**), which was elaborated to the enone **35** by the sequence shown in Scheme 3.6. Regioselective bromination [25] followed by methylation and reduction of the aldehyde function afforded the primary alcohol **39**. The alcohol **39** was dearomatized by treatment with bis(acetoxy)iodobenzene, to afford the quinone monoketal **41**. Transketalization with 1,3-propanediol followed by silylation of the primary alcohol generated the silyl ether **42** in 72 % yield over three steps.

In the next step of the sequence, the authors sought to introduce a hydroxymethylene substituent at the unsubstituted α-position of the enone. This bond construction can be carried out by conducting a Baylis–Hillman reaction with formaldehyde. In this instance, the authors used a modification of the Baylis–Hillman reaction which involves the use of a Lewis acid to activate the enone [26]. Under these conditions, the enone **42** is treated with excess paraformaldehyde in the presence of triethylphosphine (1 equiv), lanthanum triflate (5 mol%), and triethanolamine (50 mol%). It is proposed that the lanthanum triflate forms a complex with the triethanolamine. This complex is able to activate the enone toward 1,4-addition of the nucleophilic catalysts (here, triethylphosphine). In the absence of triethanolamine, the Lewis acid catalyst undergoes nonproductive complexation with the nucleophilic catalyst, leading to diminution of catalysis. Under these conditions, the hydroxymethylene derivative **37** was formed in 70 % yield. In the next step of the sequence, the authors sought to conduct a stereoselective epoxidation of the allylic

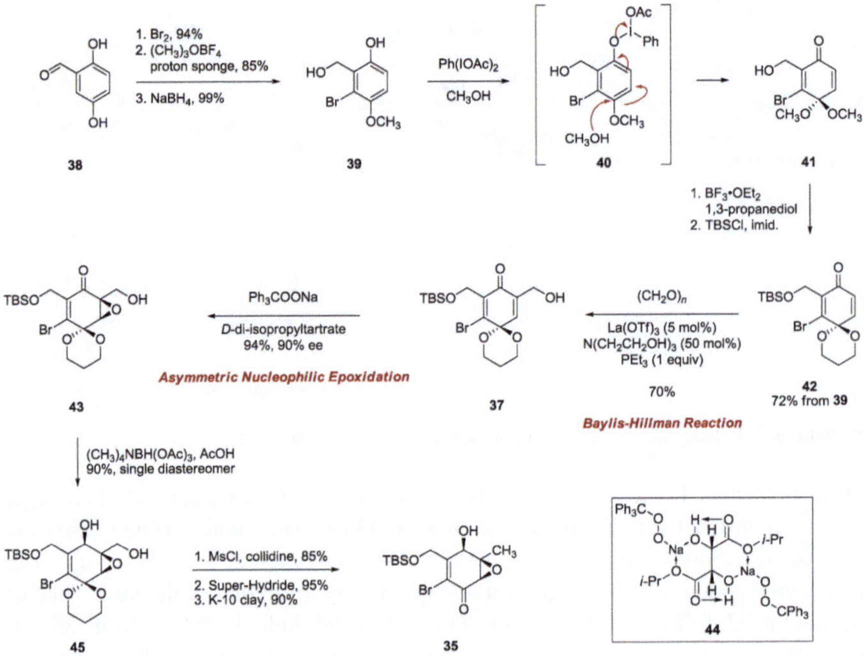

Scheme 3.6 Porco's synthesis of the enone **35**

alcohol function of **37**. Although classic Sharpless conditions afforded product, the enantiomeric excess was only 70 %. To remedy this problem, the authors applied their method of asymmetric nucleophilic epoxidation [27], which entails formation of the sodium salt of tritylperoxide and complexation with D-di-*iso*-propyl tartrate. It is proposed that under these conditions, a 2:1 complex between sodium trityl peroxide and the tartrate ester is formed (see structure **44**) [23b]. This creates an asymmetric environment that directs the stereoselectivity of the epoxidation reaction. Under these conditions, the epoxy alcohol **43** is obtained in excellent enantioselectivity (90 %) and yield (94 %). Hydroxyl-directed reduction of the ketone function of **43** then afforded the alcohol **45** as a single diastereomer (90 %). Finally, an efficient three-step sequence comprising mesylation of the primary alcohol function, reduction using lithium triethylborohydride (Super-Hydride), and cleavage of the acetal afforded the enone **35** in 73 % overall yield.

The synthesis of the second Stille coupling partner **34** was efficiently achieved in three steps. First, 2-bromojuglone (**36**) [28] was protected as its methoxymethyl ether (**46**, Scheme 3.7). The quinone was reduced using sodium thiosulfate, and the resulting hydroquinone was protected with methoxymethyl chloride to afford the arene **47**. Finally, stannylation using tetrakis-(triphenylphosphine)palladium and hexabutylditin [29] afforded the cross-coupling partner **34** in high yield.

Porco's pathway to complete the synthesis of (−)-kinamycin C (**3**) is shown in Scheme 3.8. The arylstannane **34** and the α-bromoenone **35** were efficiently coupled by a Stille reaction using tris(dibenzylideneacetone)dipalladium and triphenylarsine

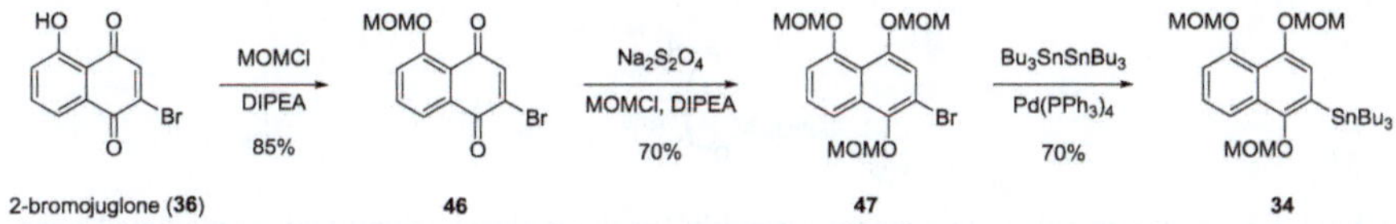

Scheme 3.7 Porco's synthesis of the arylstannane **34**

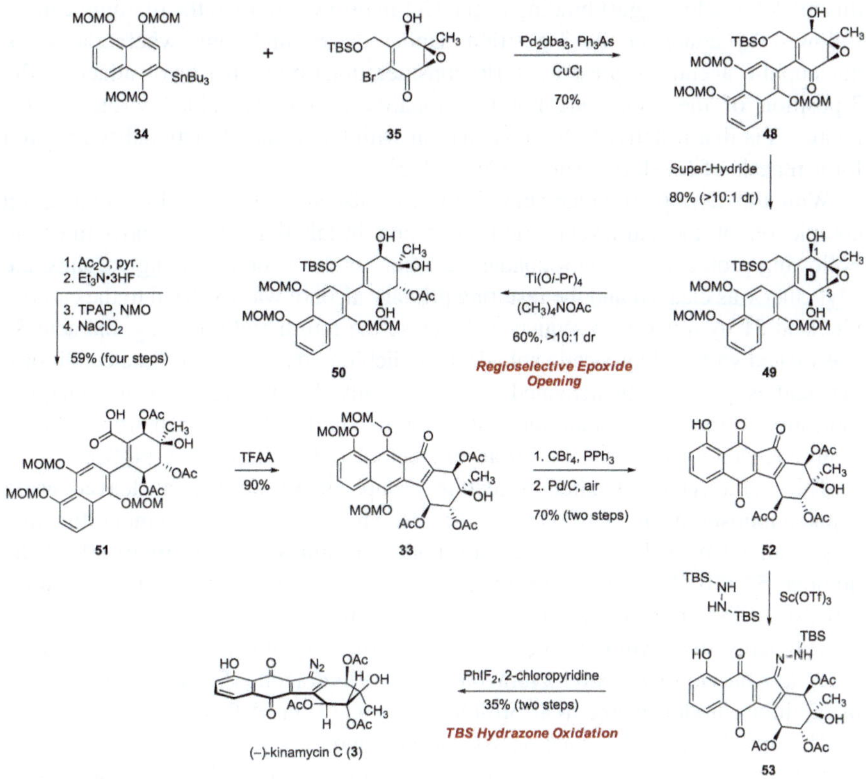

Scheme 3.8 Completion of (−)-kinamycin C (**3**) by Porco and Lei

(70 %). The coupling product (**48**) was reduced with high diastereoselectivity using Super-Hydride, to afford the diol intermediate **49** (80 %). To establish the tertiary alcohol function of the kinamycin D-ring, the epoxide ring of **49** was opened regioselectively using tetramethylammonium acetate and titanium tetraisopropoxide as mediator, to afford the acetate **50** in good yield and high diastereoselectivity.

These epoxide-opening conditions were originally developed by Sharpless and coworkers for the regiocontrolled opening of 2,3-epoxy alcohols [30]. It has been proposed that ligand exchange of the substrate with isopropoxide forms a covalently bound substrate-titanium complex (Chart 3.3). Nucleophilic attack on this complex at the 3-position is favored over attack at the 2 position. In the case of **49**,

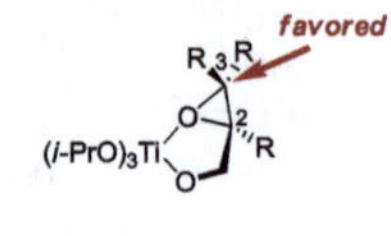

Chart 3.3 Proposed substrate–reagent assembly for the regioselective titanium-mediated opening of epoxyalcohols

this model would suggest binding of the C-1 hydroxyl group to the titanium center, followed by attack at the 3-position. Under these conditions, where the weak nucleophile acetate is present, steric considerations may also favor attack at the 3-position of the epoxy alcohol functionality of **49**. An added virtue of this approach is that it delivers the oxygenation with the acetate functionality required for synthesis of (−)-kinamycin C (**3**) in place.

With the D-ring of kinamycin C (**3**) established, Porco and Lei focused on completion of the tetracyclic framework and installation of the diazo function. Following protection of the secondary alcohols as their corresponding acetates, the silyl ether was cleaved and the resulting primary alcohol was oxidized to the carboxylic acid **51** by a two-step sequence. To close the fourth and final ring, the acid **51** was treated with trifluoroacetic anhydride in dichloroethane. Under these conditions, the acid is presumably activated to form a mixed anhydride, which undergoes ionization to form an acylium ion that is trapped by the electron-rich aromatic ring. Rearomatization then forms the tetracyclic ketone (**33**) in high yield (90 %).

The methoxymethyl ether protecting groups of **33** were then cleaved using triphenylphosphine and carbon tetrabromide. The resulting hydroquinone function was oxidized by palladium on carbon under an atmosphere of air to afford the quinone **52** (70 %). A two-step procedure was implemented to install the diazo function. First, the ketone function of **52** was condensed with *N*,*N′*-bis(*tert*-butyldimethylsilyl)hydrazine in the presence of scandium triflate, which formed the *N*-*tert*-butyldimethylsilyl hydrazone **53**. The hydrazone (**53**) was then oxidized using difluoroiodobenzene to afford kinamycin C (**3**) in 35 % yield.

This silyl hydrazone formation–oxidation sequence was originally developed as a practical alternative to the synthesis and oxidation of unsubstituted hydrazones by Myers and Furrow [31]. The formation of hydrazones directly from hydrazine and ketones is invariably complicated by azine formation. In contrast, silyl hydrazones can be formed cleanly from *N*,*N′*-bis(*tert*-butyldimethylsilyl)hydrazine and aldehydes and ketones with nearly complete exclusion of azine formation. The resulting silylhydrazones undergo many of the reactions of conventional hydrazones (Wolff–Kishner reduction, oxidation to diazo intermediate, formation of geminal and vinyl iodides) with equal or greater efficiency. It is also noteworthy that application of hydrazine in this setting may also have led to cleavage of the acetate substituents.

Porco's synthesis of (−)-kinamycin C (**3**) constituted the first reported route to any of the diazofluorene antitumor antibiotics. This synthesis invokes several powerful transformations, including a modified Baylis–Hillman reaction, a catalyst-controlled asymmetric nucleophilic epoxidation, and a regioselective epoxide opening to establish the D-ring of the kinamycins. The tetracyclic skeleton was constructed by an

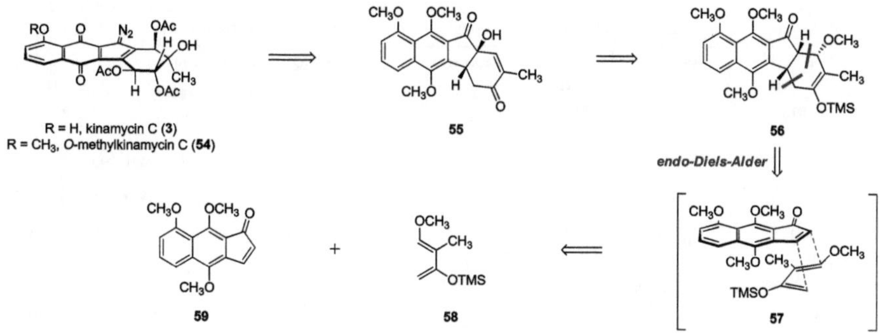

Scheme 3.9 Ishikawa's retrosynthesis of (±)-*O*-methylkinamycin C (**54**)

efficient Stille coupling–Friedel–Crafts sequence. Finally, the first successful method for synthesis of the diazofluorene function was established.

3.5.2 *Synthesis of (±)-O-Methyl-Kinamycin C [32]*

Ishikawa and coworkers reported a synthesis of (±)-*O*-methylkinamycin C (**54**) [32, 33]. Their retrosynthetic analysis is shown in Scheme 3.9. It was envisioned that **54** could be derived from the dihydroindanone **55** by D-ring oxygenation, installation of the diazo substituent, and oxidation of the protected hydroquinone function. The dihydroindanone **55** was envisioned to arise from the enol ether **56**, itself formed from an *endo* Diels–Alder reaction between the indenone **59** and the diene **58**.

The synthesis of the indenone **59** is shown in Scheme 3.10 [33]. Beginning with 1,5-dihydroxynapthalene (**60**), acylation with acetic anhydride formed 1, 5-diacetoxynapthalene (**61**) in 89 % yield. Oxidation using *N*-bromosuccinimide in a mixture of acetic acid–water formed [28, 34] *O*-acetyl-2-bromojuglone (**62**) in 84 % yield. The methyl ether **63** was then obtained by a two-step sequence involving hydrolysis of the acetate and methylation with iodomethane in the presence of silver oxide (65 %). The quinone function of **63** was reduced with tin chloride, and the resulting hydroquinone was methylated to form the aryl bromide **64** (77 %). A formyl group was then installed by lithium–halogen exchange followed by trapping with *N,N*-dimethylformamide (84 %). The aldehyde product (**66**) was then condensed with malonic acid to afford, after decarboxylation, the α,β-unsaturated carboxylic acid **67** in high yield (93 %). Finally, the olefin of **67** was reduced (H_2, Pd/C, 93 %) and the resulting product cyclized using phosphoric pentoxide in the presence of methanesulfonic acid (79 %). The resulting dihydroindanone (**69**) was oxidized by heating with IBX in DMSO (77 %) [35].

The key Diels–Alder cycloaddition was effected by treating a mixture of the indenone **59** and the diene **58** with zinc chloride in dichloromethane at −15 °C (Scheme 3.11). Under these conditions, the expected *endo*-adduct **56** was formed.

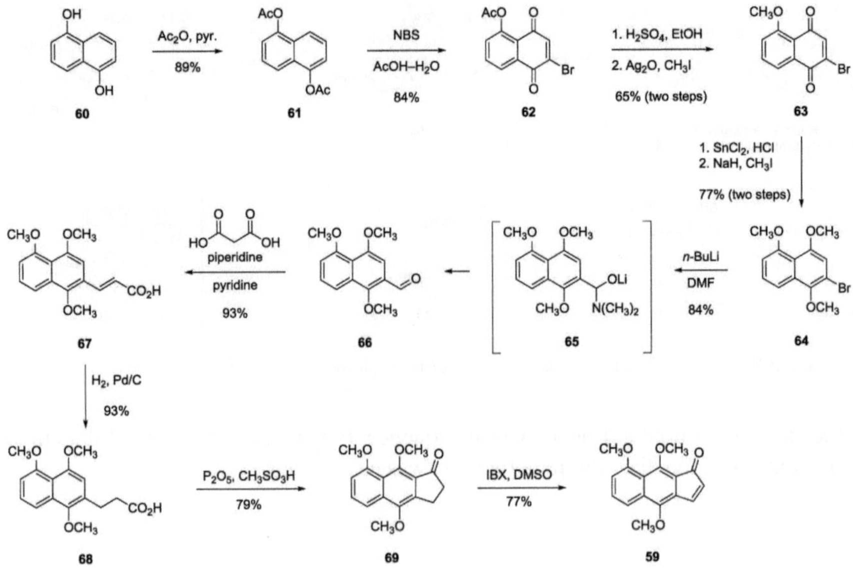

Scheme 3.10 Ishikawa's synthesis of the indenone **59**

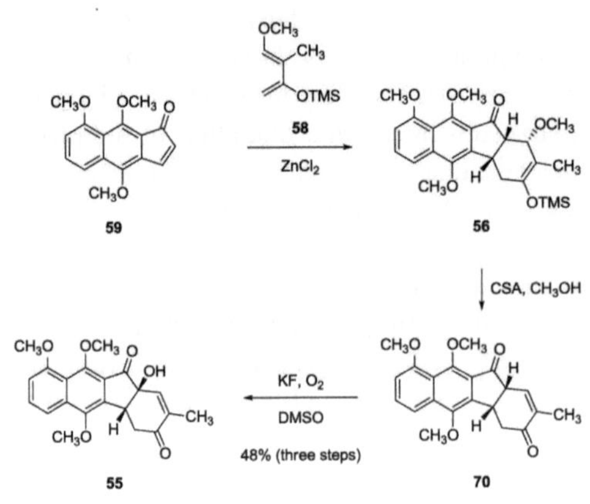

Scheme 3.11 Ishikawa's synthesis of the unsaturated ketone **55**

The adduct (**56**) was hydrolyzed (CSA, CH_3OH) to afford the unsaturated ketone **70**. Finally, the unsaturated ketone **70** was regioselectively oxygenated by treatment with potassium fluoride (0.1 equiv) in methylsulfoxide under an atmosphere of air. This sequence generated the tertiary alcohol **55** in 48 % yield over three steps.

Ishikawa's endgame toward of **54** is shown in Scheme 3.12. First, the allylic alcohol function was oxidized by a substrate-directed dihydroxylation reaction, as developed by Donohoue and coworkers (66 % yield) [36]. This reaction is conducted using 1 equiv each of osmium tetroxide and tetramethylethylene diamine (TMEDA) and provides a method to obtain the *syn*-dihydroxylation product in the

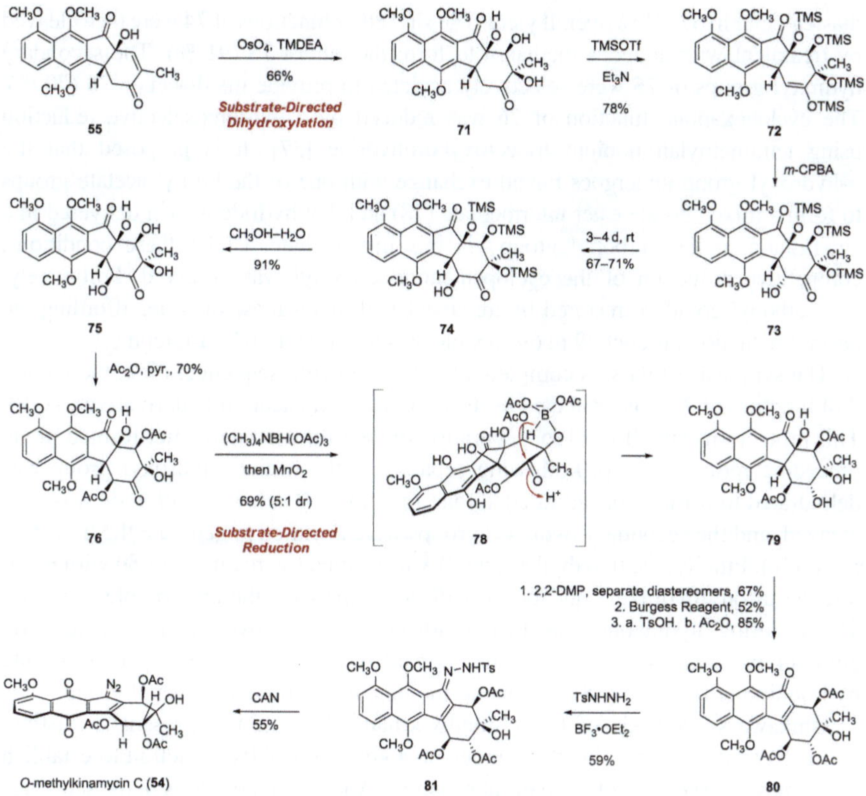

Scheme 3.12 Completion of (±)-*O*-methylkinamycin C (**54**) by Ishikawa and coworkers

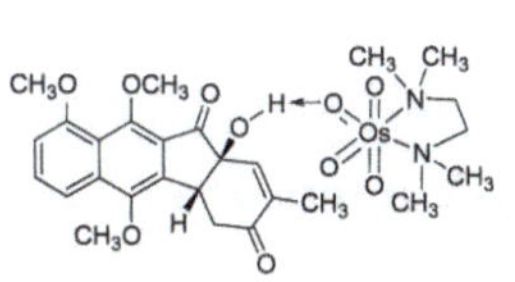

Chart 3.4 Proposed substrate–reagent assembly for the diastereoselective dihydroxylation of **55**

oxidation of cyclic allylic alcohols. Under these modified conditions, it is proposed that a 1:1 complex between osmium tetroxide and TMEDA forms (Chart 3.4). Coordination of the diamine to the osmium renders the oxo ligands more electron rich, and these ligands can hydrogen-bond to the hydroxyl group of the allylic alcohol function. A drawback of this method is that this interaction also stabilizes the product osmium complex, which prevents turnover. The dihydroxylation product **71** was then persilylated to form the enoxysilane **72** (78 %). Rubottom oxidation of **72** then generated the α-hydroxyketone **73**. Surprisingly, the oxidation was found to occur with selectivity for the undesired α-oxidation product. Fortunately, however, this diastereomer epimerized to the desired β epimer **74** on standing, to provide the desired diastereomer in 67–71 % overall yield. The silyl ether functions of **74** were then cleaved by treatment with aqueous methanol to form the tetraol **75** (91 %). The secondary hydroxyl groups of **75** were selectively acylated to provide the diacetate **76** (70 %). The cyclohexanone function of **76** was reduced by a diastereoselective reduction using tetramethylammonium triacetoxyborohydride [37]. It is proposed that the α-hydroxyl group undergoes ligand exchange with one of the bornyl acetate groups to form a mixed borate ester intermediate (**78**) and that hydride is then delivered in a *syn*-fashion to the carbonyl group via this intermediate. Under these conditions, competitive reduction of the cyclopentanone carbonyl was observed. Fortunately, this carbonyl could be restored by treatment with manganese dioxide, affording the desired reduction product **79** in 69 % yield as a 5:1 mixture of diastereomers.

The synthesis of **54** was completed by the following sequence. First, the vicinal diol function of **79** was protected as the corresponding acetonide derivative (67 %). This protection step allowed for separation of the minor diastereomer formed in the preceding reduction step. Following separation, the tertiary hydroxyl group was dehydrated to form an unsaturated ketone function (52 %). The acetonide was then cleaved, and the secondary hydroxyl groups were acylated to generate the triacetate **80** (85 %). Finally, the tosylhydrazone **81** was formed by treatment of **80** with *para*-toluenesulfonylhydrazine and boron trifluoride diethyl etherate complex (59 %). The resulting hydrazone was then oxidized to the diazo function using ceric ammonium nitrate (55 %). This also served to cleave the methyl ethers of the hydroquinone function [38] and install the naphthoquinone functional group.

Ishikawa's synthesis of (±)-*O*-methylkinamycin C (**54**) represents a distinct approach to the kinamycins that hinges on a key Diels–Alder reaction to establish the tetracyclic skeleton of the natural products. Additional key steps in the sequence include a substrate-directed dihydroxylation, substrate-directed reduction, and spontaneous epimerization of an α-hydroxyketone intermediate.

3.5.3 Syntheses of (−)-Kinamycins C, F, and J [39]

Nicolaou and coworkers reported efficient enantioselective syntheses of (−)-kinamycin C (**3**), (−)-kinamycin F (**6**), and (−)-kinamycin J (**10**) [39]. Nicolaou's retrosyntheses of these targets are shown in Scheme 3.13. The authors envisioned that all three metabolites could be accessed from the common precursor **82**. The α-hydroxyketone function of **82** was envisioned to arise from an intramolecular benzoin reaction of the ketoaldehyde **83**. This key bond disconnection would serve to forge the cyclopentyl ring of the kinamycin skeleton. The ketoaldehyde **83** was deconstructed by an Ullmann coupling of the aryl bromide **84** and the α-iodoenone **85**. The latter were anticipated to arise from the bromojuglone derivative **86** and the enantiomerically enriched enone **87**, respectively.

Nicolaou prepared the aryl aldehyde **84** by an efficient five-step sequence, as shown in Scheme 3.14. Beginning with 2-bromojuglone (**36**), radical allylation afforded the allylquinone **88** (75 %). Benzylation of the phenol group followed by

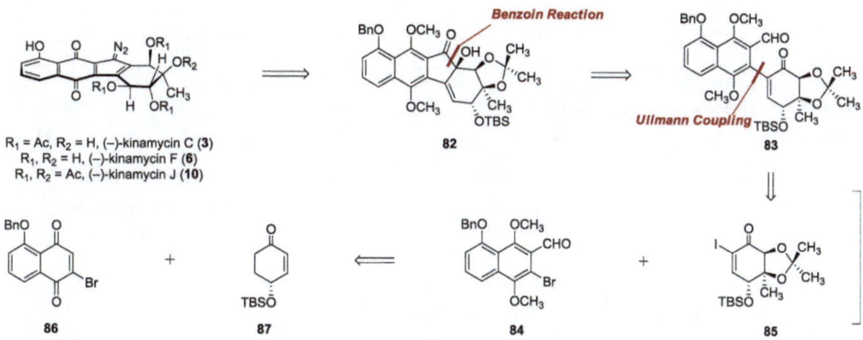

Scheme 3.13 Nicolaou's retrosynthesis of (−)-kinamycins C, F, and J (**3**, **6**, **10**)

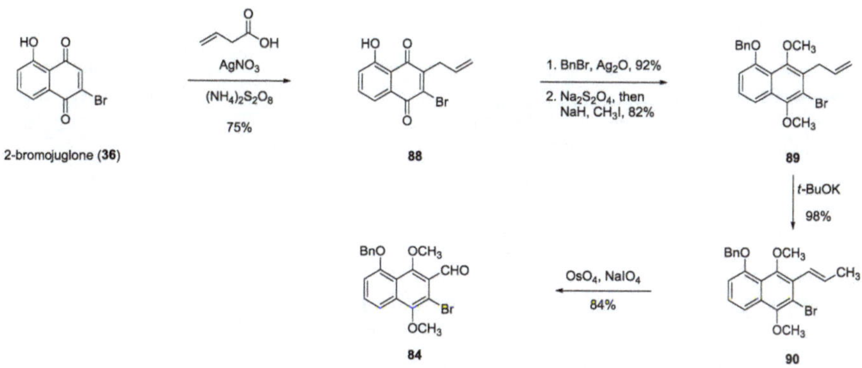

Scheme 3.14 Nicolaou's synthesis of the aryl aldehyde **84**

reduction of the quinone and subsequent methylation afforded the naphthalene **89** in excellent yield. Isomerization of the alkene to the internal position was effected by treatment with potassium *tert*-butoxide, to afford the styrene derivative **90** in nearly quantitative yield. Finally, oxidative cleavage of the olefin of **90** yielded the aryl aldehyde **84** (84 %).

Nicolaou's synthesis of the α-iodoenone **85** is shown in Scheme 3.15. The synthesis began with 4-(*tert*-butyldimethylsilyloxy)cyclohexenone (**91**). This achiral intermediate was desymmetrized by enantioselective deprotonation with a chiral amide base [40]. Trapping of the resulting enolate with chlorotrimethylsilane, followed by oxidation, afforded the enone **87** in >80 % *ee*. Copper-catalyzed 1,4-addition of methyl magnesium bromide, trapping with chlorotrimethylsilane, and reoxidation then generated the β-methylcyclohexenone **93** in excellent yield (90 %). This product (**93**) was subjected to a diastereoselective catalytic dihydroxylation to afford the diol **94** (76 %). Here, the osmium catalyst approaches the substrate from the less-hindered face, opposite the *tert*-butyldimethylsilyloxy substituent. The diol **94** was readily recrystallized to >98 % *ee*. Protection of the diol function afforded the acetonide **95** (95 %). Generation of the enoxysilane of **95** followed by oxidation then

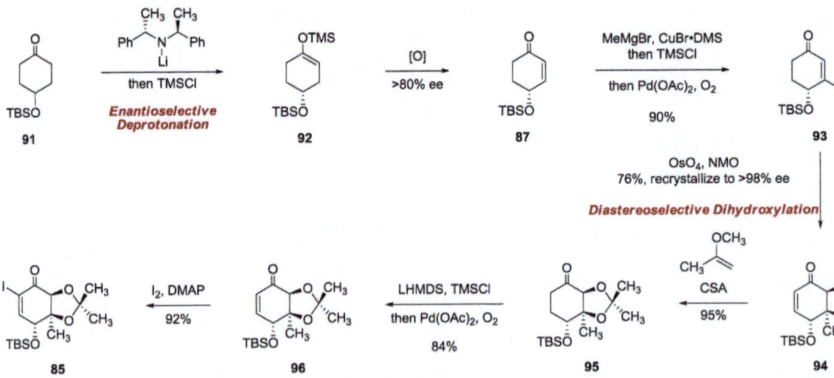

Scheme 3.15 Nicolaou's synthesis of the α-Iodoenone **85**

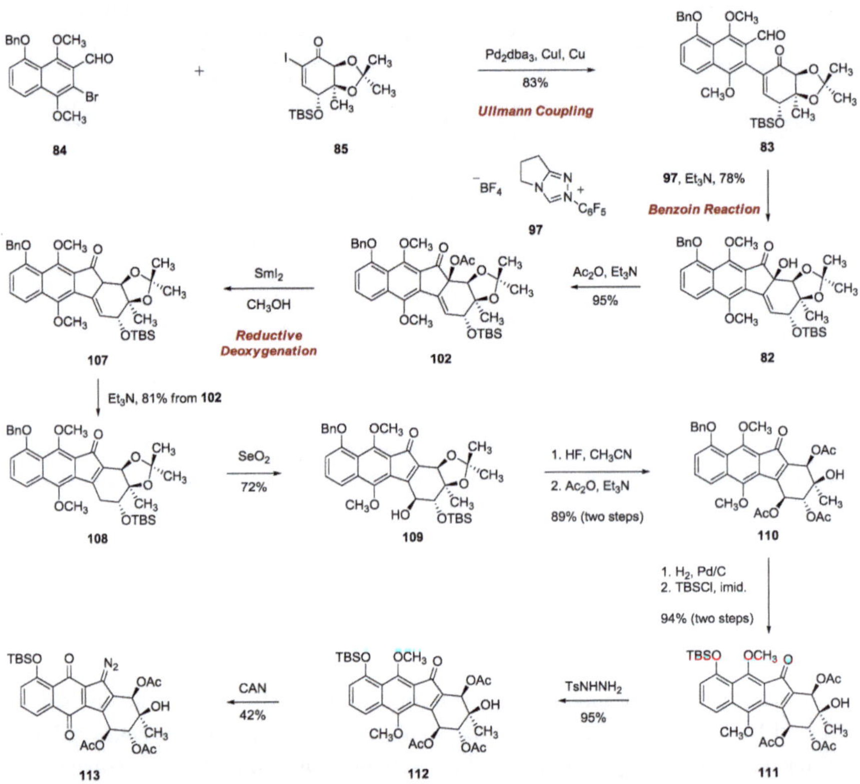

Scheme 3.16 Nicolaou's synthesis of the diazofluorene **113**

afforded the enone **96** in 84 % yield. Finally, the enone **96** was iodinated under Johnson conditions [41] to afford the target α-iodoenone **85** in 92 % yield.

The coupling of the aryl aldehyde **84** and the α-iodoenone **85** was effected by a modified Ullmann reaction (83 %, Scheme 3.16) [42]. The resulting arylation

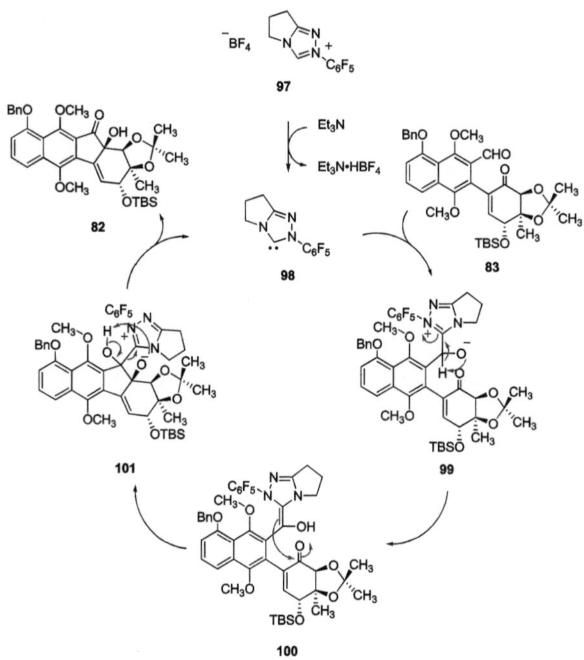

Scheme 3.17 Potential mechanism for the carbene-catalyzed benzoin reaction

product **83** was subjected to a benzoin condensation using the triazole catalyst **97** [43]. Under these conditions, a nucleophilic carbene (**98**) is generated by deprotonation of **97** with triethylamine (Scheme 3.17). This carbene undergoes addition to the aldehyde function of **83** to afford the 1,2-addition product **99**. Proton transfer affords the diaminophenol **100**, which undergoes 1,2-addition to the adjacent ketone function. The adduct (**101**) then collapses to the product **82** by proton transfer and elimination of the carbene catalyst. This reaction is highly efficient and generates the benzoin product **82** in 78 % yield. Acylation of the tertiary alcohol function of **82** then afforded the acetate **102**.

The acetate function of **98** was then cleaved by treatment with samarium diiodide in methanol in high yield (81 %) [44]. A potential mechanism for this transformation is shown in Scheme 3.18. Reduction of the ketone function forms a samarium ketyl radical (**103**). Transfer of a second electron forms a carbanion (**104**) which undergoes β-elimination of acetate to generate the samarium enolate **105**. Protonation and tautomerization then affords the observed product **107**.

Treatment of the elimination product **107** with triethylamine resulted in smooth isomerization of the olefin, to afford the α,β-unsaturated ketone **108**. Allylic oxidation of **108** then generated the secondary alcohol **109** in 72 % yield. The acetonide and silyl ether functions of **109** were cleaved in one reaction to afford a tetraol intermediate that was regioselectively acylated at the secondary alcohol functions, to provide the triacetate **110** in high yield (89 %). Hydrogenolysis of the benzyl ether

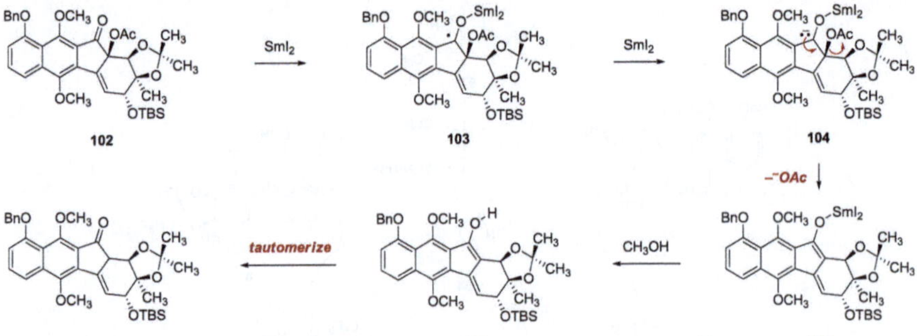

Scheme 3.18 Potential mechanism for the samarium-mediated deoxygenation reaction

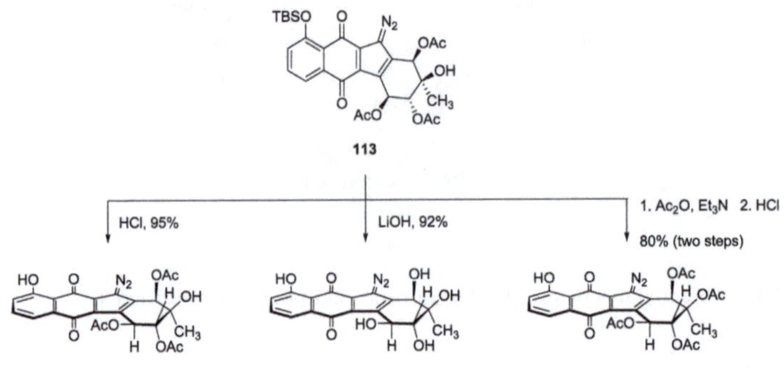

Scheme 3.19 Transformation of **113** to (−)-kinamycins C, F, and J (**3**, **6**, **10**)

of **110** followed by silylation of the resulting phenol afforded the key silyl ether **111** (94 %). Finally, condensation of **111** with *para*-toluenesulfonylhydrazine afforded a hydrazone (**112**, 95 %) that was oxidized using ceric ammonium nitrate, which installed the diazo function and concurrently created the quinone (42 %).

The elaboration of **113** to (−)-kinamycins C, F, and J, is shown in Scheme 3.19. To access (−)-kinamycin C (**3**), the silyl ether function of **113** was cleaved with aqueous hydrochloric acid (95 %). Alternatively, treatment of **113** with lithium hydroxide served to liberate the phenol function and saponify the three acetate esters, to provide (−)-kinamycin F (**6**) in 92 % yield. Finally, acylation of the tertiary hydroxyl of **113** (acetic anhydride, triethylamine) afforded a tetraacetate. Cleavage of the silyl ether then provided (−)-kinamycin J (**10**) in 80 % over two steps.

Nicolaou's work features the syntheses of three kinamycins from a common late-stage intermediate (**113**). Several noteworthy transformations are implemented in the syntheses, including a desymmetrizing deprotonation reaction to establish absolute stereochemistry, a modified Ullmann coupling to build the carbon frameworks of the targets, an intramolecular benzoin condensation to forge the tetracyclic skeleton, and a highly efficient reductive deoxygenation of an advanced intermediate.

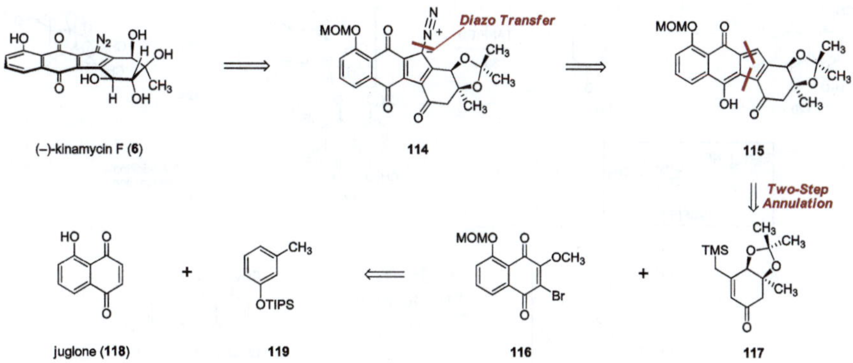

Scheme 3.20 Our retrosynthesis of (–)-kinamycin F (**6**)

3.5.4 Synthesis of (–)-Kinamycin F [45]

Our retrosynthesis of (–)-kinamycin F (**6**) is shown in Scheme 3.20 [45]. It was envisioned that (–)-kinamycin F (**6**) could be prepared from the protected diazofluorene **114** by conversion of the ketone function of **114** to a *trans*-1,2-diol, followed by deprotection of the acetonide and methoxymethyl ether protecting groups. The diazofluorene **114** was envisioned to arise from diazo transfer to the hydroxyfulvene **115**. The cyclopentadiene substructure of **115** was deconstructed by a two-step annulation sequence, to provide the bromoquinone **116** and the β-trimethylsilylmethyl unsaturated ketone **117**. The latter two intermediates were prepared from juglone (**118**) and the silyl ether **119**, respectively.

The two-step annulation procedure was first developed using a simplified model system (Scheme 3.21). The readily available dibromonaphthoquinone **120** [46] was used as a substitute for the kinamycin quinone and ultimately was expected to potentially provide access to the oxygenation pattern required for lomaiviticin synthesis. The β-trimethylsilylmethyl unsaturated ketone **121** was utilized as a model for the coupling partner **117**. β-Trimethylsilylmethyl-α,β-unsaturated ketones such as **121** are readily prepared from α,β-unsaturated ketones by a two-step sequence involving copper-mediated 1,4-addition of trimethylsilylmethyl magnesium chloride and trapping of the resulting enolate with chlorotrimethylsilane, followed by oxidation of the resulting enol ether (**122**, Scheme 3.21, bottom).

It was found that treatment of a mixture of **120** and **121** with tris(diethylaminosulfonium) trimethyldifluorosilicate [TASF(Et)] resulted in smooth addition–elimination to the naphthoquinone to form the γ-alkylation product **125** (85 %). TASF(Et) is a convenient source of soluble, anhydrous fluoride ion [47]. It is believed that exposure of **121** to TASF(Et) results in fluoride transfer to generate a hypervalent silicate anion, as depicted in structure **124**. The transfer of fluoride between TASF(Et) and **121** may be driven by stabilization of the anionic species **124** by delocalization of the carbon–silicon bond into the LUMO of the unsaturated ketone. 1,4-Addition–elimination of this species to the naphthoquinone **120** would then form the observed product.

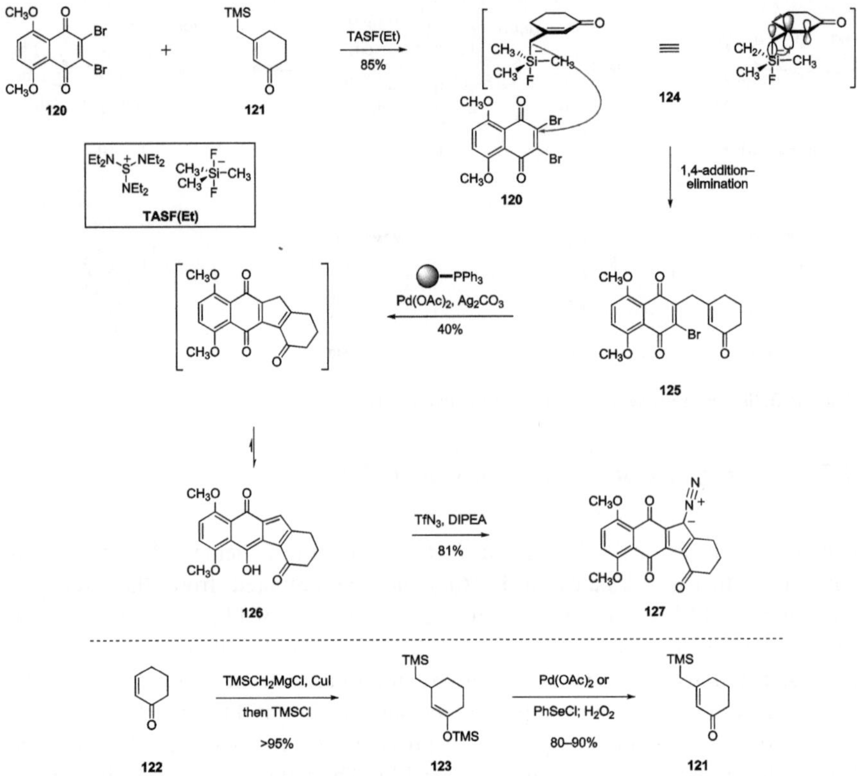

Scheme 3.21 Development of the diazofluorene synthesis

In the second step of the annulation, the γ-alkylation product **125** is heated in the presence of palladium acetate, polymer-supported triphenylphosphine, and silver carbonate. Formally, this reaction results in extrusion of hydrogen bromide from **125** with formation of the final carbon–carbon bond of the five-membered ring, to provide the cyclized product **126** in 40 % yield. It was observed that the product existed predominately in the hydroxyfulvene tautomer (**126**). The mechanism of this reaction is not clear, but several possibilities may be envisioned (Scheme 3.22). First, a Heck-type pathway involving oxidative addition of palladium to the carbon–bromine bond, followed by olefin insertion and β-hydride elimination may be operative. Alternatively, a pathway involving oxidative addition, intramolecular transmetalation of an enolate species to the vinyl palladium intermediate, and reductive elimination may be operative [48]. Finally, a sequence involving deprotonation, electrocyclic ring closure, and elimination of bromide may occur, although electrocyclic ring closures of pentadienyl anions are relatively rare [49].

To complete the synthesis of the diazofluorene, the cyclization product **126** is treated with diisopropylethylamine and triflyl azide [50], to afford the diazofluorene **127** in 81 % yield. Under these conditions, the hydroxyfulvene is likely deprotonated to generate a cyclopentadienyl anion. Attack of this anion on the trifyl azide reagent

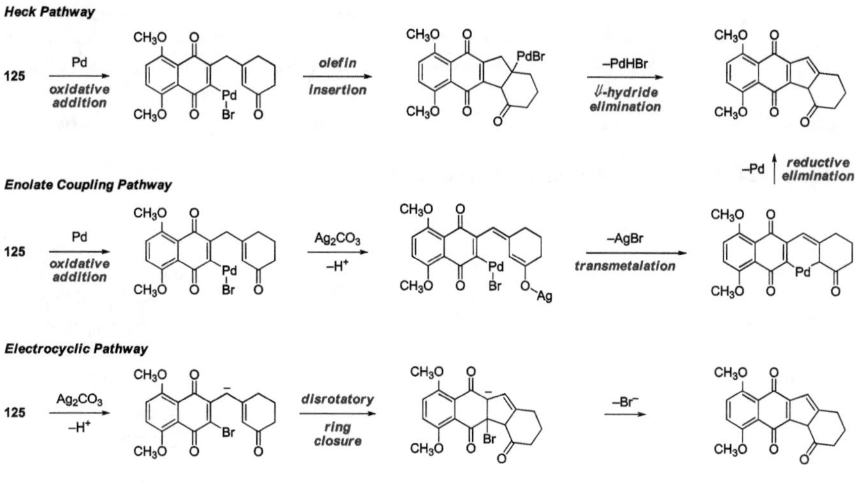

Scheme 3.22 Potential mechanisms for the palladium-mediated cyclization

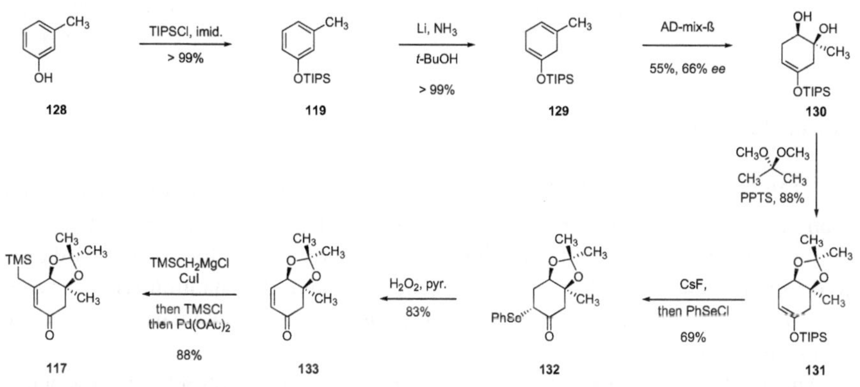

Scheme 3.23 Synthesis of the β-trimethylsilylmethyl-α,β-unsaturated ketone **117**

would provide the observed product **127**. It is noteworthy that an early application of *para*-toluenesulfonyl azide, a more common diazo transfer agent, was in the synthesis of diazocyclopentadiene itself from cyclopentadiene [51]. In the present case, *para*-toluenesulfonyl azide is not electrophilic enough to efficiently transform **126** to **127**. This is likely due to the carbonyl functional groups that surround the cyclopentadiene substructure of **126**, rendering the conjugate base less reactive.

With an effective strategy for construction of the diazofluorene established, we set out to prepare the coupling partners required for synthesis of (−)-kinamycin F (**6**). The synthesis of the enone **117** began with *meta*-cresol (**128**, Scheme 3.23). Silylation formed the silyl ether **119** in nearly quantitative yield. Birch reduction of the silyl ether **119** formed the cyclohexadiene derivative **129** in excellent yield. Asymmetric dihydroxylation [52] of **129** occurred regioselectively to afford the

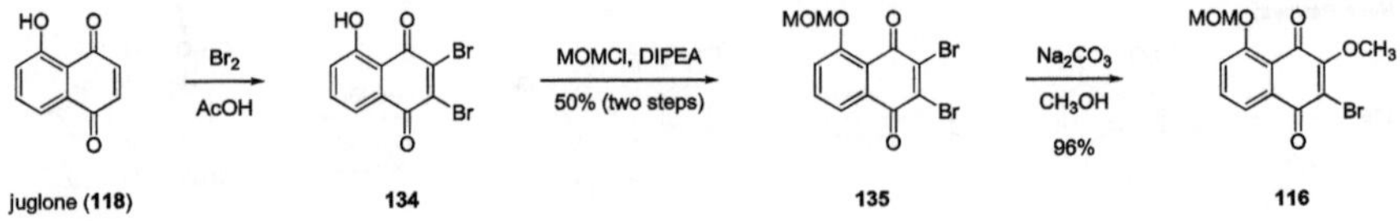

Scheme 3.24 Synthesis of the juglone derivative **116**

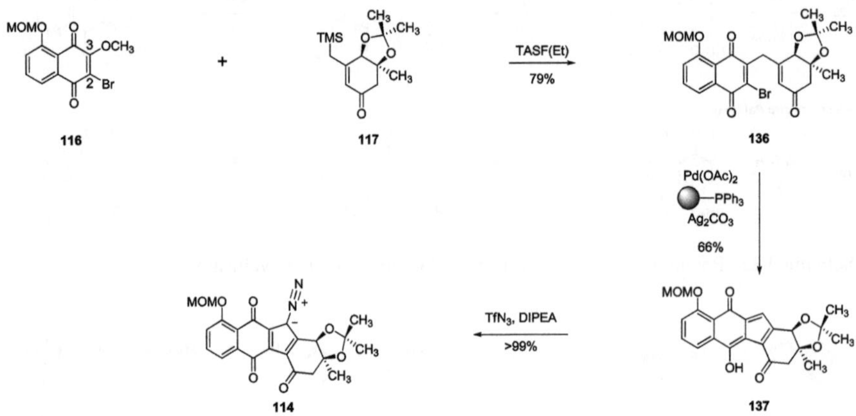

Scheme 3.25 Synthesis of the diazofluorene **114**

diol **130**. Although the dihydroxylation of electron-rich alkenes is typically faster than electron-neutral alkenes, the tri-*iso*-propylsilyl group effectively prevents the enoxysilane from entering the binding pocket of the catalyst, forcing the oxidation to occur at the normally less reactive alkene. Although the enantiomeric excess of the product was only modest (66 % *ee*), the subsequent intermediate **133** could be recrystallized to >95 % *ee*.

The diol function of **130** was protected as its acetonide **131** (88 %). Next, the enone function was installed by α-selenation of the enoxysilane, followed by peroxide oxidation and elimination (57 % over two steps). Finally, the unsaturated ketone **132** was homologated by 1,4-addition of trimethylsilylmethyl magnesium chloride, trapping with chlorotrimethylsilane, and reoxidation, to afford the target **117** (88 %).

The synthesis of the naphthoquinone **116** is shown in Scheme 3.24. Bromination of juglone (**118**) afforded the dibromojuglone derivative **134**. Protection of the phenol group as its methoxymethyl ether formed the product **135** (50 % yield over two steps). Finally, the C-3 bromide substituent was regioselectively substituted with methoxide by heating **135** in methanol in the presence of sodium carbonate (96 %). The methoxy group was installed to impart electronic bias to the naphthoquinone in the TASF(Et) coupling (vide infra).

The key intermediates **116** and **117** were coupled by the three-step sequence described above for the model system (Scheme 3.25). Thus, treatment of a mixture of **116** and **117** with TASF(Et) formed the coupling product **136** in 79 % yield. We rationalize the regioselectivity of the coupling step by considering the electronic

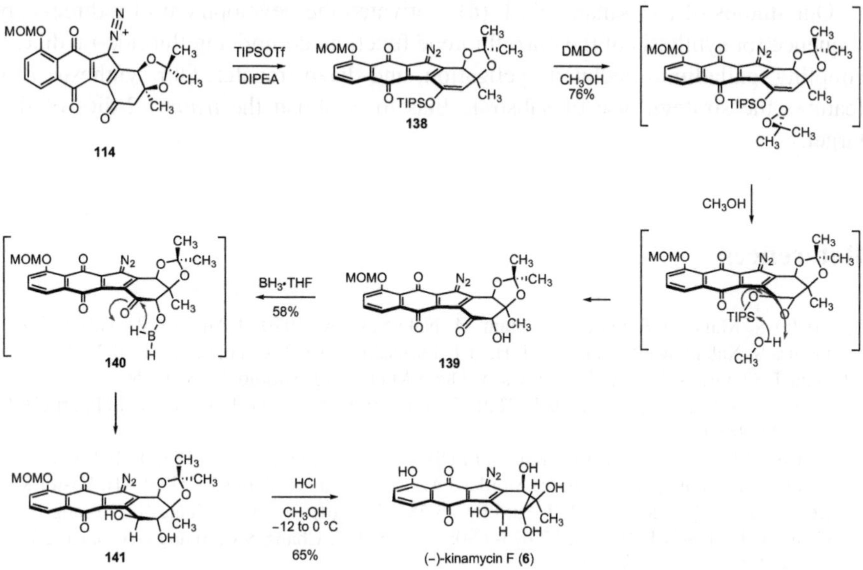

Scheme 3.26 Completion of the synthesis of (−)-kinamycin F (**6**)

bias of the naphthoquinone. Thus, the C-3 methoxy substituent donates electron density to the C-2 position, rendering this site less electrophilic and favoring *ipso*-substitution of the methoxy substituent. In support of this, attempted TASF(Et) coupling of **117** and the dibromojuglone **135** resulted in mixtures of products arising from competitive addition to the 2-position of the naphthoquinone. This strategy of controlling site-selectivity by introduction of an alkoxy substituent finds precedent in earlier mitomycin studies by Johnston and coworkers [53, 54]. Thermolysis of the coupling product **136** in the presence of palladium acetate and polymer-supported triphenylphosphine formed the hydroxy fulvene **138** (66 %). Finally, diazo transfer to the hydroxyfulvene generated the diazofluorene **114** (>99 %).

To complete the synthesis of (−)-kinamycin F (**6**), the ketone function of the diazofluorene **114** had to be elaborated to a *trans*-1,2-diol and the protecting groups needed to be removed. The steric bias of the substrate **114** was utilized to control the selectivity in the synthesis of the *trans* 1,2-diol. Thus, the enoxysilane of **114** was generated by treatment with tri-*iso*-propyl trifluoromethanesulfonate (Scheme 3.26). Exposure of **138** to dimethyldioxirane in methanol resulted in diastereoselective transfer of oxygen to the less-hindered face of the *cis*-fused 6-5 system. In the presence of methanol, the presumed silyloxyepxoide that formed was cleaved to generate the α-hydroxyketone (**139**) as a single detectable diastereomer (76 %). Reduction of the ketone function of **139** was effected by treatment with boron trifluoride etherate complex. Under these conditions, a borinate ester (**140**) may be generated in situ. Hydride is then delivered in an intramolecular and stereoselective fashion to the ketone function (58 %). Finally, global deprotection afforded (−)-kinamycin F (**6**) in 65 % yield.

Our studies of (−)-kinamycin F (**6**) motivated the development of a three-step sequence for synthesis of the diazofluorene function, comprising fluoride-mediated coupling, palladium-mediated cyclization, and diazo transfer. Our synthesis also features the strategic use of substrate bias to establish the *trans*-1,2-diol of the target.

References

1. (a) Ito S, Matsuya T, Ōmura S, Otani M, Nakagawa A (1970) J Antibiot 23:315–317; (b) Ōmura S, Nakagawa A, Yamada H, Hata T, Furusaki A (1973) Chem Pharm Bull 21:931–940
2. Hata T, Ōmura S, Iwai Y, Nakagawa A, Otani M (1971) J Antibiot 24:353–359
3. Ōmura S, Nakagawa A, Yamada H, Hata T, Furusaki A, Watanabe T (1971) Chem Pharm Bull 19:2428–2430
4. (a) Cone MC, Seaton PJ, Halley KA, Gould SJ (1989) J Antibiot 42:179–188; (b) Isshiki K et al (1989) J Antibiot 42:467–469; (c) Smitka TA, Bonjouklian R, Perun TJ, Hunt AH, Foster RS, Mynderse JS, Yao RC (1992) J Antibiot 45:581–583; (d) Lin H-C, Chang S-C, Wang N-L, Chang L-R (1994) J Antibiot 47:675–680; (e) Lin H-C, Chang S-C, Wang N-L, Chang L-R (1994) J Antibiot 47:681–687
5. For reviews, see: (a) Gould SJ (1997) Chem Rev 97:2499–2510; (b) Marco-Contelles J, Molina MT (2003) Curr Org Chem 7:1433–1442; (c) Arya DP (2006) Top Heterocycl Chem 2:129–152; (d) Nawrat CC, Moody CJ (2011) Nat Prod Rep 28:1426–1444
6. Seaton PJ, Gould SJ (1989) J Antibiot 42:189–197
7. He H, Ding WD, Bernan VS, Richardson AD, Ireland CM, Greenstein M, Ellestad GA, Carter GT (2001) J Am Chem Soc 123:5362–5363
8. Furusaki A, Matsui M, Watanabe T, Ōmura S, Nakagawa A, Hata T (1972) Isr J Chem 10:173–187
9. Seaton PJ, Gould SJ (1988) J Am Chem Soc 110:5912–5914
10. Dmitrienko GI, Nielsen KE, Steingart C, Ngai SM, Willson JM, Weeratunga G (1990) Tetrahedron Lett 31:3681–3684
11. Echavarren AM, Tamayo N, Paredes MC (1993) Tetrahedron Lett 34:4713–4716
12. Gould SJ, Tamayo N, Melville CR, Cone MC (1994) J Am Chem Soc 116:2207–2208
13. Mithani S, Weeratunga G, Taylor NJ, Dmitrienko GI (1994) J Am Chem Soc 116:2209–2210
14. Arya DP, Jebaratnam DJ (1995) J Org Chem 60:3268–3269
15. Laufer RS, Dmitrienko GI (2002) J Am Chem Soc 124:1854–1855
16. Feldman KS, Eastman KJ (2006) J Am Chem Soc 128:12562–12573
17. Khdour O, Skibo EB (2009) Org Biomol Chem 7:2140–2154
18. (a) Ballard TE, Melander C (2008) Tetrahedron Lett 49:3157–3161; (b) Heinecke CL, Melander C (2010) Tetrahedron Lett 51:1455–1458
19. Zeng W, Ballard TE, Tkachenko AG, Burns VA, Feldheim DL, Melander C (2006) Bioorg Med Chem Lett 16:5148–5151
20. O'Hara KA et al (2007) Free Radic Biol Med 43:1132–1144
21. Gould SJ, Melville CR, Cone MC, Chen J, Carney JR (1997) J Org Chem 62:320–324
22. Chen N, Carriere MB, Laufer RS, Taylor NJ, Dmitrienko GI (2008) Org Lett 10:381–384
23. (a) Nicolaou KC, Denton RM, Lenzen A, Edmonds DJ, Li A, Milburn RR, Harrison ST (2006) Angew Chem Int Ed Engl 45:2076–2081; (b) Krygowski ES, Murphy-Benenato K, Shair MD (2008) Angew Chem Int Ed Engl 47:1680–1684; (c) Morris WJ, Shair MD (2008) Org Lett 11:9–12; (d) Zhang W, Baranczak A, Sulikowski GA (2008) Org Lett 10:1939–1941; (e) Gholap SL, Woo CM, Ravikumar PC, Herzon SB (2009) Org Lett 11:4322–4325; (f) Nicolaou KC, Nold AL, Li H (2009) Angew Chem Int Ed Engl 48:5860; (g) Lee HG, Ahn JY, Lee AS, Shair MD (2010) Chem Eur J 16:13058–13062; (h) Morris WJ, Shair MD (2010)

Tetrahedron Lett 51:4310–4312; (i) Herzon SB, Lu L, Woo CM, Gholap SL (2011) J Am Chem Soc 133:7260–7263
24. Lei X, Porco JA (2006) J Am Chem Soc 128:14790–14791
25. Hu Y, Li C, Kulkarni BA, Strobel G, Lobkovsky E, Torczynski RM, Porco JA (2001) Org Lett 3:1649–1652
26. Aggarwal VK, Mereu A, Tarver GJ, McCague R (1998) J Org Chem 63:7183–7189
27. (a) Elston CL, Jackson RFW, MacDonald SJF, Murray PJ (1997) Angew Chem Int Ed Engl 36:410–412; (b) Li C, Pace EA, Liang M-C, Lobkovsky E, Gilmore TD, Porco JA (2001) J Am Chem Soc 123:11308–11309; (c) Li C, Johnson RP, Porco JA (2003) J Am Chem Soc 125:5095–5106
28. Jung ME, Hagenah JA (1983) J Org Chem 48:5359–5361
29. Azizian H, Eaborn C, Pidcock A (1981) J Organomet Chem 215:49–58
30. Caron M, Sharpless KB (1985) J Org Chem 50:1557–1560
31. (a) Furrow ME, Myers AG (2004) J Am Chem Soc 126:5436–5445; (b) Furrow ME, Myers AG (2004) J Am Chem Soc 126:12222–12223
32. Kumamoto T, Kitani Y, Tsuchiya H, Yamaguchi K, Seki H, Ishikawa T (2007) Tetrahedron 63:5189–5199
33. Kitani Y, Morita A, Kumamoto T, Ishikawa T (2002) Helv Chim Acta 85:1186–1195
34. Heinzman SW, Grunwell JR (1980) Tetrahedron Lett 21:4305–4308
35. Nicolaou KC, Montagnon T, Baran PS (2002) Angew Chem Int Ed Engl 41:993–996
36. Donohoe TJ, Blades K, Moore PR, Waring MJ, Winter JJG, Helliwell M, Newcombe NJ, Stemp G (2002) J Org Chem 67:7946–7956
37. Evans DA, Chapman KT, Carreira EM (1988) J Am Chem Soc 110:3560–3578
38. Jacob P, Callery PS, Shulgin AT, Castagnoli N (1976) J Org Chem 41:3627–3629
39. Nicolaou KC, Li H, Nold AL, Pappo D, Lenzen A (2007) J Am Chem Soc 129:10356–10357
40. Giuffredi G, Bobbio C, Gouverneur V (2006) J Org Chem 71:5361–5364
41. Johnson CR, Adams JP, Braun MP, Senanayake CBW (1992) Tetrahedron Lett 33:917–918
42. Banwell MG, Kelly BD, Kokas OJ, Lupton DW (2003) Org Lett 5:2497–2500
43. (a) Kerr MS, Read de Alaniz J, Rovis T (2002) J Am Chem Soc 124:10298–10299; (b) Kerr M, Read de Alaniz J, Rovis T (2005) J Org Chem 70:5725–5728
44. (a) Molander GA, Hahn G (1986) J Org Chem 51:1135–1138; (b) Molander GA (1992) Chem Rev 92:29–68
45. Woo CM, Lu L, Gholap SL, Smith DR, Herzon SB (2010) J Am Chem Soc 132:2540–2541
46. Huot R, Brassard P (1974) Can J Chem 52:838–842
47. (a) Middleton WJ (1976) Tris(substituted amino)sulfonium Salts. US Patent 3,940,402; (b) Fujita M, Hiyama T (1993) *erythro*-Directed reduction of a β-keto amide: erythro-1-(3-hydroxy-2-methyl-3-phenypropanoyl)piperidine. In: Organic syntheses, vol 8. Wiley, New York, p 326
48. Bellina F, Rossi R (2009) Chem Rev 110:1082–1146
49. Williams DR, Reeves JT, Nag PP, Pitcock WH Jr, Baik M-H (2006) J Am Chem Soc 128:12339–12348
50. Charette AB, Wurz RP, Ollevier T (2000) J Org Chem 65:9252–9254
51. Doering WvE, DePuy CH (1953) J Am Chem Soc 75:5955–5957
52. Kolb HC, VanNieuwenhze MS, Sharpless KB (1994) Chem Rev 94:2483–2547
53. Cox AL, Johnston JN (2001) Use of the Vicinal Element Effect for Regiochemical Control of Quinone Substitutions and Its Implication for Convergent Mitomycin Construction. Org Lett 3:3695
54. Williams AL, Srinivasan JM, Johnston JN (2006) Synthesis of an Advanced Intermediate en Route to the Mitomycin Natural Products. Org Lett 8:6047

Chapter 4
A Short Synthesis of Strychnine from Pyridine

David B. C. Martin and Christopher D. Vanderwal

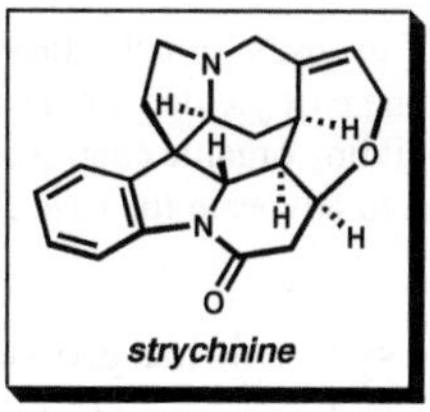

4.1 Introduction

The synthesis of complex natural products serves as a source of great inspiration and challenge for organic chemists. The synthesis of strychnine is no exception. With six contiguous stereocenters, five of which adorn the central cyclohexane ring, a stereodefined trisubstituted olefin, and seven rings, strychnine (**1**, Fig. 4.1) presents a significant challenge for synthesis. In 1952, Sir Robert Robinson said "for its molecular size strychnine is the most complex substance known" [1].

Clearly, our increasingly powerful methods for chemical synthesis, as well as improved methods for natural product isolation and structural elucidation, have changed our definitions of complexity, and Robinson's statement no longer holds true. Still, beginning as early as the 1830s, it took the efforts of countless chemists terminating with a "decades-long chemical degradative assault" to elucidate the full structure of strychnine [2–5]. That some of organic chemistry's greatest minds of the time were engaged in this pursuit is as much a testament to the challenge of structural determination in that era as it is to the possibility for new discoveries along the way. With the advent of modern spectroscopic and X-ray crystallographic

D.B.C. Martin • C.D. Vanderwal (✉)
Department of Chemistry, University of California, Irvine, CA 92697, USA

J.J. Li and E.J. Corey (eds.), *Total Synthesis of Natural Products*,
DOI 10.1007/978-3-642-34065-9_4, © Springer-Verlag Berlin Heidelberg 2012

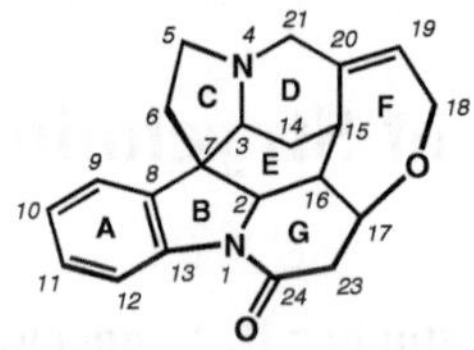

Fig. 4.1 The biogenetic numbering system used throughout this chapter (C22 has been biochemically excised)

methods in the decades following strychnine's structural elucidation, clever degradation experiments were gradually marginalized. As Woodward foresaw and elegantly stated when describing in detail his group's inaugural synthesis of strychnine, the field of natural product synthesis would effectively replace structure elucidation as the major source of serendipitous discovery in organic chemistry:

> Of course, men make much use of excuses for activities which lead to discovery, and the lure of unknown structures has in the past yielded a huge dividend of unsought fact, which has been of major importance in building organic chemistry as a science. Should a surrogate now be needed, we do not hesitate to advocate the case for synthesis.
> R.B. Woodward, 1963 [6].

The field of complex molecule synthesis has grown tremendously since Woodward prophetically wrote these words, and one cannot overstate the contributions in terms of new methods, new strategies, and new reactivity that have arisen either deliberately or fortuitously through the exercise of natural product synthesis [7]. In the relatively short time that our laboratory has been engaged in the synthesis of natural products, we, like many other research groups, have uncovered unexpected reactivity and unusual molecular rearrangements [8]. These serendipitous discoveries have led to useful new reactions and inspired approaches to new natural product targets. In the narrative that follows, we describe our efforts directed toward the *Strychnos* family of indole alkaloids using Zincke aldehydes (5-amino-2,4-pentadienals, available from the ring opening of pyridinium salts) as key intermediates, and we highlight some of the unforeseen adventures that have kept things interesting along the way [9].

4.2 Synthesis of Strychnine: A Historical Perspective

The most famous and recognizable member of the *Strychnos* family, strychnine is thought to be the first indole alkaloid ever isolated in pure form, as reported by Pelletier and Caventou in 1818 [2]. Its elemental composition would be determined by Regnault a mere 20 years later [3], but the correct structure would not be put forth in the literature until 1946 by Robinson, and confirmed 2 years later by Woodward [4, 5]. With its moderate level of structural complexity, diverse possibilities for constructing its polycyclic ring system and an intriguing history, it is no wonder that strychnine has attracted substantial interest as a target for synthesis despite its lack of important biological activity. Over the past 60 years, 18 groups have reported

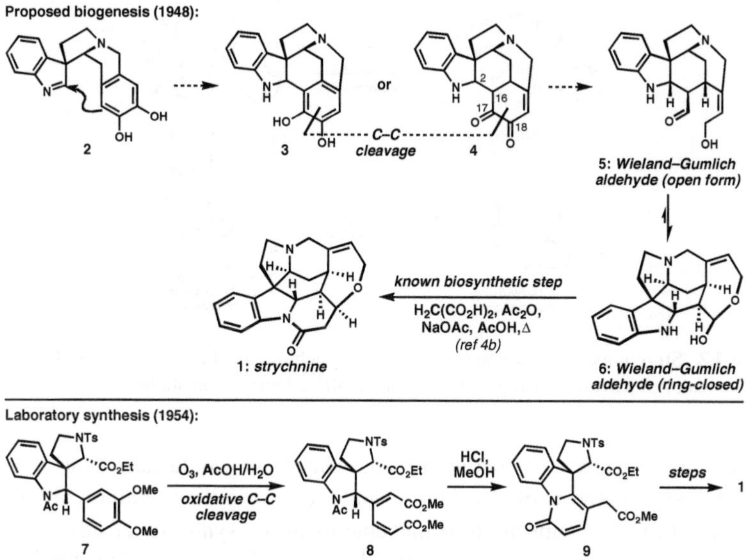

Scheme 4.1 Oxidative cleavage of an electron-rich aromatic ring was a key component of Woodward's proposal for a biogenesis of the *Strychnos* alkaloids and a key step in his subsequent synthesis of strychnine

syntheses of this heptacyclic alkaloid, each reporting a unique approach to its construction [10]. The first such achievement was reported by Woodward and co-workers in 1954, a mere 6 years after the correct structure of strychnine had been rigorously established. Woodward took advantage of many lessons learned from degradation studies in the planning and execution of this synthesis and was equally inspired by his early thoughts on the biogenesis of the *Strychnos* alkaloids, incorporating elements of his proposal into the synthesis. For example, an early biogenetic proposal is outlined in Scheme 4.1, wherein intermediate **2** might be formed by successive Pictet–Spengler reactions of tryptamine with two different aldehydes (3,4-dihydroxyphenylacetaldehyde and formaldehyde) [11]. One further nucleophilic attack of the electron rich arene could forge the C16–C2 bond [12], providing a structure (**3** or **4**) that "contains striking similarities to many of the features of the strychnine molecule." Cleavage of the C17–C18 bond and minor redox adjustments would yield the Wieland–Gumlich aldehyde (**5**/**6**), which could be converted to strychnine by incorporation of an N_a–Ac group, followed by another simple condensation reaction. With respect to this proposal, Woodward wrote: "On the whole, the possibility of building up so complicated a structure as (strychnine) by a series of simple reactions from plausible starting materials *is so striking that it is difficult to believe that the scheme lacks significance*." While these early musings were mistaken with regard to the origins of the Wieland–Gumlich aldehyde [13], the thought process of successive, simple reactions leading to a rapid buildup of complexity was compelling, and a similar oxidative cleavage of a catechol derivative (**7** → **8**) was used in Woodward's successful synthesis [10a].

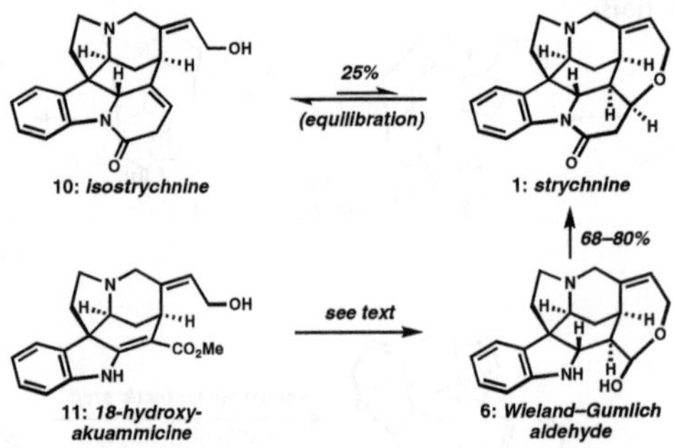

Scheme 4.2 Strychnine is readily accessed via isostrychnine or the Wieland–Gumlich aldehyde using chemistry discovered during structural elucidation/degradation studies

Following Woodward's inaugural achievement, it would take nearly 40 years before the next synthesis of strychnine would appear in the literature, although the intervening time saw substantial achievements in the synthesis of other *Strychnos* alkaloids and members of the related *Aspidosperma*, *Iboga*, and *Vinca* families [14]. The reported syntheses of strychnine since 1990 have included milestones such as the first asymmetric synthesis by Overman [10c], the first biomimetic synthesis by Martin [10h], and particularly concise syntheses by Kuehne [10d], Rawal [10e], Bodwell [10k], Reissig [10p], and MacMillan [10r]. The reported syntheses to date are all creative and instructive, many relying on new technologies that have arisen since the days of Woodward and some of which were clearly inspired by the challenge of the target alkaloid itself. In this context, strychnine continues to serve as a benchmark target against which organic chemists may measure the state-of-the-art of natural product synthesis strategy.

Although each synthesis of strychnine has been accomplished using unique methods or strategic disconnections, they share certain elements. All reported syntheses have proceeded via either isostrychnine (**10**) or the Wieland–Gumlich aldehyde (**6**) as the penultimate intermediate (Scheme 4.2) [15]. Both of these molecules are themselves naturally occurring alkaloids; isostrychnine is, as the name suggests, a constitutional isomer of strychnine, and the Wieland–Gumlich aldehyde is a biogenetic precursor of strychnine [4]. Part of the reason for this commonality is that there exist well-established procedures for the conversion of each precursor to strychnine. The equilibration of isostrychnine and strychnine favors the former, providing a 70:30 mixture that can be resolved by chromatography [16]. The conversion of the Wieland–Gumlich aldehyde to strychnine was reported by Robinson and Anet and is reported to proceed in a much higher 68–80 % yield [4c]. In addition, many of the syntheses proceeding via the intermediacy of the Wieland–Gumlich aldehyde have also employed 18-hydroxyakuammicine (**11**) as an earlier precursor, as first described by Overman and co-workers [10c, f, h, k, m, o, p]. The reliable introduction of the final ring and three stereocenters are attractive reasons for this endgame, and within 18-hydroxyakuammicine and isostrychnine remain most of the structural challenges that attract chemists to this class of alkaloids.

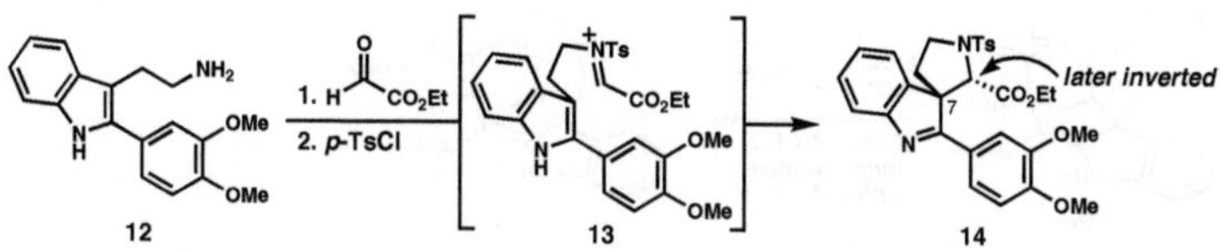

Scheme 4.3 Formation of the C7 quaternary center by C7–C3 bond formation by Woodward and co-workers

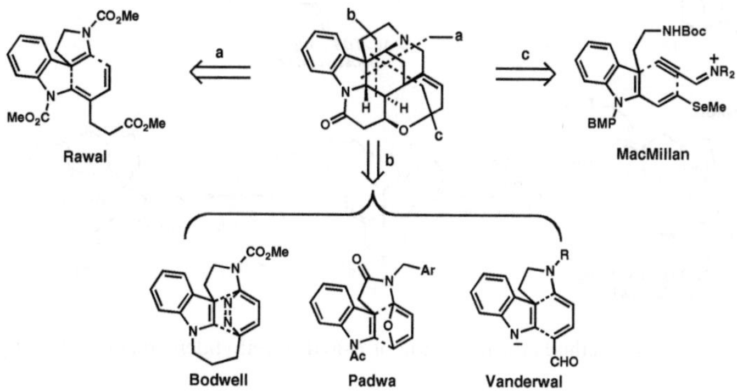

Scheme 4.4 Diels–Alder disconnections used in the synthesis of strychnine

4.3 Structural Challenges

Among the many rings and stereogenic elements that contribute to strychnine's challenge as a target for synthesis, two key structural features have inspired unique chemical solutions and deserve some comment. One important challenge is the C7 quaternary stereogenic center (see Fig. 4.1 for atom and ring labeling). At the junction of three separate rings, this center has inspired a number of clever strategies for its construction, including sigmatropic rearrangements, cycloadditions, and transition metal-catalyzed reactions. The most widely adopted strategy involves cyclization of a C7 nucleophile onto an iminium ion to forge the C7–C3 bond, often taking advantage of the inherent nucleophilicity of the indole group's 3-position. This strategy was used at an early stage in Woodward's inaugural synthesis (Scheme 4.3), and figured prominently in the syntheses of Magnus, Overman (as part of the key aza-Cope/Mannich sequence), and others [10a–c, d, f, m].

Another prominent solution makes use of the Diels–Alder cycloaddition, a reliable reaction for the construction of quaternary centers. Rawal pioneered this strategy in an intramolecular Diels–Alder reaction that constructed both the indoline B-ring and the E-ring (Scheme 4.4, disconnection a), allowing for the controlled formation of three contiguous stereocenters [10e]. Two approaches previous to ours have enlisted the indole as the 2π component in an intramolecular Diels–Alder reaction with either a pyridazine [10k], or an amidofuran [10n], as the

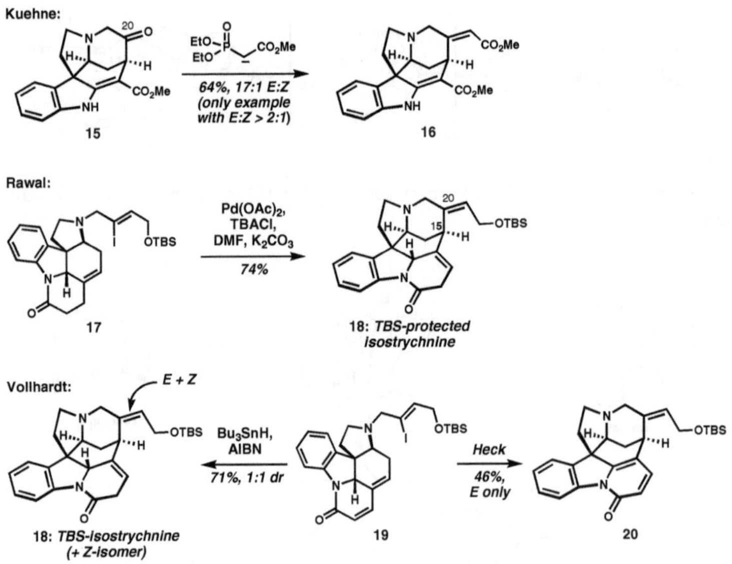

Scheme 4.5 Previous methods for formation of E-hydroxyethylidene functional group

4π component (disconnection b, for details, see section below). Finally, the recent synthesis reported by MacMillan and co-workers established the viability of incorporating a vinyl indole as the 4π component in an application of their enantioselective catalysis via chiral iminium ions (disconnection c) [10r]. Notably, three of the four distinct Diels–Alder disconnections that allow direct construction of the C7 quaternary center have been successfully employed in the synthesis of strychnine.

Another challenge associated with the synthesis of strychnine, as well as other *Strychnos* alkaloids, is the stereoselective construction of the E-hydroxyethylidene side chain (Scheme 4.5). Strategies that rely on a late-stage olefination reaction of a C20 ketone often suffered from low diasterocontrol for the newly formed alkene, with the best exception provided by Kuehne (**15** → **16**) [10a, b, d, f]. Overman, Martin, and Shibasaki took advantage of highly selective β-elimination reactions of β-hydroxy esters or amides at earlier stages of the synthesis, the products of which would be later elaborated into the Wieland–Gumlich aldehyde (not shown) [10c, h, l]. Arguably the most general of the approaches is the stereospecific closure of the D-ring piperidine using a vinylmetal species derived from a precursor vinyl halide (C20–C15 bond formation). This strategy has taken the form of a Cu/Mn-, Ni- or Pd-mediated conjugate addition, a Pd-catalyzed enolate vinylation, and an intramolecular Heck reaction [10g, e, i, j, n, r][1]. The Heck approach first reported by Rawal in

[1] A synthesis of strychnine by the Stork group was apparently disclosed in lecture form, and some of the details are provided in the review by Bonjoch and Solé from 2000 [15a]. While we prefer not

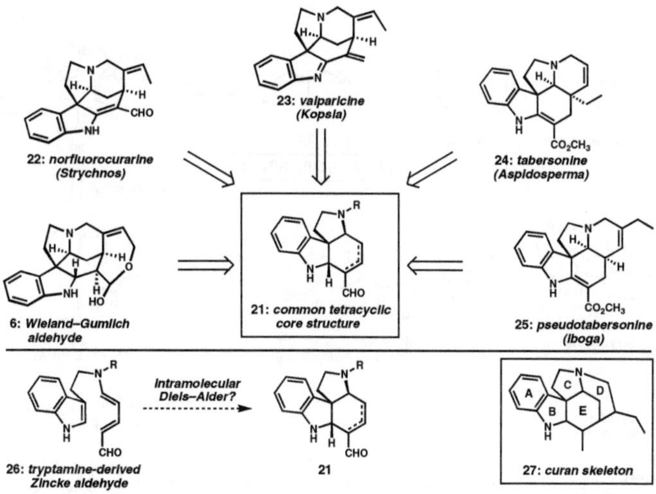

Scheme 4.6 The common core structure of many indole monoterpene alkaloids might be accessed through a tetracyclic Zincke aldehyde-derived building block

1994 (**17** → **18**) was so successful that it has been applied to closely related substrates by Vollhardt, Mori, Andrade, and MacMillan and has featured in many other alkaloid syntheses [10e, i, j, k, o, p, r, 17–21]. This versatile approach has also been used to great effect for the synthesis of simpler Strychnos alkaloids, as we will elaborate later. Vollhardt and co-workers reported an effective radical-mediated ring closure (**19** → **18**), but the product was isolated as a 1:1 mixture of alkene isomers, owing to the known rapid isomerization of vinyl radical intermediates; these workers also eventually used a Heck cyclization for this bond formation (**19** → **20**) [10i].

In our efforts toward strychnine, we have integrated an intramolecular Diels–Alder approach to its tetracyclic core with an organometallic D-ring closure. Both of these key reactions were certainly inspired by the previous accomplishments of others as described above.

4.4 Background: Zincke Aldehydes

We became particularly interested in strychnine when we noticed that the tetracycle **21** (Scheme 4.6), which might be readily available by an intramolecular Diels–Alder cycloaddition of a tryptamine-derived aminodiene, contains much of the complexity of this popular alkaloid target. In fact, this tetracycle is common to many indole monoterpene alkaloids including members of the *Strychnos*, *Aspidosperma*, and

to cite a synthesis that is not available in the primary literature, based on the account provided in [15a], the Stork group utilized a vinyl carbanionic D-ring-closure by C15–C20 bond formation that presumably proceeded via conjugate addition onto an α,β-unsaturated ester.

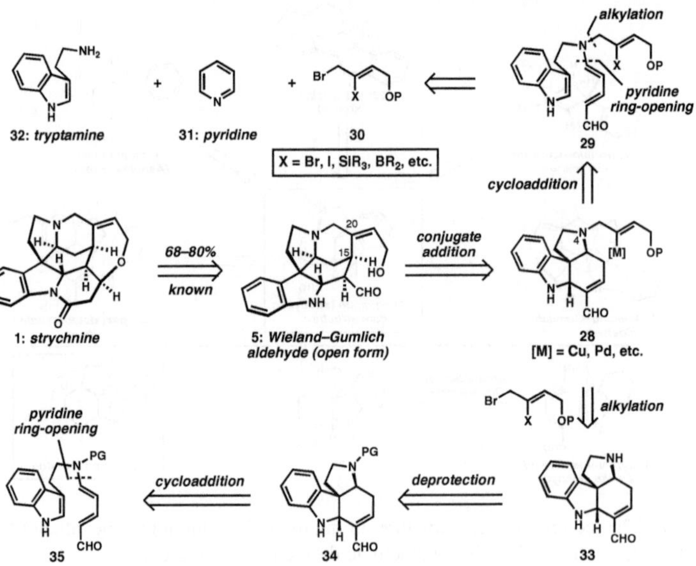

Scheme 4.7 Retrosynthetic options considered for strychnine

Iboga families (see **22–25**), comprising the ABCE rings of the curan skeleton **27**. It was our interest in the particular class of 1-amino-2,4-dienes called Zincke aldehydes (5-amino-2,4-pentadienals) [22], which derive readily from the ring-opening reactions of pyridinium salts, that caused us to recognize an opportunity for a particularly direct approach to strychnine and related alkaloids. Intramolecular cycloaddition of a tryptamine-derived Zincke aldehyde such as **26** could generate tetracycle **21** directly with probable conjugation of the unsaturated aldehyde. This line of thinking encouraged our in depth studies of Zincke aldehydes and inspired other efforts within our research group to engage Zincke aldehydes in Diels–Alder reactions (see below) [8].

Our retrosynthesis is presented in more detail in Scheme 4.7. Our objective was the Wieland–Gumlich aldehyde (shown in its hydroxy-aldehyde tautomeric form, **5**) because it can be efficiently converted to strychnine (68–80 % yields reported). Furthermore, the C17 aldehyde would be derived without redox manipulations from the Zincke aldehyde, further streamlining the synthesis and highlighting the utility of Zincke aldehydes. No previous synthesis of the Wieland–Gumlich has made direct use of a C17 aldehyde; instead previous efforts have relied on more robust esters or other surrogates, necessitating an eventual oxidation state adjustment. We envisioned that the Wieland–Gumlich aldehyde could be obtained by a conjugate addition of a vinyl nucleophile onto an α,β-unsaturated aldehyde to forge the C20–C15 bond and close the D-ring (**28** → **5**). This disconnection was pioneered by Rawal in his synthesis of dehydrotubifoline where he used an intramolecular Heck reaction (as discussed in Scheme 4.5) [18], which has proven to be a robust and strategically advantageous disconnection in several subsequent syntheses. α,β-Unsaturated aldehyde **28** (or its unconjugated isomer) would be the product of our proposed intramolecular cycloaddition of a fully functionalized Zincke aldehyde **29** or could potentially be derived from an *N*-protected tetracycle **34**. The exact nature of intermediate **28** would depend on the precise method of D-ring closure, the functional group tolerance of the cycloaddition reaction, and the success of accessing unsubstituted tetracycle **33**. In addition to providing the framework for a particularly concise approach to strychnine, we envisioned that judicious choice of the N4-substituent and manipulation of the aldehyde group would enable the synthesis of many alkaloids of this class by a unified strategy.

4.5 Background: Intramolecular Cycloadditions of Indoles

There are two major challenges in the proposed intramolecular Diels–Alder reaction: first, the C2–C3 indole π bond is part of a stable aromatic system and is a notoriously poor dienophilic partner for [4 + 2] cycloadditions. As a result, most of the reported examples, although they are intermolecular reactions, involve electronic perturbation of the indole system. For example, triplet photosensitization [23], stoichiometric generation of an indole radical cation [24], or deprotonation to give the indole anion [25] are methods that successfully led to formal cycloaddition outcomes. For selected examples, see Eqs. (4.1) and (4.2). In many cases, the reaction proceeds in a stepwise fashion with radical and/or charged intermediates rather than via concerted cycloaddition. The lack of stereospecificity [e.g., Eq. (4.1)] is a telltale sign of a stepwise process, in this case an anionic Michael/Mannich sequence. Intramolecular examples of indole Diels–Alder reactions include the work of Bodwell and Padwa, both used in approaches to strychnine, where an *N*-acyl or *N*-alkyl indole reacts with a tethered amidofuran or pyridazine (see disconnection b, Scheme 4.4, specific reactions not shown) [10k, n]. Despite the lowered entropy of activation afforded by tethering the two reaction partners, both of these reactions occur only at elevated temperature (150–215 °C) and benefit from a subsequent irreversible fragmentation of the bicyclic [4 + 2] product.

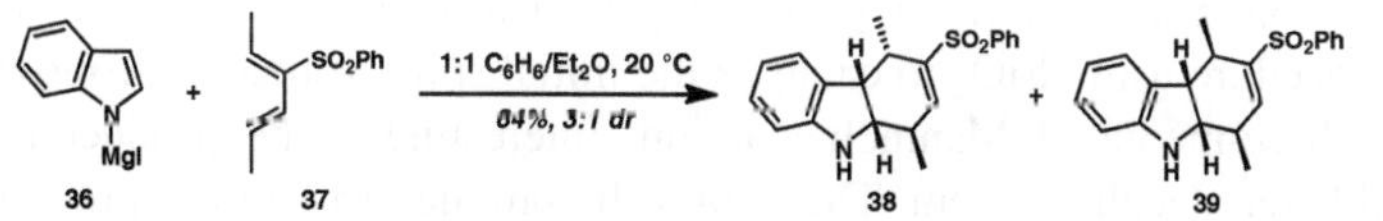

(4.1)

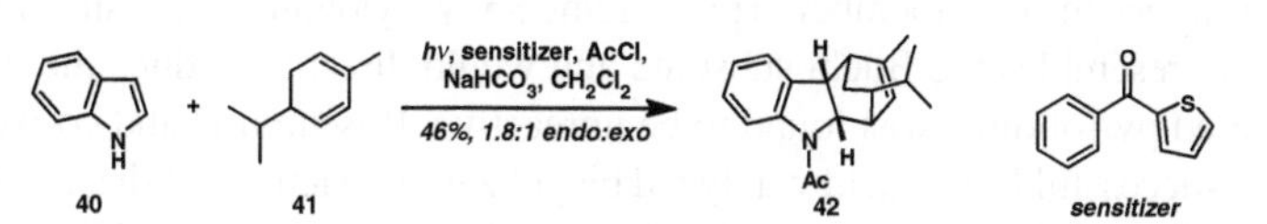

(4.2)

In contrast to the failure of Diels–Alder reactions, dipolar cycloadditions of indoles are much more successful, and the Boger group has reported a fascinating [4+2]/1,3-dipolar cycloaddition cascade involving indole as the dipolarophile in their impressive synthesis of vindoline (Scheme 4.8) [26]. After the initial

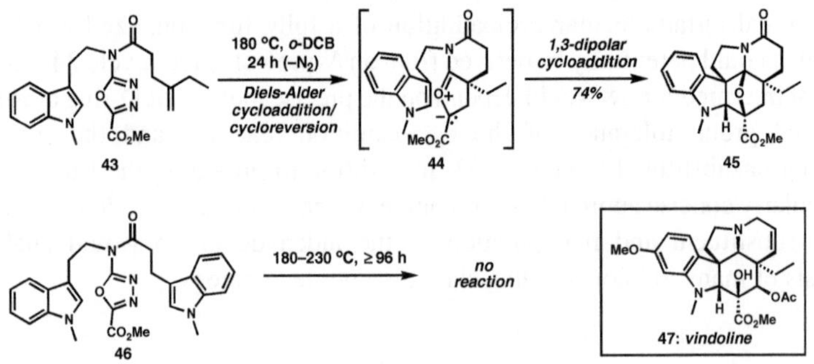

Scheme 4.8 Boger's cascade cycloaddition chemistry to access indole alkaloids features dipolar cycloaddition to indole, and not Diels–Alder cycloaddition

Diels–Alder reaction of the 1,3,4-oxadiazole with the pendant olefin and loss of N_2, the C2–C3 π bond participates in a subsequent 1,3-dipolar cycloaddition with the carbonyl ylide to generate complex polycycles such as **45** as single diastereomers with up to six new stereocenters. That the cascade reaction is initiated by a Diels–Alder reaction with the alkene rather than with the indole is supported by the lack of reaction even under forcing conditions with substrate **46**, in which a Diels–Alder reaction with the indole C2–C3 π bond would be required [26a].

A number of stepwise indole annulations of tryptamine derivatives can be found in the literature that bear a resemblance to our proposed transformation. Early reports from Büchi and co-workers detail the synthesis of **48** and treatment with $BF_3{\cdot}OEt_2$ at elevated temperatures to give tetracycle **49**, which has become known as the Büchi ketone (Scheme 4.9) [27]. This reaction presumably arises from conjugate addition of indole C3 onto the Lewis-acid-activated electron-deficient vinylogous amide, followed by enolate equilibration and final Mannich cyclization. Tetracycle **49** was carried on to complete a concise synthesis of vindorosine (**50**). A related reaction catalyzed by $TiCl_4$ was reported by Pandit and co-workers, directly incorporating a C17 ester (not shown) [28]. More recently, Markó and co-workers reported a different method to access a similar tetracyclic ketone [29]. They found that exposure of tryptamine derivative **51** to KO*t*-Bu at low temperature gave rise to tetracyclic ketone **52**, albeit in low yield [29c]. They also reported a higher-yielding stepwise procedure using SiO_2 to catalyze the initial spirocyclization, followed by subsequent base-mediated Mannich reaction. Interestingly, the product bears a trans-fused heterodecalin system. The tetracyclic product was further processed to eventually produce the five-membered pyrrolidine **53**. Vinylogous amide substrate **54**, which closely resembles the Büchi substrate and would directly produce the desired five-membered pyrrolidine, was found to be unreactive. Rosenmund and co-workers reported the successful Lewis acid-catalyzed bicyclization reactions of diesters **55a/b** which also deliver the desired pyrrolidine C-rings, but these harsh conditions led to mixtures of products in only modest yields [30]. Nonetheless, reactions of these doubly activated diene substrates directly yield full-fledged ABCE tetracycles, with C17 in an appropriately oxidized form relevant to most curan alkaloids. That the

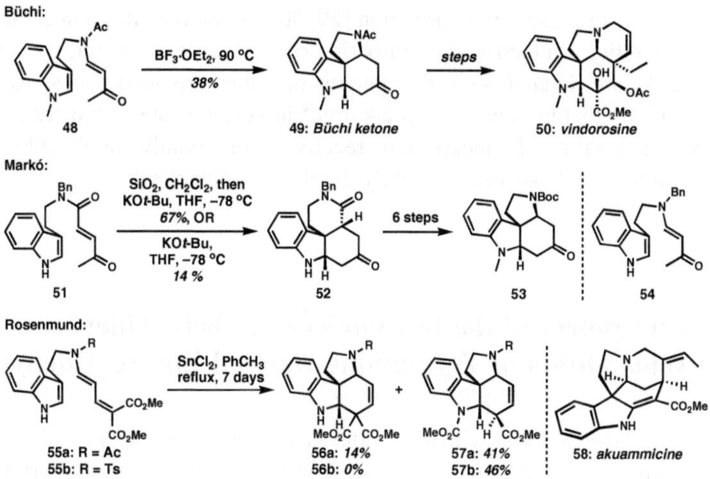

Scheme 4.9 Some bicyclization reactions relevant to our desired transformation

Rosenmund group never applied this reaction to the *Strychnos* alkaloids (e.g., akuammicine, **58**) is somewhat surprising.

Beyond the reluctance of indoles to participate in Diels–Alder reaction, the second challenge of our proposed intramolecular cycloaddition is that Zincke aldehydes are also known to be poor dienes in [4 + 2] cycloadditions [31]. To the best of our knowledge, there are no reported examples of successful Diels–Alder reactions with Zincke aldehydes serving as the 4π component, almost certainly owing to their high donor–acceptor stabilization. In the most closely related example that we have found, Zincke aldehyde-like compound **59a** [Eq. (4.3)] was found to be unreactive as the 2π component in an enantioselective organocatalytic Diels–Alder cycloaddition until the donor nature of the indole nitrogen was attenuated by *N*-tosylation (see **59b**) [32]. While excellent enantioselectivity was achieved, this intermolecular reaction remained inefficient, although it did serve as a key step in Kerr's synthesis of (+)-hapalindole Q.

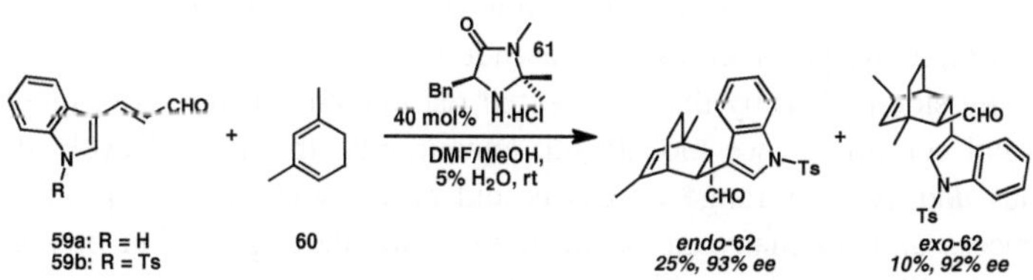

(4.3)

While we were concerned by the potential problems with our desired reaction, particularly the poor dienophilicity of indoles and the failure of **54** to cyclize under stepwise cyclization conditions, we were nonetheless inspired to pursue this potentially direct strategy. The successes of Markó and Rosenmund in related systems (**51** → **52** and **55** → **56**, respectively), the ease of substrate synthesis, and the significant utility of the reaction products compelled us to evaluate Zincke aldehydes in the proposed transformation [29, 30]. Given the intramolecular setting and the possibility of electronic perturbation of the indole fragment, we remained optimistic. As we detail below, this pursuit not only provided access to several *Strychnos* alkaloids (and one *Kopsia* alkaloid) in very few steps, but we have also uncovered a wealth of unexpected reactivity potentially applicable to the stereocontrolled synthesis of completely unrelated molecules.

4.6 Development of the Intramolecular Diels–Alder Cycloaddition of Tryptamine-Derived Zincke Aldehydes

We began our investigations with the synthesis of model substrate **66** (Scheme 4.10), which is readily available on multigram scale in 83 % overall yield from tryptamine, benzaldehyde, and pyridinium salt **65** via Zincke pyridinium ring opening [9]. We first attempted thermal Diels–Alder reactions in a variety of solvents. At lower temperatures, the starting material remained unchanged, but above 150 °C in aromatic solvents, we observed the clean formation of a new product. Examination of ^{1}H-NMR spectra of the crude products showed that the indole resonances remained, while the salient resonances of the Zincke aldehyde were replaced with a new set of signals in the alkene region. The product was determined to be *Z*-α,β,γ,δ-unsaturated amide **67** [8b]. This completely unexpected reaction turned out to be quite general (see **68** → **69**) and has been observed in all of our attempts to initiate thermal Diels–Alder reactions with other Zincke aldehydes. This interesting chemistry has been further explored in our group in terms of scope and mechanism and has resulted in a new synthesis of polycyclic lactams from allyl-substituted Zincke aldehydes (**70** → **71** → **72**) [8b, d, f]. Examination of the structure of these products led us to ponder a new synthesis of gelsemine (**73**), another polycyclic alkaloid that has been a popular target over the last 20 years [33]. A Zincke aldehyde precursor **70** with suitable substitution (A = vinyl, B = H, etc.) should lead to a lactam that maps nicely with respect to substitution and stereochemistry onto three of the six rings of gelsemine. This ongoing project aims to highlight the rapid buildup of complexity provided by this cascade reaction.

After being sidetracked by the unexpected rearrangement of Zincke aldehydes under thermal conditions, we looked into encouraging the desired cycloaddition with various additives. A range of Lewis and protic acids did not promote the desired process and, in many cases, resulted in indole degradation or apparent Pictet–Spengler-like reactivity. Nuhant, Marazano, and co-workers have since reported a method for performing Pictet–Spengler-type reactions of similar Zincke aldehydes using TFAA [Eq. (4.4)], giving access to tetrahydro-β-carboline products such as **75** which could serve as useful intermediates for the synthesis of natural products such as corynantheal (**76**) [34]. We also briefly examined the use of aminium catalysis through the use of the commercially available radical cation

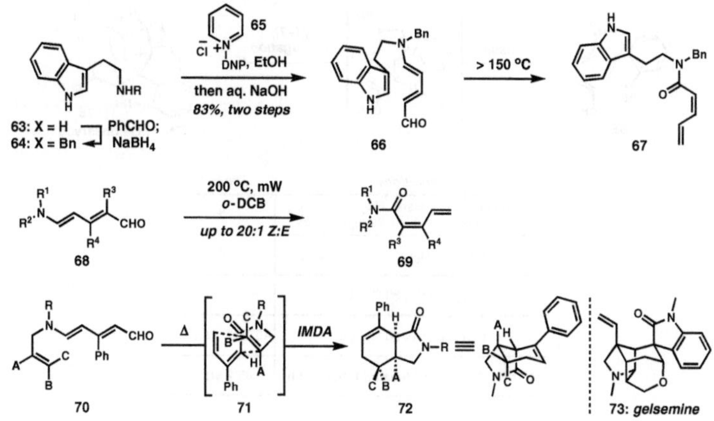

Scheme 4.10 Unexpected thermal rearrangement of Zincke aldehydes leads to some interesting reactivity (DNP = 2,4-dinitrophenyl)

salt $N(4\text{-}BrC_6H_4)_3SbF_6$ [24]. We anticipated that an indole radical cation might engage the Zincke aldehyde; however, productive reactivity was never observed.

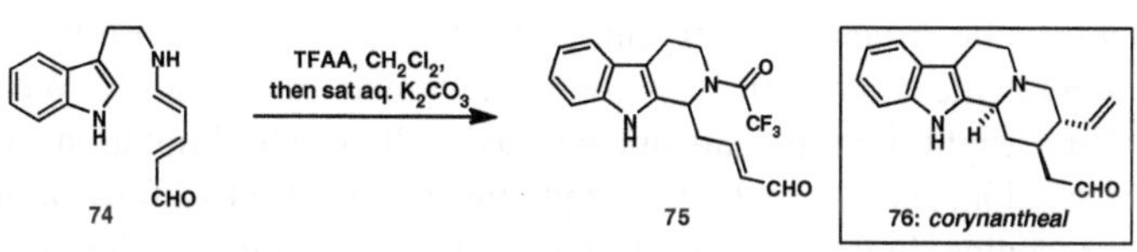

(4.4)

The final strategy that we explored to induce cycloaddition was indole metallation. Inspired by the anionic bicyclization of Markó that eventually led to tetracycles related to the Büchi ketone (**49** in Scheme 4.9), we examined a variety of organic and inorganic bases, and promising results were quickly obtained. Exposure of Zincke aldehyde **66** to a stoichiometric quantity of KO*t*-Bu in THF at 50 °C for 4 h led to 40 % conversion to a new isomeric product (Scheme 4.11). We identified the product as tetracyclic α,β-unsaturated aldehyde **78**; not surprisingly, the basic conditions of the reaction resulted in conjugation of the alkene with the aldehyde. The relative configuration of the product was readily discerned by nOe experiments and matched that found in the ABCE ring systems of the majority of known indole alkaloids that share that skeleton. Optimization led to a simple and reliable protocol: treatment of Zincke aldehyde **66** in THF (0.06 M) with KO*t*-Bu (commercially available 1 M solution in THF) at 80 °C in a sealed tube routinely afforded tetracycle **78** in 85 % yield.

A particularly interesting aspect of the reaction is its apparent counter-cation dependence. Our initial screen of bases included inorganic bases with a variety of counterions, including Li^+, Na^+, K^+, and MgX^+, but only potassium bases were effective. The precise role of the counterion is not completely understood, but it is critical for success. To date, the use of KO*t*-Bu or KH have consistently provided the best results, with KHMDS used as a lower-yielding alternative if absolutely necessary (in cases where the substrate was unstable to alkoxide bases).

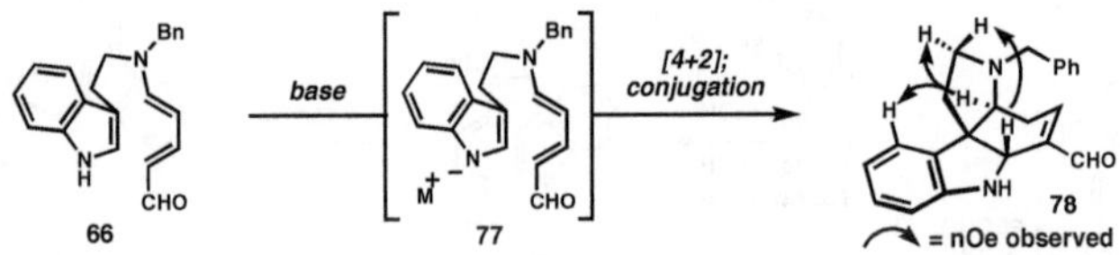

Conditions	*Result*
DBU, phosphazine bases, Na_2CO_3, K_2CO_3, Cs_2CO_3, variety of solvents/temperatures	NR
MeLi, NaH or EtMgBr, THF, 0 to 50 °C	NR
EtMgBr, THF, 80 °C μW	NR
KO*t*-Bu, THF, 50 °C, 4 h	40% conversion
KO*t*-Bu, THF, 80 °C (sealed tube), < 2 h	**85% yield**

Scheme 4.11 Selected results in the base-mediated cycloaddition reaction

4.7 Synthesis of Norfluorocurarine

With the success of our new stereoselective cycloaddition, we turned our attention to its application in the synthesis of *Strychnos* alkaloids. Rather than immediately tackle the heptacyclic structure of strychnine itself, we initially chose to validate our methodology in the context of a less complex member of the family. We hoped to gain further insight into the advantages as well as the limitations of our bicyclization reaction, and we could then apply this knowledge to a fully optimized approach to strychnine. In this context, we targeted the *Strychnos* alkaloid norfluorocurarine (**22**, Schemes 4.6 and 4.12) that possesses the curan skeleton and includes a C17 aldehyde group, providing an ideal target for our studies. This relatively simple *Strychnos* alkaloid was first reported in 1961 by Stauffacher and was later isolated from a second source as a racemate [35].

Norfluorocurarine has been previously synthesized by the groups of Harley-Mason, Bonjoch, and Rawal [36]. The first synthesis by Crawley and Harley-Mason (1971) begins with the synthesis of macrocyclic ketone **79** (Scheme 4.12) [36a]. Closure of the D-ring and formation of the exocyclic C19–C20 alkene occurs in a base-mediated, one-pot process of **79**; presumably, attack of an enolate onto the alkyl bromide forges the C15–C20 bond, which is followed by elimination of the β-methoxyamide (giving a 1.4:1 mixture of alkene isomers). Incorporation of C17 is accomplished by a Wittig reaction to give **80**. A late-stage oxidation (Pt and O_2) forms a C3–N4 iminium ion which undergoes transannular attack by the indole (C7) to give **81**. This oxidative approach was later applied by Magnus in his synthesis of strychnine [10b].

The synthesis reported by Bonjoch and co-workers utilizes a Ni-mediated reductive cyclization of vinyl iodide **82** with concomitant reductive indolenine formation; quenching of the presumed intermediate (dehydrotubifoline, **83**) with the Vilsmeier reagent affords *N*-formyl derivative **84** [36b]. Photoisomerization gives norfluorocurarine (**22**) in low yield. Similarly, Rawal and He produce

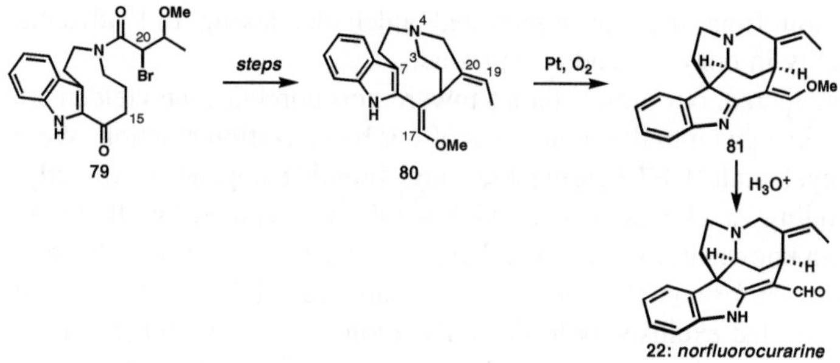

Scheme 4.12 First synthesis of norfluorocurarine by Harley-Mason and Crawley

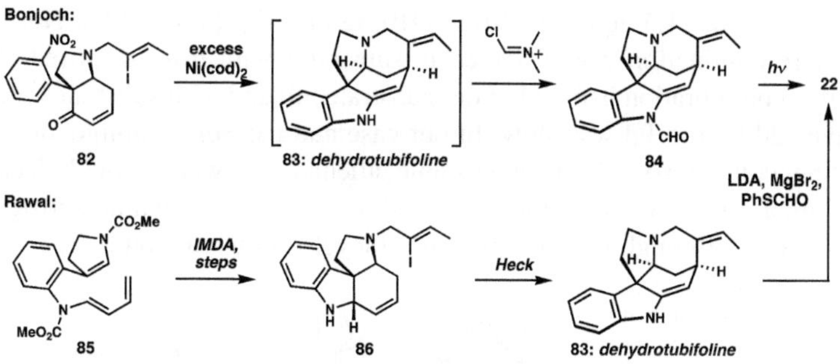

Scheme 4.13 Norfluorocurarine syntheses by Bonjoch and Rawal groups both proceed via dehydrotubifoline

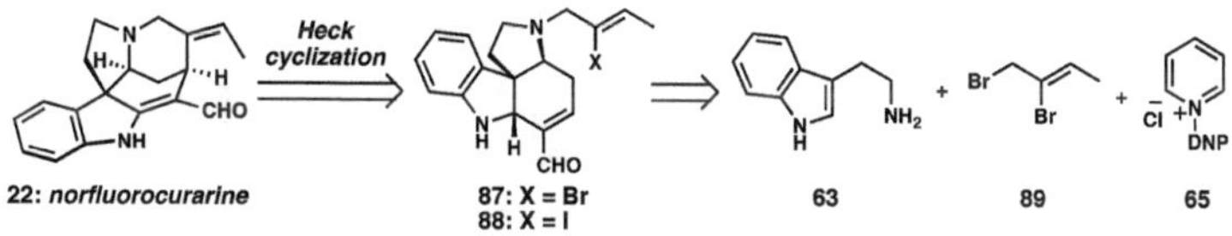

Scheme 4.14 Plan for the synthesis of norfluorocurarine

norfluorocurarine via the intermediacy of dehydrotubifoline [36c], but they reported new conditions that favor a direct *C*-formylation in preference to *N*-formylation [36b, 37]. This synthesis of dehydrotubifoline involved a Diels–Alder reaction of triene **85** and closure of the D-ring using a Heck reaction of vinyl iodide **86** (Scheme 4.13).

We reasoned that an intramolecular Heck reaction of vinyl halide **87/88** (Scheme 4.14), inspired by the work of Rawal, would serve as the final key step to close the D ring and deliver the natural product [18]. Based on previous reports outlined above, we anticipated that this strategy could efficiently effect closure of the D-ring with complete control of alkene geometry. This Heck-based strategy would also directly generate the vinylogous amide of norfluorocurarine by alkene transposition from the α,β-unsaturated aldehyde, taking full advantage of the functionality in our cycloaddition products.

As anticipated, our initial studies toward norfluorocurarine yielded a number of important insights into the limitations of our bicyclization reaction. We planned to prepare cycloadduct **87** bearing the vinyl bromide appendage directly from the corresponding Zincke aldehyde, which would be prepared by alkylation of tryptamine with known dibromide **89** (Scheme 4.14) [38a], followed by Zincke aldehyde formation. Unfortunately, but the standard reaction conditions for anionic bicyclization led exclusively to decomposition of substrate **90** [Eq. (4.5)] and we were unable to isolate any of the desired tetracyclic product. The major product isolated was the corresponding alkyne **91**, presumably formed via base-mediated dehydrohalogenation. This process was observed by Solé and co-workers in a Pd-catalyzed enolate vinylation to give bridged, bicyclic products that proceeded under similar conditions [1.5 equiv KO*t*-Bu, THF, reflux; Eq. (4.6)] [20a]. Not surprisingly, more elimination was observed in substrates with *trans*-disposed H–I and H–Br (this configuration is found in our substrate **90**) and with substrates for which oxidative addition to Pd was slow. In our case and the Solé example, dehydrohalogenation could also afford an unstable allenamine, which could decompose further. As a result, we were forced to find a suitable vinyl halide surrogate that could be carried through the strongly basic bicyclization reaction.

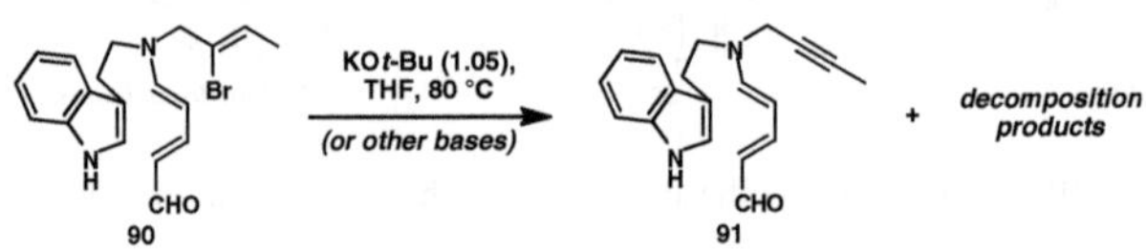

(4.5)

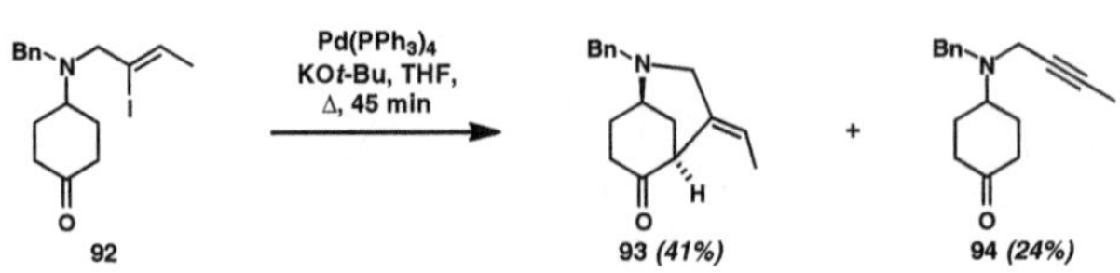

(4.6)

The most obvious vinyl halide surrogate, and one that ultimately proved useful, was a vinylsilane. Several methods are available for the stereocontrolled synthesis of vinylsilanes, including alkyne hydrosilylation, addition of vinyl nucleophiles to chlorosilanes, and reduction/functionalization of alkynylsilanes. Using the latter method, and according to the protocol of Metz and Linz [39], 1-(trimethylsilyl)-propyne was converted in one step into vinylsilane **95** via hydroalumination and reaction with paraformaldehyde. The allylic alcohol was converted to a leaving group and subsequently treated with tryptamine to afford **96** (Scheme 4.15). Zincke aldehyde **97** was produced by treatment of two equivalents of **96** with pyridinium salt **65** followed by hydrolysis of the resulting conjugated iminium ion. The second equivalent of amine liberated in the hydrolysis step was easily recovered in high yield. As we anticipated, the standard reaction conditions for cycloaddition produced tetracycle **98** in 84 % yield, once again as a single diastereomer, and without any decomposition of the vinylsilane. At this point, stereospecific conversion of the

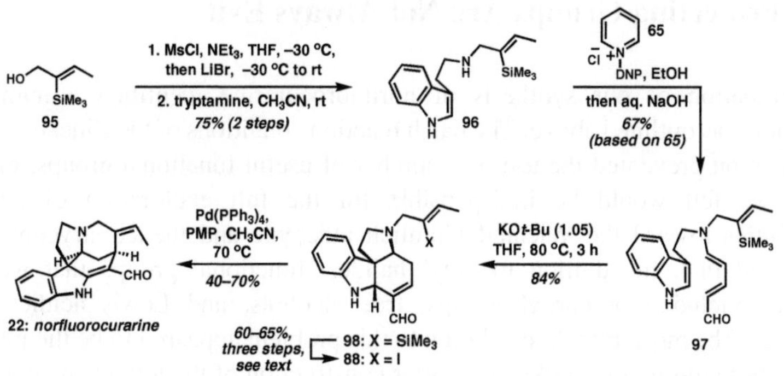

Scheme 4.15 Synthesis of norfluorocurarine (PMP = 1,2,2,6,6-pentamethylpiperidine)

vinylsilane to the corresponding vinyl iodide was required to provide Heck substrate **88**. Unfortunately, many attempts using known iododesilylation conditions led to complex reaction mixtures in which only small quantities of the desired vinyl iodide could be observed [40][2]. This reaction was primarily complicated by undesired halogenation of the electron-rich aromatic ring. This problem could be circumvented using a three-step sequence that featured transitory *N*-trifluoroacetylation of the indoline, and the desired iodide **88** was obtained in good yield. Exposure of this substrate to a catalytic quantity of $Pd(PPh_3)_4$ and a hindered amine base instigated an intramolecular Heck reaction to yield norfluorocurarine (**22**) directly. The successful route proceeds via a longest linear sequence of seven steps from 1-(trimethylsilyl)propyne using the poorly efficient direct iodination protocol, and nine steps using the optimal three-step iodination sequence. Our synthesis of norfluorocurarine served to validate our Zincke aldehyde strategy as an efficient method to access tetracyclic intermediates with well-positioned functionality for subsequent D-ring formation [9a].

In the course of our successful synthesis, we identified several limitations of our new method and associated strategy: (1) the harsh conditions of the bicyclization reaction do not tolerate base-sensitive functionality such as vinyl halides; (2) post-cyclization manipulations such as iododesilylation reactions are complicated by the sensitive/reactive functionality of the products (α,β-unsaturated aldehyde, indoline, etc.); and (3) the incorporation of the required functionality into the Zincke aldehyde requires the synthesis of a complex tryptamine derivative, resulting in a lengthy, non-convergent route. In order to develop a concise route to strychnine, we would have to address each of these issues, and a straightforward solution to obviate all of these is described below.

[2] Although bromide **87**, could also be suitable, we spent most of our efforts on accessing iodide **88**, which would presumably be more reactive in the final Heck cyclization.

4.8 Protecting Groups Are Not Always Evil

Reexamination of our synthesis of norfluorocurarine identified a number of limitations, as outlined above. The harsh reaction conditions of the Zincke aldehyde bicyclization prevented the use of a number of useful functional groups; many of which we felt would be indispensible for the full exploration of endgame possibilities toward the Wieland–Gumlich aldehyde and thence strychnine. We discovered that, in addition to vinyl halides, functional groups that were not tolerated included propargyl groups, free alcohols, and Lewis acidic silanes (Fig. 4.2). The most attractive solution to this problem appeared to be the introduction of these motifs by *N*-alkylation after construction of the tetracyclic core. This strategy would allow for the incorporation of virtually any group and would also enable a more convergent approach. Synthesis of the tetracyclic core and the key D-ring precursor reagent in parallel followed by convergent coupling of the two pieces would result in a shorter linear sequence and more importantly would allow for late-stage divergent synthesis of many substrates via a common intermediate. The pursuit of this common tetracyclic intermediate therefore became our primary goal.

The common intermediate we sought was tetracycle **33**, which bears a free secondary amine. This goal structure could not be synthesized directly by our cycloaddition chemistry, owing to the limited stability of Zincke aldehydes derived from primary amines, which are known to readily convert to pyridinium salts [22a]. What we required was a precursor incorporating a removable substituent—in essence, a protecting group. The virtues of protecting-group free synthesis hardly need to be emphasized. It is clear that unnecessary protection and deprotection steps can lengthen a synthetic sequence and detract from the underlying chemistry. Nonetheless, protecting groups have a long-standing history in organic chemistry, and the synthesis of many classes of complex natural product (e.g., polyketides, polypeptides, carbohydrates) is often not feasible without the incorporation of strategic blocking groups. Given the advantages that a protecting group would provide in our case (greater convergency and divergent access to more complex substrates), we began to evaluate Zincke aldehydes bearing commonly used protecting groups on the nitrogen atom.

The century-old method for the synthesis of Zincke aldehydes placed certain limitations on our choice of substrates: strongly electron-withdrawing groups (Boc, Ts, Ac, etc.) are precluded because formation of Zincke aldehydes is only efficient for relatively electron-rich secondary amines. Furthermore, as vinylogous imides, these particularly electron-deficient systems were expected to be labile to basic conditions. Having established efficient three-step access to *N*-benzyl-type substrates such as **78** (Scheme 4.11), we initially aimed to convert these substrates to our desired common intermediate via amine deprotection. A survey of reaction conditions typically used for the removal of these groups led largely to undesired side reactions of the other functional groups. Cleavage of benzyl-type protecting groups under reductive conditions was precluded by the reactive α,β-unsaturated aldehyde group. Oxidative removal of the electron-rich PMB and DMB

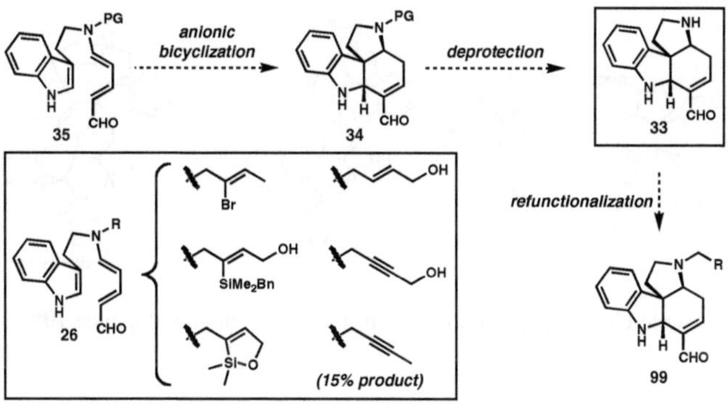

Fig. 4.2 Problematic functional groups (*inset*) and protecting group strategy to alleviate problems caused by the harsh cycloaddition conditions

(2,4-dimethoxybenzyl) groups was similarly unsuccessful owing to complications from the electron-rich indoline. Treatment of DMB-protected amines with a strong acid such as TFA in the presence of a sacrificial nucleophile such as anisole is also known to cleave the C–N bond [41]. Under these conditions, we observed clean dimerization of our starting material to give **101** (Scheme 4.16) containing a central eight-membered diazocine ring. This product is produced by the mutual condensation of the secondary amino and aldehyde groups to give a single diastereomer of product, which we presume is the homodimer[3]. Our identification of the product was guided by the knowledge that dimers such as toxiferine I (**102**) are themselves natural products. Natural and semisynthetic analogs of toxiferine I have been investigated because of their allosteric modulation of the muscarinic acetylcholine receptor [42]. A similar biomimetic dimerization of *Strychnos* alkaloids has been reported to occur under acidic conditions (AcOH, NaOAc, 70 °C); it is interesting to note that in our case a single diastereomer of the dimer is produced from the racemic mixture of starting material.

After evaluation of several other potential protecting groups (including methyl, trimethylsilylethyl, phenylselenenylethyl) and multiple different means for their removal [9c], we turned our attention to the allyl group [41]. We began with N_b-allyl tryptamine, a known compound that is most efficiently prepared by reaction of tryptophyl bromide with allylamine, which was converted to the corresponding Zincke aldehyde **103** in good yield (Scheme 4.17) [43]. The cycloaddition reaction of **103** was particularly sensitive to concentration, giving the highest yield (64 %) of

[3] Our efforts to concretely determine the relative stereochemistry of this dimer have been met by failure. We have made attempts to resolve several of the monomeric tetracyclic aminoaldehydes of type **100** by HPLC using chiral stationary phase, in order to know for sure the structure of the homodimer. The poor solubility of these compounds in typical HPLC solvents hampered these efforts to access enantiopure monomer. A few attempts at diastereomeric salt formation from compounds of type **101** using chiral carboxylic acids were also unsuccessful. Computational analysis corroborates the assumption that the homodimer should be formed preferentially.

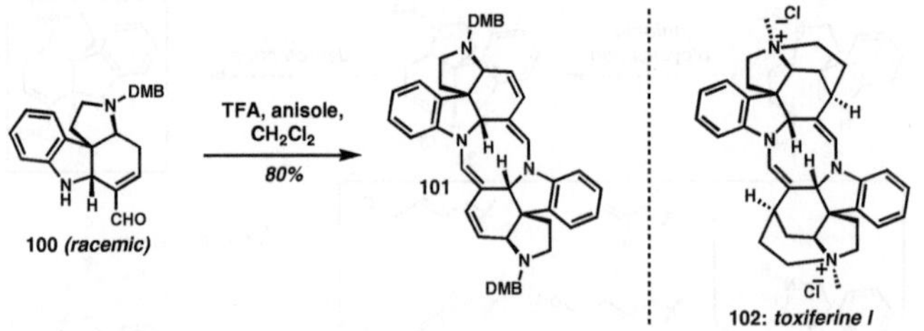

Scheme 4.16 Facile biomimetic dimerization of tetracyclic amino aldehyde **100**

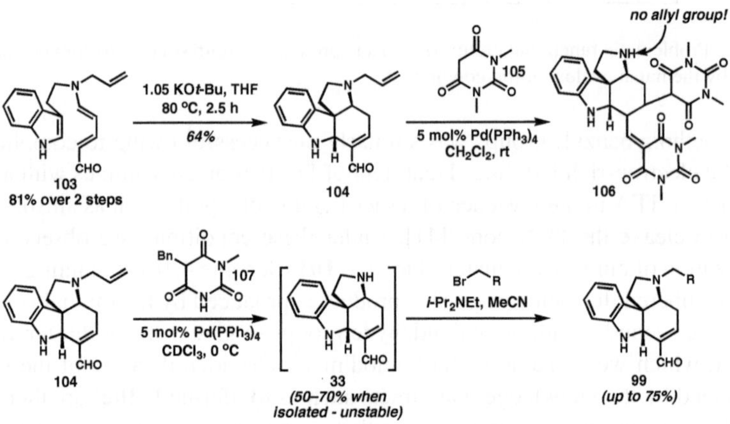

Scheme 4.17 Generation of secondary amine **33** by deprotection of *N*-allyl group enables access to functionalized core structures

tetracycle **104** at 0.02 M. Pd-catalyzed deallylation was effective under very mild conditions (0 °C, <1 h) using *N,N'*-dimethylbarbituric acid as the nucleophilic allyl scavenger [44]; however, this deallylation was accompanied by a Knoevenagel condensation of the barbituric acid derivative with the aldehyde, followed by Michael addition of a second equivalent of this nucleophile. This undesired process could be suppressed by incorporation of an alkyl group on the barbituric acid derivative to prevent the dehydration step of the Knoevenagel reaction. *C*-Benzylated derivative **107** or commercially available 5-methyl Meldrum's acid were used for all our subsequent investigations, and either reagent enabled the isolation of tetracycle **33** in reasonable yield. However, the presence of two nucleophilic amines and two electrophilic carbon atoms resulted in poor stability upon purification and storage. We soon found that in situ realkylation of the liberated secondary amine provided the desired products in good yields (up to 75 %). Fortunately, and surprisingly, the residual palladium did not appear to cause side reactions of the allylic, propargylic, and vinyl halides present in the electrophiles we explored. The excess barbituric acid/ Meldrum's acid derivatives likewise did not interfere. Using this approach, we explored a number of endgame strategies, some of which are detailed in the sections that follow.

4.9 Strategies for D-Ring Formation for Strychnine

At this stage of our investigation, we had only begun to consider the multitude of possibilities for completion of the synthesis of strychnine. Based on their established success in closely related systems, we considered different strategies involving the cyclization of vinylmetal species onto the α,β-unsaturated aldehyde. Inspired by the previous work of others, we contemplated Pd-catalyzed Heck reactions, Ni-mediated reductive Heck/conjugate additions, and Cu-mediated conjugate additions, all of which required a vinyl halide precursor [15]. A Heck reaction, as we and others had demonstrated, would lead to β-hydride elimination toward nitrogen, installing an unwanted alkene at C2–C16 (**108** → **109**, Scheme 4.18). While used to advantage in our synthesis of norfluorocurarine and Rawal's synthesis of dehydrotubifoline (among others) [9a, 15, 18, 32c], such a result is not advantageous for the synthesis of strychnine via the Wieland–Gumlich aldehyde. As mentioned above, several previous syntheses of the Wieland–Gumlich aldehyde have proceeded via the analogous unsaturated ester 18-hydroxyakuammicine (**11**). Reduction of the alkene of this vinylogous carbamate has been accomplished by Zn/H_2SO_4 or $NaBH_3CN$ (**11** → **110**), leaving the ester group intact for further reduction to give the Wieland–Gumlich aldehyde [10c, f, h, k, m, o, p]. At the outset of our studies, it seemed unlikely that the potentially delicate functional group arrangements in **109** would allow for a selective reduction of the C–C π bond. We later learned that Rawal and He had in fact accomplished this transformation using Li/NH_3 in reasonable yield, completing a second synthesis of strychnine by this group that was only published in thesis form [36c]. At the time, however, we were attracted to an alternative possibility that would potentially avoid this post-cyclization redox adjustment and would probe an interesting question of reactivity.

Our hypothesis centered on a question of inherent selectivity in the β-hydride elimination step of the Heck reaction. Allylic alcohol **111** was readily obtained by reduction of norfluorocurarine precursor **88** (Scheme 4.19) and was envisioned as a model system for this approach. A Heck reaction of this substrate could have two possible outcomes: β-hydride elimination from intermediate **112** could occur in the direction of N1 (as it did in the case of norfluorocurarine) to give an enamine that should tautomerize to imine **113**, a natural product named dehydrodesacetylretuline [45]. Alternatively, β-hydride elimination could take place toward the hydroxyl group to give, after tautomerization, aldehyde **114** (deshydroxy-Wieland–Gumlich aldehyde, also a natural product) [46]. Some related systems had been reported to give variable selectivity and could in at least one case be tuned by reaction conditions (see below) [18a]. We saw this as an opportunity to probe such selectivity further, and as it turned out, either result would prove advantageous for alkaloid synthesis. If the aldehyde product predominated, then extension to a C18-hydroxylated vinyl iodide could provide a means to access the Wieland–Gumlich aldehyde. If the allylic alcohol product predominated, we envisioned a short synthesis of the recently reported alkaloid valparicine. This curan alkaloid was isolated from *Kopsia arborea* by Kam and co-workers and exhibits pronounced

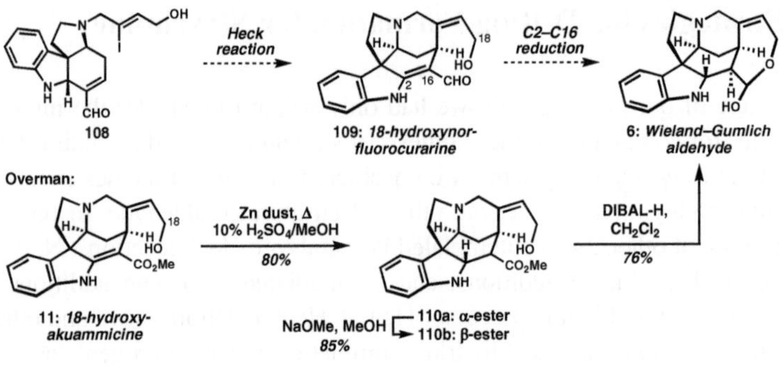

Scheme 4.18 Potential Heck-based endgames for strychnine

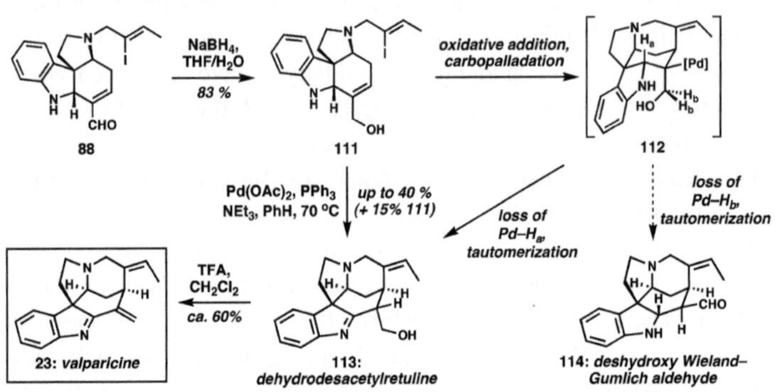

Scheme 4.19 Synthesis of dehydrodesacetylretuline and valparicine via Heck reaction

cytotoxicity toward drug-sensitive and drug-resistant KB cells as well as Jurkat cells [47, 48]. Under all Heck conditions examined with **111**, we never observed any aldehyde products of type **114**; dehydrodesacetylretuline **113** was the major product of successful cyclizations. We later learned that allylic alcohol **111** had been previously synthesized by Rawal and He via a different route and that they observed similar selectivity in the Heck reaction [36c]. When we treated **113** with trifluoroacetic acid, it was converted to valparicine in good yield by dehydration. The spectral data of synthetic valparicine were identical to those reported by Kam and co-workers [47].

The selectivity in the Heck reaction of allylic alcohol **111** is interesting, and the factors that lead to the observed preference for β-hydride elimination toward nitrogen in this system are unclear, although a combination of steric effects and stereoelectronic factors (i.e., alignment of C–H and C–Pd bonds, $n_N \rightarrow \sigma^*_{C\text{–}H}$ interactions) is likely involved. Examination of related examples from the literature (Scheme 4.20) reveals no clear trend. Rawal and Michoud examined substrate **115**, which lacks the influence of both the amine and hydroxyl substituents and also seems to favor β-hydride elimination within the six-membered ring over formation of the exocyclic olefin under standard Heck conditions [18a]. However, under

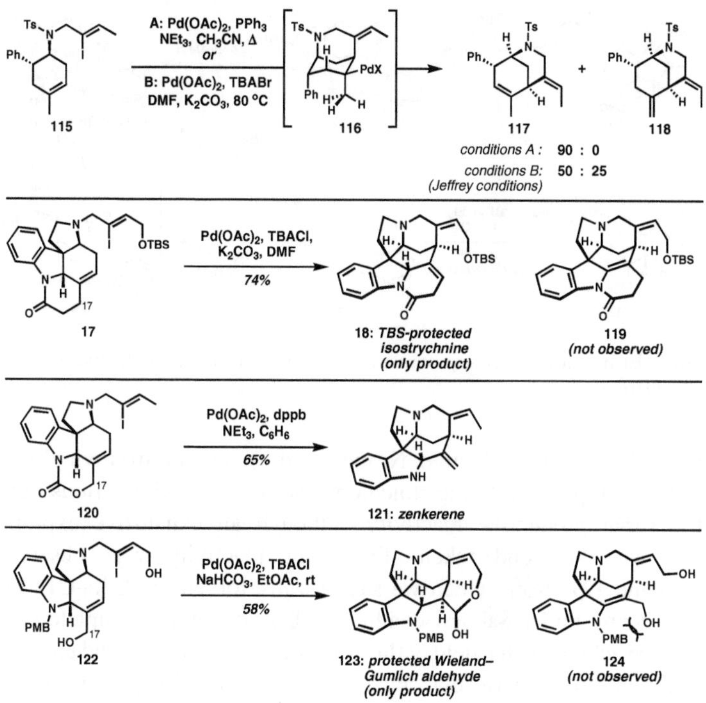

Scheme 4.20 β-H elimination selectivity in relevant Heck cyclizations by the Rawal group and in the MacMillan strychnine synthesis

Jeffery's conditions (with a tetrabutylammonium halide additive) a change in selectivity is observed and a substantial amount of the exocyclic olefin is produced [49]. Rawal and Iwasa reported that pentacyclic substrate **17** undergoes selective β-hydride elimination toward C17 using Jeffery's conditions to give TBS-protected isostrychnine (**18**) in 74 % yield [10e]. Attempts by the Rawal group to favor β-hydride elimination toward a hydroxylated C17 to access the Wieland–Gumlich aldehyde included a Heck reaction of cyclic carbamate **120**, which underwent an unexpected Pd(II)-carboxylate elimination to give zenkerene (**121**) [36c]. To our knowledge, the only example of selective β-hydride elimination toward a C17-hydroxyl group to give an aldehyde product is found in MacMillan's recent synthesis of strychnine (**122** → **123**) [10r]. They report the use of an N_a-PMB group to disfavor formation of a C2–C16 alkene, which would introduce substantial $A_{1,3}$-strain between C17 and the substituent on nitrogen (see **124**).

Having established that the inherent selectivity of the Heck reaction would not readily allow us to access aldehyde products related to the Wieland–Gumlich aldehyde, we considered a number of other possible reaction manifolds for C20–C15 bond formation. Each substrate was accessed by deallylation of cycloadduct **104** followed by alkylation with the appropriate allylic or propargylic halide. Inspired by the extensive studies of Bonjoch, Bosch, Solé, and co-workers,

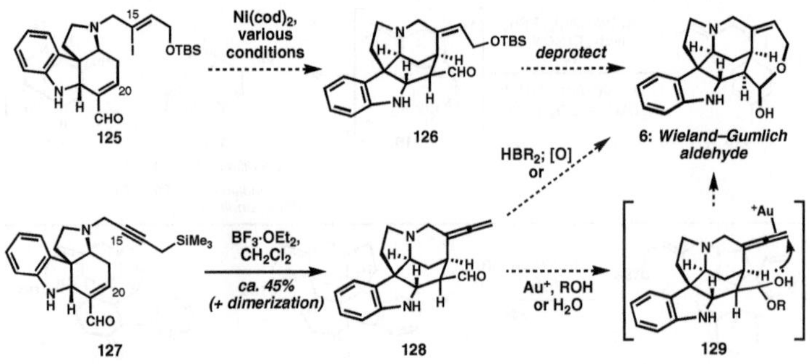

Scheme 4.21 Efforts toward D-ring formation via nickel-mediated reductive cyclization and Sakurai allenylation

we explored the use of a reductive, Ni-mediated cyclization from vinyl iodide **125** (Scheme 4.21) [10g, 36b]. Using stoichiometric Ni(cod)$_2$ with various ligands, we did not observe productive reactivity, although an exhaustive exploration of conditions was not undertaken. Promising reactivity was observed with propargylsilane **127**, from which we successfully forged the C20–C15 bond and closed the D-ring under Sakurai conditions. Unfortunately, complete conversion could never be achieved, the yield of this transformation was low and unreliable, and varying amounts of dimeric diazocine products were also produced. A similar Sakurai product to **128** had been selectively hydrogenated by Bonjoch and co-workers in a synthesis of akuammicine; however, rather than reduction, we required a hydration of the allene terminal double bond [36b, 50]. Initially we envisioned hydration via hydroboration/oxidation; this hydroboration would need to be chemo-, regio-, and stereoselective and would need to leave the aldehyde untouched. An alternative method would be a gold-catalyzed hydration or 7-endo cyclization event of the allene and aldehyde (or hemiacetal/hydrate thereof). Given our limited access to allene **128**, these two conceptually intriguing strategies were only briefly examined with no obvious success, and we soon turned to alternative strategies for D-ring formation.

Another option that presented itself was cyclization by the direct conjugate addition of our previously synthesized vinylsilane substrate **98** to form 18-deshydroxy-Wieland–Gumlich aldehyde [Eq. (4.7)]. Again, this reaction was meant to serve as a model for the eventual approach to the Wieland–Gumlich aldehyde itself. Vinylsilanes are not particularly nucleophilic; however, they do benefit from stereospecificity in electrophilic desilylation reactions [51]. We began by simply heating vinylsilane **98** but no reaction was observed, except for decomposition at very high temperature (>200 °C). We next treated **98** with Lewis acids to activate the α,β-unsaturated aldehyde toward nucleophilic attack. A variety of Lewis acids ($TiCl_4$, $Sc(OTf)_3$, etc.) were ineffective in generating any cyclization products, although in some cases dimerization occurred to give diazocine products, as we had observed earlier with protic acids.

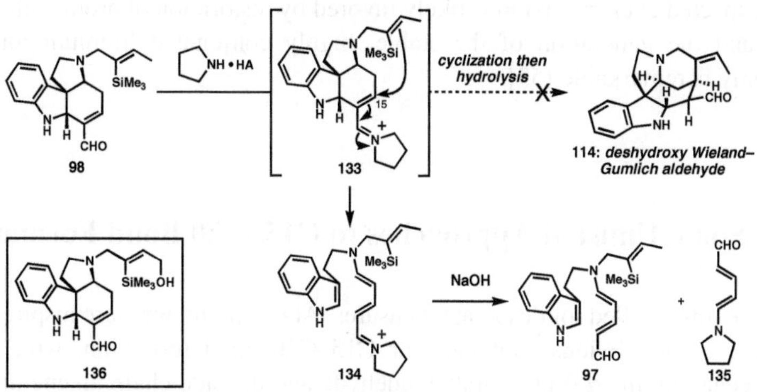

Scheme 4.22 An unexpected cycloreversion reaction

Another method of activation we considered was the use of a secondary amine to generate a more electrophilic iminium species, examples of which have been used in vinylsilane-terminated cyclization reactions, particularly by Overman and co-workers [Eq. (4.8)] [51]. In our case, the unsaturated iminium ion would be activated for

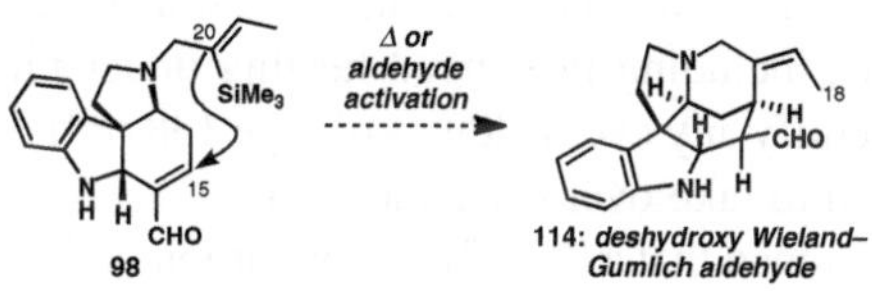

(4.7)

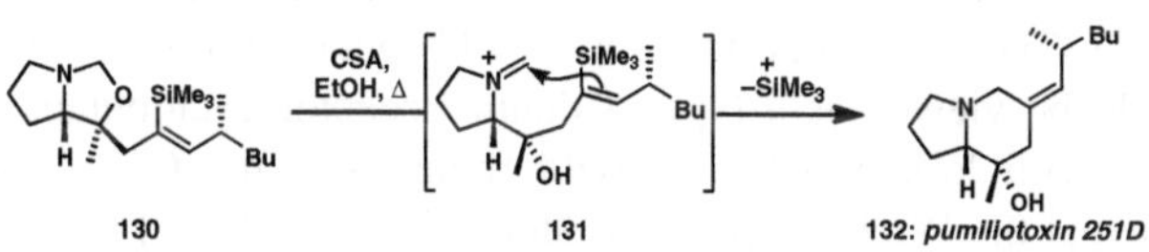

(4.8)

intramolecular attack at C15, analogous to the iminium activation popularized by MacMillan and others [52]. Using pyrrolidine with acid catalysis, we observed the formation of the iminium species **133** from **98** by mass spectrometry (Scheme 4.22). Even after substantial heating (up to 140 °C), no cyclization was observed, but a new product with the same mass as the iminium species was eventually observed. After aqueous work-up, we isolated a familiar product: Zincke aldehyde **97**. We also observed Zincke aldehyde **135**, as well as the secondary amines derived from hydrolysis of the intermediate iminium ion. Similar results were obtained using different secondary amines, in different solvents, and using the C18-hydroxylated silane **136** (see below for its synthesis). Apparently, iminium species **133** undergoes cycloreversion by a stepwise mechanism [53] to give the fully conjugated iminium species **134** rather than engaging the pendant vinylsilane as a nucleophile and leading to **114**. Contrasting with the successful cycloaddition, in which the formation of the tetracyclic product—as its corresponding enolate—is clearly favored, this unexpected cycloreversion is likely favored by restoration of aromaticity of the indole and the generation of the stable, highly conjugated iminium ion. This dichotomy is remarkable [53].

4.10 Some Unusual Approaches to C15–C20 Bond Formation

Desperate times called for desperate measures. At this point, we were inspired to try a number of nonobvious strategies for C15–C20 bond formation, with an eye toward engaging the α,β-unsaturated aldehyde and the side-chain alkene in cycloaddition reactions. Cycloaddition reactions are particularly well suited to forming hindered bonds in complex systems, and the lower entropy of activation for intramolecular processes provides an additional advantage [54]. First, we imagined that if the α,β-unsaturated aldehyde could act as a heterodiene in a [4 + 2] reaction with the side-chain alkene, for example, the vinylsiloxane in **137**, the product would contain the required C15–C20 bond. This product would likely exhibit strain and might undergo a subsequent fragmentation to reintroduce the C19–C20 alkene with loss of silicon. Although the alignment of the orbitals in question did not appear ideal for a concerted fragmentation, stepwise processes involving either a silicon-stabilized carbocation or a siliconate complex could also be envisioned. Under a number of conditions, including prolonged heating under microwave irradiation, we were unable to identify any cycloadducts of type **138** or fragmentation products, and degradation began to take over (Scheme 4.23).

Similarly, we envisioned that a [2 + 2] cycloaddition of the C15–C16 olefin with a pendant alkene could deliver a similar product containing the required C15–C20 bond (**136** → **139**) [55]. This cycloadduct would contain a strained four-membered ring that could be viewed as a type of donor–acceptor cyclobutane. Subsequent fragmentation with loss of silicon would reintroduce the C19–C20 olefin and potentially reveal the Wieland–Gumlich aldehyde. The [2 + 2] cycloaddition was explored under a variety of conditions including photoirradiation with UV light or full-spectrum light using a mercury-vapor lamp. Reactions were performed at various temperatures and in some cases with catalytic copper(II) salts, which are known to catalyze [2 + 2] cycloadditions of unactivated systems [56]. Once again, we were unable to identify any desired products, even using the less-hindered allyl substrate that would have provided **142**. The strain in the desired hexacyclic products might be sufficiently high that the barrier to cycloaddition cannot be overcome, and decomposition becomes the dominant process under forcing conditions. We also note the paucity of literature examples of photochemical [2 + 2] cycloadditions of α,β-unsaturated aldehydes. Although decarbonylation might appear to be a concern, one of the most detailed studies on the intramolecular enal–olefin photocycloaddition suggests that post-cyclization rearrangements represent the bigger problem [57].

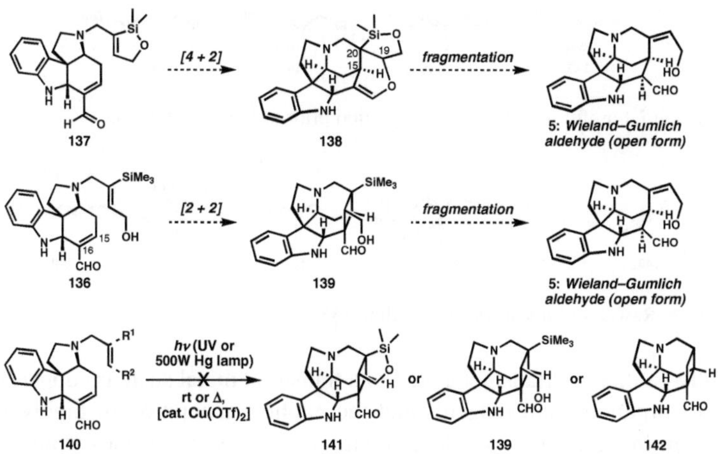

Scheme 4.23 Some unusual cycloaddition approaches to forge the C15–C20 bond

4.11 A Successful Route to Strychnine

Given our failure to engage the α,β-unsaturated aldehyde using the C19–C20 π bond in a direct cyclization or a cycloaddition reaction, we reexamined the opportunities presented by reactive vinylmetal intermediates. A conjugate addition reaction of a suitable vinylmetal species was the most attractive option to directly deliver the Wieland–Gumlich aldehyde without the need for other redox manipulations. Most often, vinylmetal species are generated by lithium–halogen exchange of the precursor vinyl halides, and reactivity can be tempered with catalytic or stoichiometric quantities of copper salts [58]. A specific example of this approach is found in Stork's unpublished synthesis of strychnine, some details of which are available in the literature [15a, 17]. Exposure of protected vinyl iodide **143** to *t*-BuLi followed by a mixture of $CuCl_2$ and $MnCl_2$ at −30 °C is reported to give rise to pentacyclic ester **144** in moderate yield [Eq. (4.9)]. This reaction demonstrated the feasibility of a copper-mediated ring closure via conjugate addition.

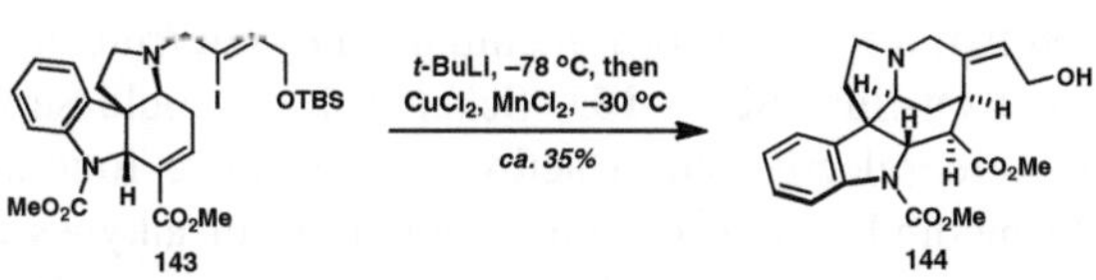

(4.9)

The conjugate addition substrate that we were considering bore an obvious resemblance to Stork's; however, it would incorporate a free NH and OH, as well as a more sensitive aldehyde. In this light, an approach involving reactive organolithium species seemed unlikely to be successful, and a few experiments quickly validated this surmise. We therefore sought a precursor that could

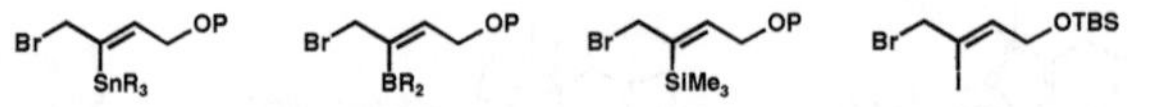

Fig. 4.3 Potential useful reagents for *N*-alkylation prior to D-ring formation

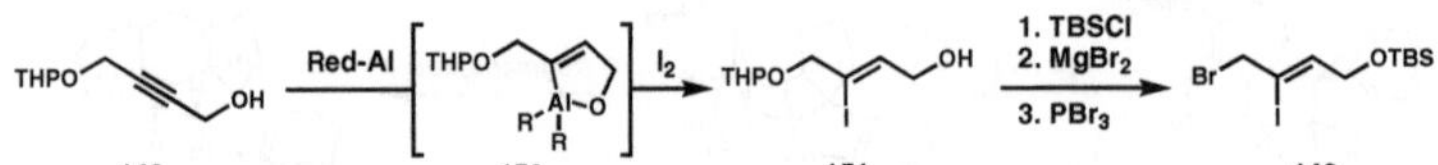

Scheme 4.24 Rawal's synthesis of vinyl iodide **155**

potentially be directly transmetallated to a transition metal such as copper, without the use of organolithium reagents and the explicit generation of highly reactive vinyllithium intermediates. While a vinyltin or vinylboron species would be ideal for this purpose, the stereoselective synthesis of vinylstannane **145** or vinylborane **146** (Fig. 4.3), to be incorporated by *N*-alkylation of tetracycle **33**, appeared challenging. In addition, β-bromostannanes related to **145** were found to be unstable with respect to allene formation and loss of R_3SnBr [59]. Based on our previous experience, we expected that the corresponding silane would be quite stable. The challenge became to find a concise and stereoselective synthesis of vinylsilane **147**.

Vinyl iodide **148**, used by Rawal and others, was made using a *trans*-hydroalumination of propargylic alcohol **149** with Red-Al (sodium bis(2-methoxyethoxy)-aluminum hydride) followed by an iodine quench to afford stereodefined, trisubstituted alkene **151** (Scheme 4.24) [10e]. This method is one of the reliable ways to access a trisubstituted olefin of this type with Z stereochemistry, but this reaction depends upon the presence of the propargylic alcohol. Most alkyne hydrometallations occur in a syn manner (see, e.g., the synthesis of vinylsilane **95**), and we are not aware of any syn-difunctionalization (or hydrometallation/functionalization) reactions that would provide any of the substrates shown in Fig. 4.3 from readily available materials in very few steps. Regardless of the methods used, we sought to uncover a more direct synthesis to maximize convergency. For comparison, iodide **148** is made in a total of six steps from propargyl alcohol, including three protecting group manipulations. Because we could gain access to our pivotal *N*-allyl substrate **104** in only three steps, we felt compelled to advance a new and more direct solution.

We recognized that a KO*t*-Bu- or ruthenium-catalyzed *trans*-hydrosilylation of alkynes (e.g., for KO*t*-Bu: **152** → **153**, Scheme 4.25) could serve as a pivotal reaction for the stereoselective construction of a suitable four-carbon fragment [60]. These mild methods for *trans*-hydrometallation of alkynes are directed by nearby hydroxyl groups and tolerate a broader range of functionality than the *trans*-hydroalumination reaction. Each reaction manifold can be used to give access to siloxacyclopentenes such as **153**, and this functional group has previously been directly transmetallated to copper and used for C–C bond formation (**153** → **154**) [60a]. With complex propargylic alcohol **155**, we found the most success with the method of Trost and Ball that uses $[Cp^*Ru(NCCH_3)_3]PF_6$ as the catalyst and $Me_2SiH(OEt)$ as the silane. The presumed initial product undergoes loss of EtOH

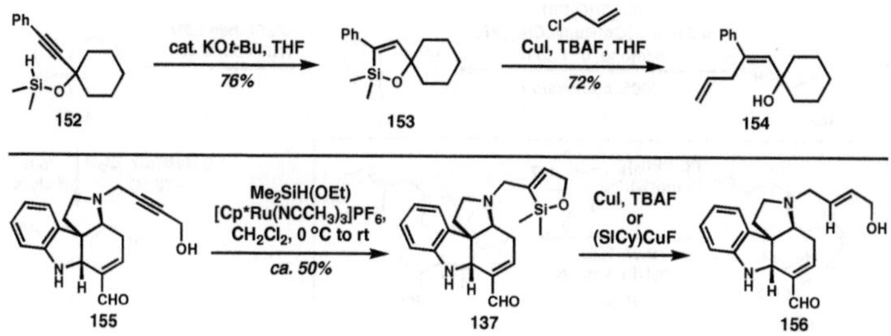

Scheme 4.25 Cyclic siloxane as a potential precursor to C15–C20 bond formation

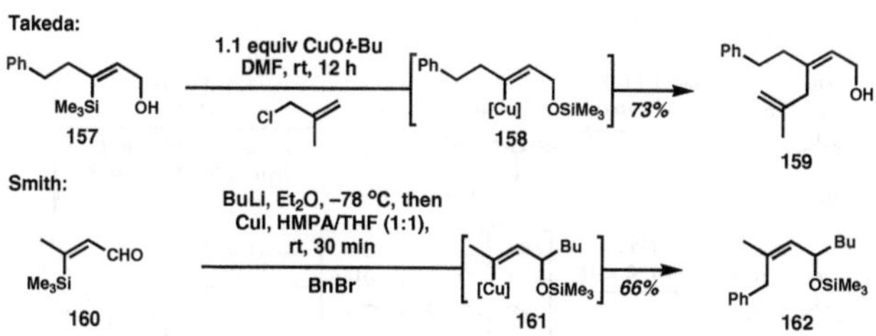

Scheme 4.26 Precedent for Brook rearrangement/transmetallation of vinylsilanes bearing allylic alcohols

after the hydrosilylation event to give rise to siloxacyclopentene **137** [60b]. Unfortunately, when we attempted to effect the transmetallation/conjugate addition reaction in this challenging setting, we observed only protodesilylation product **156**. The protodesilylation of silacycle **137** occurs in the presence of TBAF without a copper source, suggesting that under rigorously anhydrous conditions, protodemetallation might be avoided. Using anhydrous conditions, including an NHC·CuF complex (SICy)CuF designed and supplied to us by Ball and Herron [61], we still observed only protodemetallation; the rate of cyclization was simply too slow to compete with premature quenching of the reactive organocopper intermediate under these conditions.

Further examination of the literature led us to a report by Takeda and co-workers that demonstrated the generation of vinylcopper species from vinyltrimethylsilanes, followed by C–C bond formation (**157** → **158** → **159**, Scheme 4.26) [62]. Although tetraalkylsilanes generally transmetallate quite slowly, the *cis*-disposed alcohol can participate in activation of the silicon toward transmetallation to copper in what might be described as a copper-assisted Brook rearrangement [63]. This general reaction type has been further expanded by Smith and co-workers to include different preparations of the key alkoxide (**160** → **161**) and a variety of uses of the resultant vinyl copper, including transmetallation to Pd and subsequent cross-coupling (not shown) [64].

As a result of the excellent precedent from the Takeda and Smith groups, we targeted vinylsilane **136** as a key intermediate in a Brook rearrangement/conjugate

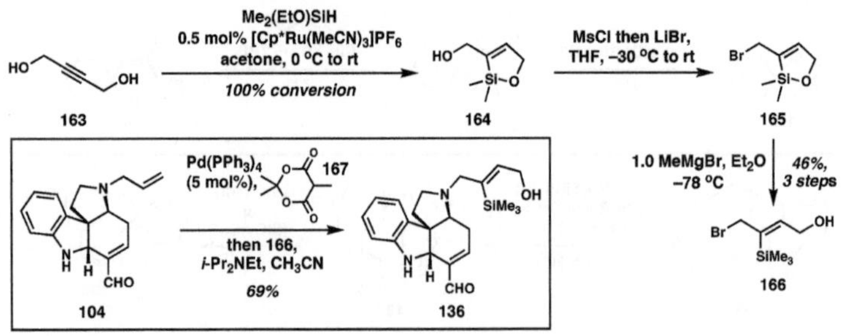

Scheme 4.27 Synthesis of Brook rearrangement substrate **136** featuring a three-step synthesis of polyfunctional building block **166**

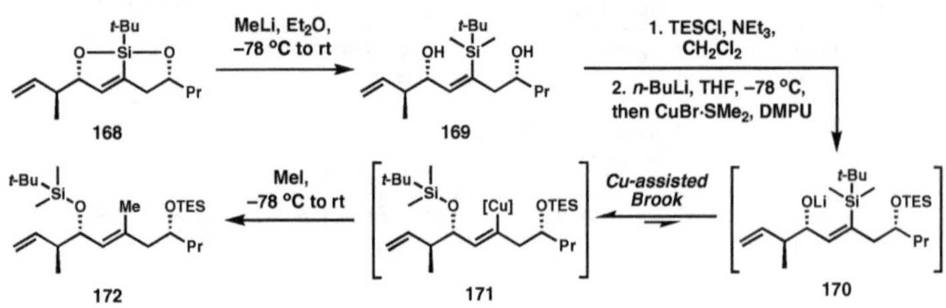

Scheme 4.28 Leighton's precedent for siloxacycle ring opening and Brook rearrangement/C–C bond formation

addition strategy to build the D-ring of strychnine. The synthesis of **136** (Scheme 4.27) begins with the ruthenium-catalyzed *trans*-hydrosilylation of 1,4-butynediol using the method of Trost and Ball, providing silane **164** in high yield with complete control of alkene geometry [60b]. This reaction is scalable, proceeds with low catalyst loadings, and converts inexpensive 1,4-butynediol into the required Z-vinylsilane, while at the same time differentiating the two hydroxyl groups. With one of the hydroxyl groups internally protected as part of the siloxacycle, the other could be converted to the bromide (**165**). Subsequent treatment with methylmagnesium bromide chemoselectively opened the siloxacycle and provided the desired **166** in 46 % yield over three steps. A related example of ring opening of a siloxacycle with methyllithium is found in Leighton's synthesis of dolabelide D, revealing a vinyl-TBS group (see Scheme 4.28) [65]. The efficient sequence leading to silane **166** provides rapid access to significant quantities of this polyfunctional, stereodefined trisubstituted alkene and may prove useful for the synthesis of related vinylsilanes. Fortunately, the free alcohol in **166** did not readily suffer alkylation, and it could be stored neat (at −20 °C), allowing for easy handling of this important intermediate. When bromide **166** was added to the reaction mixture after complete deallylation of **104** using methyl Meldrum's acid (**167**), refunctionalized core **136** was isolated in 69 % yield. The alkylation was chemoselective for N_b-alkylation, and the ability to work with the free hydroxyl group eliminated the need for further protecting group manipulations.

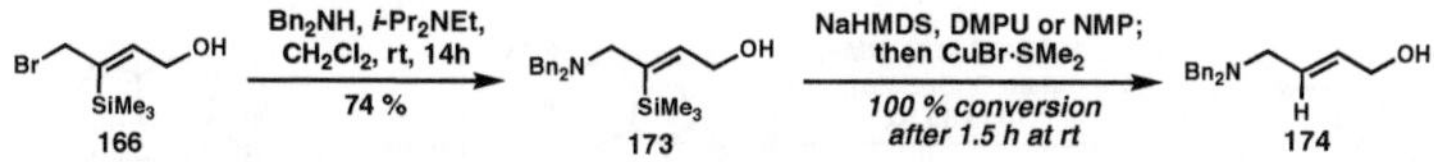

Scheme 4.29 Synthesis and Brook rearrangement of model system **173**

With convenient access to vinylsilane **136**, we again investigated the C15–C20 bond formation. Application of conditions reported by Takeda (CuO*t*-Bu, THF or DMF) [62] and co-workers resulted in recovery of starting materials at lower temperature and decomposition at elevated temperature. Similarly, conditions inspired by Smith and co-workers (*n*-BuLi or NaHMDS, then CuI, THF/HMPA or THF/DMPU) [64] did not result in C–C bond formation in the real system or model system **173** (Scheme 4.29) which also incorporates the potentially problematic allylic tertiary amine. Smith had demonstrated that the addition of a polar cosolvent was necessary to trigger the Brook rearrangement of the alkoxide (see Scheme 4.26). Similarly, a Brook rearrangement in Leighton's synthesis of dolabelide D proceeded with addition of $CuBr{\cdot}SMe_2$ in DMPU to trigger the migration event (Scheme 4.28) [65]. In our case also, it was not until DMPU (or NMP) was used as the sole solvent that the Brook rearrangement/protonolysis product was observed in the model system, cleanly generating **174**.

When these optimized conditions were applied to **136**, Brook rearrangement could be triggered in the presence of excess base, but again cyclization did not occur. When the reaction mixture was heated to 40 °C, partial cyclization was observed, and <5 % of the Wieland–Gumlich aldehyde was painstakingly isolated from the mixture of reaction products. Apparently, the free alcohol present in **136** had indeed enabled a Brook rearrangement with transmetallation to copper, followed by intramolecular conjugate addition to afford the Wieland–Gumlich aldehyde, albeit in low yield. Further optimization identified NMP as a better solvent, NaHMDS and KHMDS as preferred bases (Li prevents Brook rearrangement), and 65 °C as the best temperature for cyclization, yielding the Wieland–Gumlich aldehyde in up to 10 % yield. While the yield of this final bond construction is certainly lower than desired, there are several other reports that highlight the difficulty in forging this type of bond in related settings[4]. With the known conversion of the Wieland–Gumlich aldehyde to strychnine [4c], the adoption of this Brook rearrangement-based strategy enabled the completion of the synthesis in only six linear steps from commercially available starting materials because of the rapid, convergent assembly of vinylsilane **136**. Although the brevity of our synthesis of strychnine is certainly noteworthy, the rapid buildup of complexity using the reactivity options provided by the Zincke aldehyde is at the heart of this accomplishment. In the course of only four chemical steps, four new

[4] (a) For a relevant discussion of the difficulty of C15–C20 bond construction, see [101] and [65a]. (b) For a discussion of D-ring formation in related pyrroloindoline natural products, see [20d].

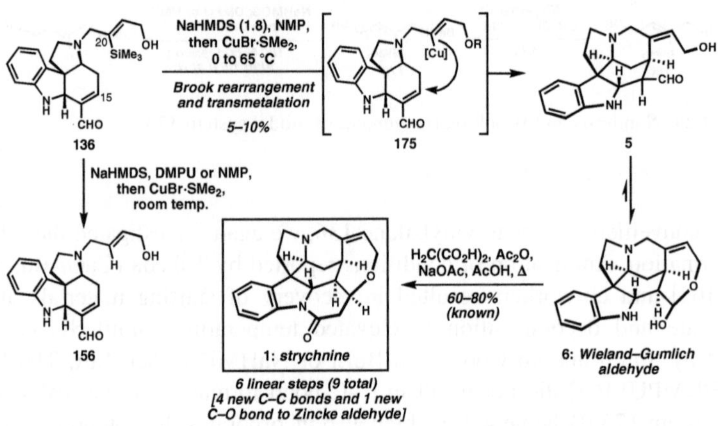

Scheme 4.30 Completion of a six-step linear synthesis of strychnine using a Brook rearrangement/conjugate addition to generate the Wieland–Gumlich aldehyde

carbon–carbon bonds and one carbon–oxygen bond are forged to the five carbons of the Zincke aldehyde, exploiting much of the latent functionality present in this type of donor–acceptor diene (Scheme 4.30).

4.12 Conclusions

The synthesis of natural products continues to be an exciting and tremendously fulfilling pursuit and one that regularly uncovers unexpected and interesting reactivity. Our goal of synthesizing members of the *Strychnos* alkaloids was initiated by our interest in using Zincke aldehydes as starting points for the rapid generation of molecular complexity, and the wealth of interesting reactivity encountered, both anticipated and unexpected, surpassed our expectations. In our early efforts toward these alkaloids, an unusual pericyclic cascade rearrangement was found that has inspired studies toward unrelated alkaloids. Eventually, a base-mediated cycloaddition reaction of tryptamine-derived Zincke aldehydes was developed, and the synthesis of several *Strychnos* alkaloids could be achieved, including a short synthesis of strychnine. With nature's seemingly limitless capacity to create complex molecules that challenge our understanding of what can be made from a few simple elements, natural product synthesis will continue to be a source of inspiration for many years to come, especially as the emphasis continues to shift from syntheses that are minimally productive to those that are short and efficient.

Acknowledgment We dedicate this account to the memory of David Gin, a wonderful person and truly inspiring scientist. We thank the NSF (CAREER Award CHE-0847061) for support of our work. This project also benefited from support through graduate fellowships to D.B.C.M. from Eli Lilly, the NSERC of Canada, and Bristol–Myers Squibb, as well as an Amgen Young Investigator

Award, an AstraZeneca Award for Excellence in Chemistry, and an Eli Lilly Grantee Award to C. D.V. D.B.C.M is a recipient of a Roche Excellence in Chemistry Award, and C.D.V. is a fellow of the A.P. Sloan Foundation and is a University of California, Irvine Chancellor's Faculty Fellow.

References

1. Robinson R (1952) In: Cook JW (ed) Progress in organic chemistry. Butterworths, London, p 2
2. (a) Pelletier PJ, Caventou JB (1818) Ann Chim Phys 8:323; (b) Pelletier PJ, Caventou JB (1819) Ann Chim Phys 10:142
3. Regnault V (1838) Ann 26:17–35
4. (a) For a historical review of the chemistry of strychnine, see: Smith GF (1965) In: Manske RH (ed) The alkaloids, vol 8. Academic, New York, NY, pp 591–671; (b) For a lead ref that includes the synthesis of the Wieland–Gumlich aldehyde from strychnine, see: Anet FAL, Robinson R (1955) J Chem Soc:2253–2262; (c) For the original conversion of the Wieland–Gumlich aldehyde into strychnine, see: Anet FAL, Robinson R (1953) Chem Ind:245
5. (a) Briggs LH, Openshaw HT, Robinson R (1946) J Chem Soc:903–908; (b) Holmes HL, Openshaw HT, Robinson R (1946) J Chem Soc:910–912; (c) Woodward RB, Brehm WJ (1948) J Am Chem Soc 70:2107–2115
6. Woodward RB, Cava MP, Ollis WD, Hunger A, Daeniker HU, Schenker K (1963) Tetrahedron 19:247–288
7. (a) Hofmann AW (1851) Ann 78:253–286; (b) Hofmann AW (1881) Ber 14:659–669; (c) Polonovski M, Polonovski M (1927) Bull Soc Chim Fr 41:1190–1208; (d) Polonovski M, Polonovski M (1930) Bull Soc Chim Belg 39:1–39; (e) von Braun J (1900) Ber 33:1438; (f) Emde H (1909) Ber Deutsch Chem Ges 42:2590
8. For our previous work with Zincke aldehydes, see: (a) Kearney AM, Vanderwal CD (2006) Angew Chem Int Ed 45:7803–7806; (b) Steinhardt SE, Silverston JS, Vanderwal CD (2008) J Am Chem Soc 130:7560–7561; (c) Michels TD, Rhee JU, Vanderwal CD (2008) Org Lett 10:4787–4790; (d) Steinhardt SE, Vanderwal CD (2009) J Am Chem Soc 131:7546–7547; (e) Michels TD, Kier MJ, Kearney AM, Vanderwal CD (2010) Org Lett 12:3093–3095; (f) Paton RS, Steinhardt SE, Vanderwal CD, Houk KN (2011) J Am Chem Soc 133:3895–3905; (g) Vanderwal CD. J Org Chem published online 30 August (2011) J Org Chem 76:9555–9567
9. For preliminary accounts of our research in this area, see: Martin DBC, Vanderwal CD (2009) J Am Chem Soc 131:3472–3473; (b) Martin DBC, Vanderwal CD (2011) Chem Sci 2:649–651; (c) Martin DBC, Nguyen LQ, Vanderwal CD (submitted for publication) (2012) J Org Chem 77:17–46
10. (a) Woodward RB, Cava MP, Ollis WD, Hunger A, Daeniker HU, Schenker K (1954) J Am Chem Soc 76;4749–4751; (b) Magnus P, Giles M, Bonnert R, Kim CS, McQuire L, Merritt A, Vicker N (1992) J Am Chem Soc 114:4403–4405; (c) Knight SD, Overman LE, Pairaudeau G (1993) J Am Chem Soc 115:9293–9294; (d) Kuehne ME, Xu F (1993) J Org Chem 58:7490–7497; (e) Rawal VH, Iwasa S (1994) J Org Chem 59:2685–2686; (f) Kuehne ME, Xu F (1998) J Org Chem 63:9427–9433; (g) Solé D, Bonjoch J, García-Rubio S, Peidró E, Bosch J (1999) Angew Chem Int Ed 38:395–397; (h) Ito M, Clark CW, Mortimore M, Goh JB, Martin SF (2001) J Am Chem Soc 123:8003–8010; (i) Eichberg MJ, Dorta RL, Lamottke K, Vollhardt KPC (2000) Org Lett 2:2479–2481; (j) Nakanishi M, Mori M (2002) Angew Chem Int Ed 41:1934–1936; (k) Bodwell GJ, Li J (2002) Angew Chem Int Ed 41:3261–3262; (l) Ohshima T, Xu Y, Takita R, Shimuzu S, Zhong D, Shibasaki M (2002) J Am Chem Soc 124:14546–14547; (m) Kaburagi Y, Tokuyama H, Fukuyama T (2004) J Am Chem Soc 126:10246–10247; (n) Zhang H, Boonsombat J, Padwa A (2007) Org Lett 9:279–282; (o) Sirasani G, Paul T, Dougherty W Jr, Kassel S, Andrade RB (2010) J Org Chem 75:3529–3532;

(p) Beemelmanns C, Reissig H-U (2010) Angew Chem Int Ed 49: 8021–8025; (q) Ref. [9b]; (r) Jones SB, Simmons B, Mastracchio A, MacMillan DWC (2011) Nature 475:185–188
11. Woodward RB (1948) Nature 168:155–156
12. Le Men J, Taylor WI (1965) Experientia 21:508–510
13. For biogenetic studies of the Strychnos alkaloids, see: (a) Bisset NG (1980) In: Phillipson JD, Zenk MH (eds) Indole and biogenetically related alkaloids, Academic, London, pp 27–61; (b) Kisakürek MV, Leeuwenberg AJM, Hesse M (1983) In: Pelletier SW (ed) Alkaloids: chemical and biological perspectives, vol 1. Wiley, New York, NY, pp 211–376
14. (a) Dadson BA, Harley-Mason J, Foster GH (1968) J Chem Soc Chem Commun:1233; (b) Harley-Mason J (1975) Pure Appl Chem 34:167–174; (c) Ando M, Büchi G, Ohnuma T (1975) J Am Chem Soc 97:6880–6881; (d) Overman LE, Sworin M, Burk RM (1983) J Org Chem 48:2685–2690; (e) Kuehne ME, Podhorez DE, Mulamba T, Bornmann WG (1987) J Org Chem 52:347–353
15. For a review of strychnine syntheses prior to 2000, see: (a) Bonjoch J, Solé D (2000) Chem Rev 100:3455–3482; (b) For a more recent review, see: Mori M (2010) Heterocycles 81:259–292; (c) Cannon JS, Overman LE (2012) Angew Chem Int Ed 51:4288–4311
16. Prelog V, Battegay J, Taylor WI (1948) Helv Chim Acta 31:2244–2246
17. Link JT (2002) Organic reactions, vol 60. Wiley, Hoboken, NJ, Chapter 2
18. For seminal work on the use of an intramolecular Heck reaction for the synthesis of Strychnos alkaloids, see: (a) Rawal VH, Michoud C (1991) Tetrahedron Lett 32:1695–1698; (b) Rawal VH, Michoud C, Monestel RF (1993) J Am Chem Soc 115: 3030–3031; (c) Ref. [10e]
19. (a) Geissoschizal: Birman VB, Rawal VH (1998) Tetrahedron Lett 39:7219–7222; (b) Minfiensine: Dounay AB, Humphreys PG, Overman LE, Wrobleski AD (2008) J Am Chem Soc 130:5368–5377; (c) Apparicine: Bennasar M-L, Zulaica E, Solé D, Alonso S (2009) Chem Commun:3372–3374; (d) Aspidophylline: Zu L, Boal BW, Garg NK (2011) J Am Chem Soc 133:8877–8879
20. For the related Pd-catalyzed enolate vinylation, see: (a) Solé D, Diaba F, Bonjoch J (2003) J Org Chem 68:5746–5749; (b) Boonsombat J, Zhang H, Chughtai MJ, Hartung J, Padwa A (2007) J Org Chem 73:3539–3550; (c) Ref. [19b]; (d) Zhang D, Song H, Qin Y (2011) Acc Chem Res 44:447–457
21. For a review of heterocycle synthesis using various Pd-catalyzed methods, see: Zeni G, Larock RC (2006) Chem Rev 106:4644–4680
22. Seminal papers: (a) Zincke T (1903) Liebigs Ann Chem 330:361–374; (b) Zincke T (1904) Liebigs Ann Chem 333:296–345; (c) Zincke T, Wurker W (1905) Liebigs Ann Chem 338:107–141; (d) König W (1904) J Prakt Chem 69:105–137
23. González-Béjar M, Stiriba S-E, Domingo LR, Pérez-Prieto J, Miranda MA (2006) J Org Chem 71:6932–6941
24. (a) Haberl U, Steckhan E, Blechert S, Wiest O (1999) Chem Eur J 5:2859–2865; (b) Peglow T, Blechert S, Steckhanb E (1999) Chem Commun:433–434
25. (a) Bäckvall J-E, Plobeck NA, Juntunen SK (1989) Tetrahedron Lett 30:2589–2592; (b) Sato S, Fujino T, Isobe H, Nakamura E (2006) Bull Chem Soc Jpn 79:1288–1292
26. (a) Elliott GI, Fuchs JR, Blagg BSJ, Ishikawa H, Tao H, Yuan Z-Q, Boger DL (2006) J Am Chem Soc 128:10589–10595; (b) Ishikawa H, Elliott GI, Velcicky J, Choi Y, Boger DL (2006) J Am Chem Soc 128:10596–10612; (c) Sasaki Y, Kato D, Boger DL (2010) J Am Chem Soc 132:13533–13544
27. (a) Büchi G, Matsumoto K, Nishimura H (1971) J Am Chem Soc 93:3299–3301; (b) Ref. [14c]
28. For a Lewis-acid-mediated cyclization reaction of a tryptamine-derived vinylogous amide that affords the Aspidosperma skeleton, see: Huizenga RH, Pandit UK (1991) Tetrahedron 47:4155–4164
29. (a) Markó IE, Southern JM, Adams H (1992) Tetrahedron Lett 33:4657–4660; (b) Turet L, Markó IE, Tinant B, Declercq J-P, Touillaux R (2002) Tetrahedron Lett 43:6591–6595; (c) Heureux N, Wouters J, Markó IE (2005) Org Lett 7:5245–5248
30. Rosenmund P, Hosseini-Merescht M, Bub C (1994) Liebigs Ann Chem:151–158

31. Baldwin JE, Claridge TDW, Culshaw AJ, Heupel FA, Lee V, Spring DR, Whitehead RC (1999) Chem Eur J 5:3154–3161, and references therein
32. Kinsman AC, Kerr MA (2003) J Am Chem Soc 125:14120–14125
33. Lin H, Danishefsky SJ (2003) Angew Chem Int Ed 42:36–51
34. Nuhant P, Raikar SB, Wypych J-C, Delpech B, Mazarano C (2009) J Org Chem 74:9413–9421
35. (a) Stauffacher D (1961) Helv Chim Acta 44:2006–2015; (b) Rakhimov DA, Malikov VM, Yusunov CY (1969) Khim Prir Soedin 5:461–462; (c) For NMR data, see: Clivio P, Richard B, Deverre J-R, Sevenet T, Zeches M, Le Men-Oliver L (1991) Phytochemistry 30:3785–3792
36. (a) Crawley GC, Harley-Mason J (1971) Chem Commun:685–686; (b) Bonjoch J, Solé D, García-Rubio S, Bosch J (1997) J Am Chem Soc 119:7230–7240; (c) He S (2001) Ph.D. Dissertation (Rawal), University of Chicago, Chicago, IL
37. (a) Overman reported a similar inability to effect N-formylation of dehydrotubifoline: Angle SR, Fevig JM, Knight SD, Marquis RW Jr, Overman LE (1993) J Am Chem Soc 115:3966–3976; (b) In contrast, C-functionalization could be effected in the related Aspidosperma series (32 % C vs. 22 % N); see Ref. [14d]
38. (a) For Z-1,2-dibromo-2-butene, see: Miyaura N, Ishikawa M, Suzuki A (1992) Tetrahedron Lett 33:2571–2574; (b) For Z-1-bromo-2-iodo-2-butene, see Ref. [18b]
39. Metz P, Linz C (1994) Tetrahedron 50:3951–3966
40. Ilardi EA, Stivala CE, Zakarian A (2008) Org Lett 10:1727–1730
41. Wuts PGM, Greene TW (2007) Greene's protective groups in organic synthesis, 4th edn. Wiley, Hoboken, NJ, Chapter 7
42. Zlotos DP (2004) Eur J Org Chem:2375–2380; Zlotos DP, Buller S, Stiefl N, Baumann K, Mohr K (2004) J Med Chem 47:3561–3571
43. For a related preparation, see: Martin SF, Williamson SA, Gist RP, Smith KM (1983) J Org Chem 48:5170–5180
44. Garro-Helion F, Merzouk A, Guibé F (1993) J Org Chem 58:6109–6113
45. Massiot G, Thépenier P, Jacquier M-J, Lounkokobi J, Mirand C, Zèches M, Le Men-Olivier L, Delaude C (1983) Tetrahedron 39:3645–3656
46. Massiot G, Massousa B, Jacquier M-J, Thepenier P, Le Men-Olivier L, Delaude C, Verpoorte R (1988) Phytochemistry 27:3293–3304
47. (a) Lim K-H, Low Y-Y, Kam T-S (2006) Tetrahedron Lett 47: 5037–5039; (b) Lim K-H, Hiraku O, Komiyama K, Koyano T, Hayashi M, Kam T-S (2007) J Nat Prod 70:1302–1307
48. In 2007, Padwa reported the first synthesis of valparicine using his Diels–Alder/fragmentation methodology. See: Boonsombat J, Zhang H, Chughtai MJ, Hartung J, Padwa A (2007) J Org Chem 73:3539–3550
49. Jeffery T (1996) Tetrahedron 52:10113–10130
50. Solé D, Bonjoch J, García-Rubio S, Suriol R, Bosch J (1996) Tetrahedron Lett 37:5213–5216
51. (a) Blumenkopf TA, Overman LE (1986) Chem Rev 86:857–873; (b) Overman LE, Bell K, Ito F (1984) J Am Chem Soc 106:4192–4201
52. Iminium catalysis: Lelais G, MacMillan DWC (2006) Aldrichimica Acta 39:79–87
53. Pham HV, Martin DBC, Vanderwal CD, Houk KN (2012) Chem Sci 3:1650–1655
54. Shea KJ, Wise S, Burke LD, Davis PD, Gilman JW, Greeley AC (1982) J Am Chem Soc 104:5708, and references therein
55. Crimmins MT, Reinhold TL (1993) Org Reactions 44:297–588
56. (a) Salomon RG, Kochi JK (1972) J Am Chem Soc 94:1889–1897; (b) Salomon RG, Ghosh S, Raychaudhuri SR, Miranti TS (1984) Tetrahedron Lett 25:3167–3170
57. Wolff S, Barany F, Agosta WC (1980) J Am Chem Soc 102:2378–2386
58. Knochel P, Betzemeier B (2002) In: Krause N (ed) Modern organocopper chemistry. Wiley-VCH, Weinheim, Chapter 2
59. For the elimination of R_3 SnBr to give allenes, see: (a): Eichberg MJ, Dorta RL, Grotjahn DB, Lamottke K, Schmidt M, Vollhardt KPC (2001) J Am Chem Soc 123:9324–9337; (b) Overman LE, personal communication
60. (a) Maifield SV, Lee D (2005) Org Lett 7:4995–4998; (b) Trost BM, Ball ZT (2005) J Am Chem Soc 127:17644–17655

61. (a) Herron JR, Ball ZT (2008) J Am Chem Soc 130:16486–16487; (b) Herron JR, Russo V, Valente EJ, Ball ZT (2009) Chem Eur J 15:8713–8716; (c) Russo V, Herron JR, Ball ZT (2010) Org Lett 12:220–223
62. (a) Taguchi H, Ghoroku K, Tadaki M, Tsubouchi A, Takeda T (2002) J Org Chem 67:8450–8456; (b) Taguchi H, Ghoroku K, Tadaki M, Tsubouchi A, Takeda T (2001) Org Lett 3:3811–3814
63. Brook AG (1974) Acc Chem Res 7:77–84
64. (a) Smith AB III, Kim W-S, Tong R (2010) Org Lett 12:588–591; (b) Smith AB III, Tong R, Kim W-S, Maio WA (2011) Angew Chem Int Ed 50:8904–8907
65. Park PK, O'Malley SJ, Schmidt DR, Leighton JL (2006) J Am Chem Soc 128:2796–2797

Chapter 5
Bryostatin 7

Yu Lu and Michael J. Krische

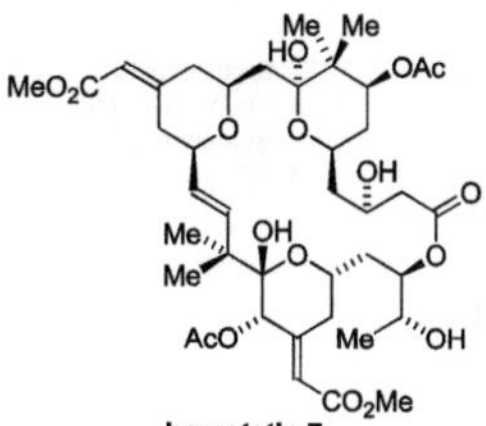

bryostatin 7

5.1 Introduction

The bryostatins are a family of marine natural products originally isolated from the bryozoan *Bugula neritina* by Pettit and coworkers in the course of their search for new antineoplastic agents derived from marine organisms [1]. Later, during a large-scale isolation, 18 g of bryostatin 1, the first discovered and most naturally abundant member of this family, was obtained from a collection of 10,000 gallons of wet bryozoan [2]. Following the discovery of bryostatin 1 in the early 1980s, nineteen other structurally related congeners (bryostatin 2 to bryostatin 20) were isolated over the years from the Gulf of Mexico, Gulf of California, and Sagami Bay of Japan. Presently, it is believed that the bryostatins are not produced by *Bugula neritina* itself, but instead by a bacterial symbiont "*Candidatus* Endobugula sertula" of the bryozoan [3].

The structures of the bryostatins were determined by a combination of single-crystal X-ray diffraction analysis and/or a series of detailed spectroscopic analyses.

Y. Lu • M.J. Krische (✉)
Department of Chemistry and Biochemistry, University of Texas, Austin, Austin, TX 78712, USA
e-mail: mkrische@mail.utexas.edu

J.J. Li and E.J. Corey (eds.), *Total Synthesis of Natural Products*,
DOI 10.1007/978-3-642-34065-9_5, © Springer-Verlag Berlin Heidelberg 2012

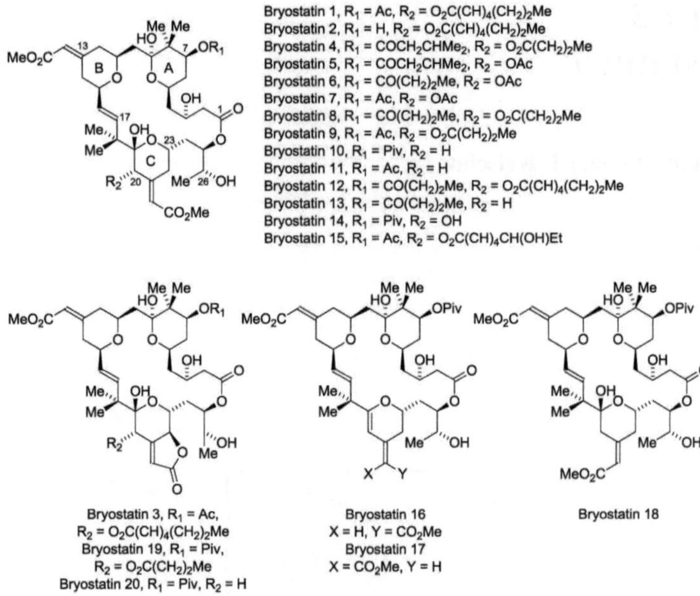

Fig. 5.1 Structures of bryostatins 1–20

All bryostatins share several distinctive architectural features, including a 20-membered macrolactone with three highly functionalized pyran rings, *gem*-dimethyl groups at C8 and C18, exocyclic methyl enoates at C13 and C21, and a *trans*-olefin at C16–C17. In most cases, members of the bryostatin family are differentiated structurally through the variation or absence of acyloxy substituents at the C7 and C20 positions. Bryostatins 3, 19, and 20 are further differentiated by the presence of oxygen at C22, which takes the form of a butenolide incorporating the C21 exocyclic enoate. Finally, bryostatins 16 and 17 embody a C-ring dihydropyran rather than *trans*-vicinal oxygenation at C19 and C20 found in other members (Fig. 5.1).

5.2 Pharmacology

As described in several monographs [4], bryostatin 1 exhibits significant in vitro and in vivo antineoplastic activity against a range of tumor cell lines including murine leukemia, B-cell lymphoma, reticulum cell sarcoma, ovarian carcinoma, and melanoma. It is also effective in the modulation of apoptotic function [5], the reversal of multidrug resistance [6], and stimulation of the immune system [7]. These unique features displayed by bryostatin 1 are attributed to its high affinity for protein kinase C (PKC) isozymes and its ability to selectively modulate their functions [8]. PKCs are a type of intracellular serine and threonine kinase that control *O*-phosphorylation of substrate proteins. Accordingly, expression of the PKCs is linked with the signal transduction pathway and regulation of cell growth [9]. Common exogenous PKC activators, such as phorbol esters, hyper-activate PKCs and trigger cell proliferation through induction of a conformational change in the PKC C1 binding domain [10]. Therefore, most phorbol esters (e.g., phorbol 12-myristate-13-acetate, PMA) act as tumor promoters. It was postulated that although bryostatin 1 interacts with the same region, the resulting complex adopts a stabilizing conformation and, therefore, does not promote tumor growth [11]. Because of its promising effects in early animal studies, bryostatin 1 has been investigated as a drug candidate for human cancer therapy in numerous clinical trials. Subsequently, it was found that bryostatin 1 acts synergistically with established therapeutic agents, enhancing their anticancer activity at exceptionally low doses (~50 $\mu g/m^2$) [12]. However, clinical studies have not yet shown bryostatin 1 is effective at curing cancer, either as a single-agent drug or in combination with other antitumor drugs.

Beyond its antineoplastic activity, bryostatin 1 displays pharmacological activity in the induction of synaptogenesis [13] and the enhancement of learning and memory in animal models [14], offering potential utility in the treatment of Alzheimer's disease (AD) and other neurological disorders [15]. A phase II clinical trial using bryostatin 1 as the single anti-AD drug has been initiated[1]. Bryostatin 1 also has been suggested as a potential therapeutic agent for the treatment of dementia and depression [16] and for the reversal of neural damage incurred during a stroke by stimulating neural growth and repair [17]. More recently, preliminary studies have shown bryostatin 1 can modulate latent HIV-1 infection, suggesting a potential means of clearing HIV from those infected [18].

A pharmacophore model has been proposed by Wender and Blumberg through computer-based structure–activity relationship (SAR) studies [19]. The southern hemisphere of bryostatin appears to serve as a recognition domain, whereas the northern hemisphere serves to orient the recognition domain into the PKC binding pocket. The collective data suggest that the C1, C19, and C26 oxygens in bryostatins are pharmacophoric elements. The C3 and C11 oxygens participate in transannular hydrogen bonding, which stabilizes the molecular conformation favorable for PKC binding. Accordingly, the presence and stereochemistry of the C3 and C11 oxygens are crucial in promoting high PKC binding affinity. Based on this model, a series of simplified bryostatin analogues were prepared by Wender, and their PKCα binding affinities were evaluated [20]. A parallel effort led by Keck reveals the importance of bryostatin A-ring functionality, particularly the C8 *gem*-dimethyl group, in retaining a good PKC binding affinity and the suppression of tumor cell growth [21]. Substituents on C7 and C20 appear to be tunable, providing a means to adjust function and physical properties without significantly affecting the PKC binding affinity (Fig. 5.2).

[1] For current information, see: http://clinicaltrials.gov

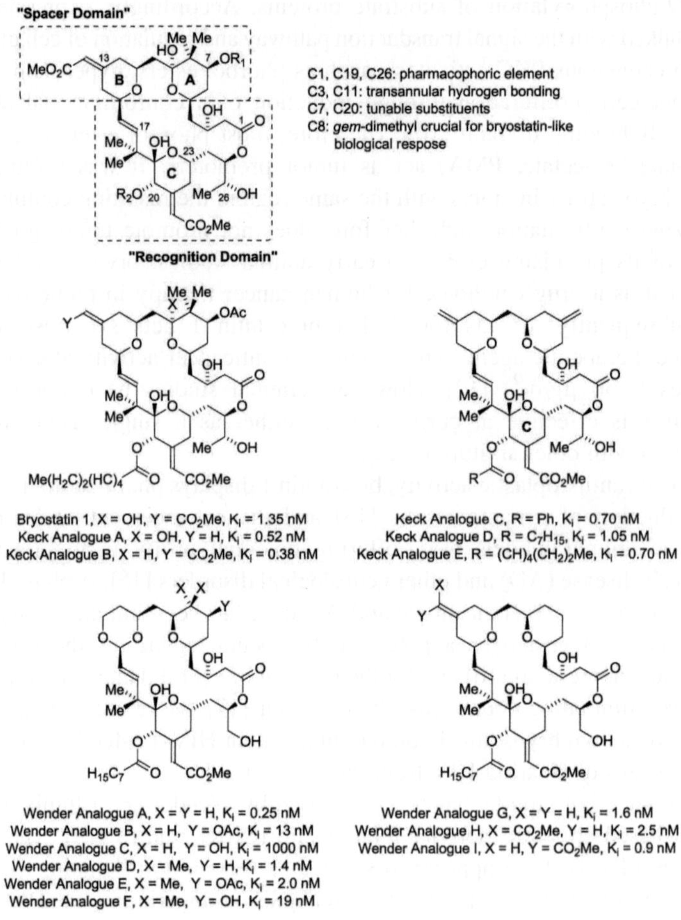

Fig. 5.2 PKC binding affinities for bryostatin 1 and selected analogues

5.3 Biosynthesis

The bryostatins are produced by modular polyketide synthases (PKS) found in a bacterial symbiont of *B. neritina* [3]. A putative genomic region, the *bry* cluster, has been identified, and a pathway for the biosynthesis of the bryostatins was proposed (Scheme 5.1) [22]. BryA, a PKC gene with four modules (L and M1–3), is believed to be responsible for the synthesis of the C19–C27 carbon chain found in the bryostatin southern hemisphere. Following installation of the C21 β-ketone by the ketone synthase (KS) embedded in M3, the HMG-CoA synthase condenses acetyl-CoA onto the β-keto group. After dehydration, the resulting carboxylic acid undergoes *O*-methylation by *O*-methyltransferase (*O*-MT) to produce the C21 exocyclic enoate. Then bryB (M4–7) elongates the chain to C11–C27. The MT

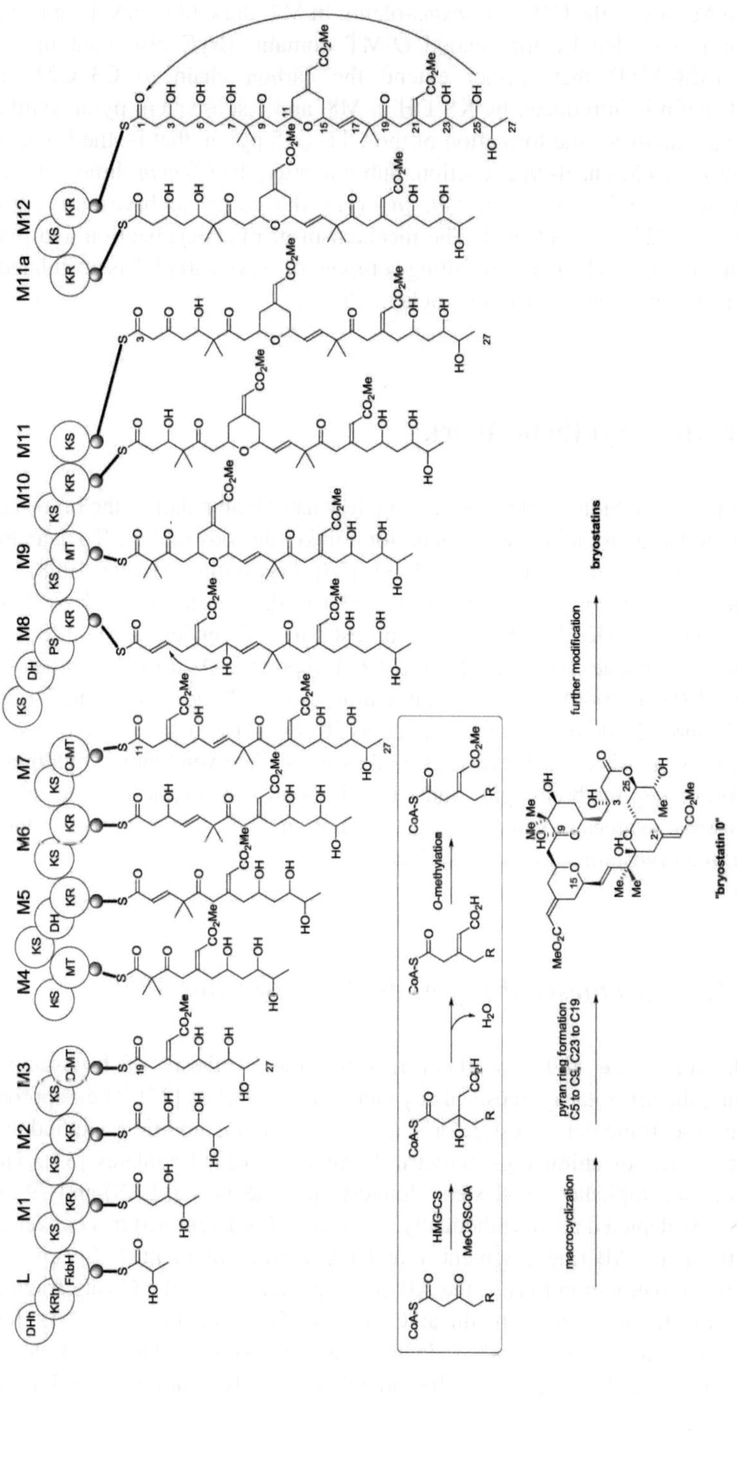

L, loading module; DH, dehydratase; KS, β-ketosynthase; KR, ketoreductase; MT: methyltransferase; PS, pyran synthase; DHh and KRh are DH and KR-like sequences, together with the FkbH domain, they are involved in the formation of D-lactate starter unit; HMG-CS, hydroxy-methyl-glutaryl CoA synthase. Acyl-carrier-protein domains are shown as small filled balls with chain attached by the thiol group. The box shows the HMG-CS pathway for the formation of exocyclic enoate.

Scheme 5.1 Proposed biosynthesis of the bryostatins

domain in M4 introduces the C18 *gem*-dimethyl group, and the dehydratase (DH) domain in M5 forms the C16–C17 *trans*-olefin. In M7, the C13 exocyclic enoate in bryostatin is installed by the second *O*-MT domain. BryC also contains four modules (M8–M11) that further extend the carbon chain to C3–C27. The C10–C11 olefin is introduced by KS-DH in M8, and a subsequent pyran synthase (PS) domain catalyzes the formation of the C11–C15 pyran, that is, the bryostatin B-ring, through a Michael-type reaction. Subsequently, the C8 *gem*-dimethyl group is added by the MT domain in M9. In bryD, the complete bryostatin carbon backbone C1–C27 is established. The mechanism of macrocyclization and pyran formation remains unclear. The resulting compound, "bryostatin 0," is postulated to be the progenitor of the bryostatins (Scheme 5.1).

5.4 Previous Synthetic Work

Given their diverse biological properties and low natural abundance, the bryostatins have emerged as a vibrant testing ground for polyketide construction. To date, total syntheses of bryostatin 7 (Masamune 1990) [23], bryostatin 2 (Evans 1998) [24], bryostatin 3 (Nishiyama and Yamamura 2000) [25], bryostatin 16 (Trost 2008) [26], bryostatin 1 (Keck 2011) [27], bryostatin 9 (Wender 2011) [28], and bryostatin 7 (Krische 2011) [29] have been reported. A formal synthesis of bryostatin 7 (Hale 2006) [30] and total syntheses of C20-*epi*-bryostatin 7 (Trost 2010) [31] and C20-deoxybryostatin (Thomas 2011) [32] also have been disclosed. A very brief synopsis of each previous total synthesis is given below, highlighting application of key methodologies and strategic bond formations. For the sake of brevity, formal syntheses, syntheses of structural analogues, and synthetic studies related to the bryostatins are not described.

5.4.1 Total Synthesis of Bryostatin 7 (Masamune 1990)

In 1990, Masamune and coworkers reported the synthesis of bryostatin 7, constituting the first total synthesis of any member of this class [23]. Their approach showcases the principal investigator's asymmetric aldol addition methodology based on the use of chiral C_2-symmetric borolane-modified enolates [33]. Their synthesis is accomplished in 41 steps (longest linear sequence, LLS) and 79 total steps (TS). As depicted retrosynthetically, bryostatin 7 is assembled in a convergent fashion from the AB-ring fragment **1** and the C-ring fragment **2**. Specifically, Julia–Lythgoe olefination [34] of the AB-ring fragment **1** and the C-ring fragment **2** installs the hindered *trans*-olefin at C16–C17. The product of Julia–Lythgoe olefination is then converted to a C3-aldehyde, which is subjected to stereoselective aldol addition (3:1 dr at C3) using the indicated chiral borolane-modified acetate

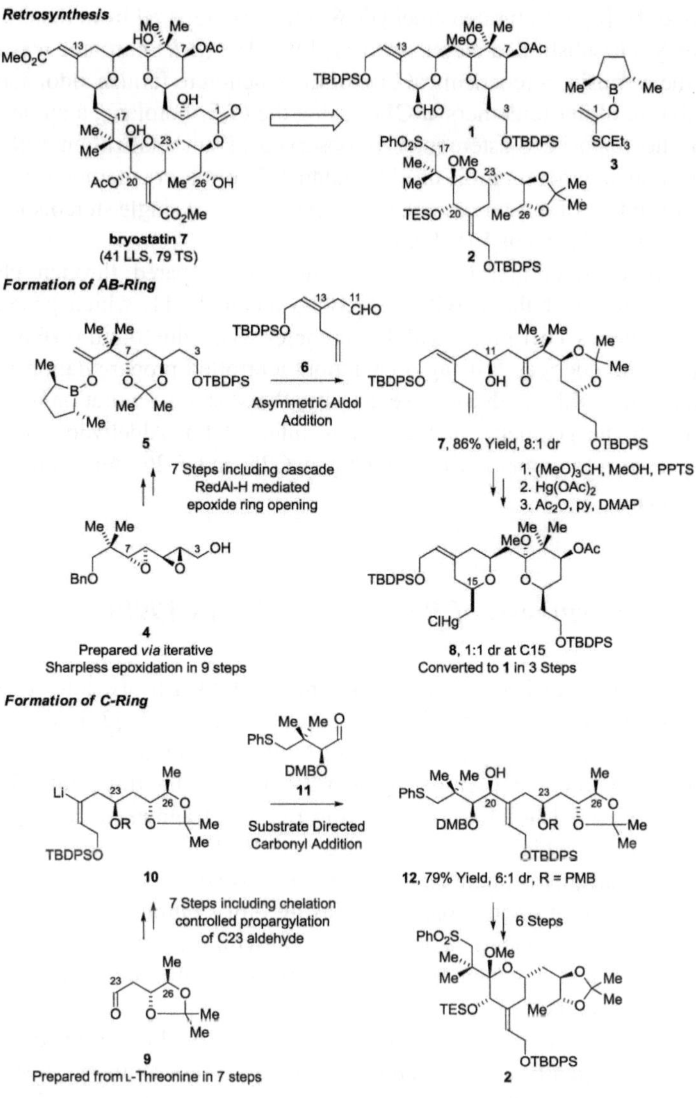

Scheme 5.2 Total synthesis of bryostatin 7 by Masamune

enolate **3** to provide the corresponding *seco*-acid and, ultimately, through DCC-mediated macrolactonization, the 20-membered macrolide (Scheme 5.2).

Construction of the AB-ring fragment **1** is achieved through iterative Sharpless asymmetric epoxidation [35] of neopentyl glycol to furnish the *bis*-epoxide **4**, which upon cascading regioselective RedAl-H-mediated reductive ring opening installs the C5 and C7 hydroxyl moieties. A key step in the synthesis of AB-ring fragment **1** is the asymmetric aldol addition of the methyl ketone-derived boron enolate **5** to the β,γ-unsaturated aldehyde **6**, which is prepared from the C13 alkyne using Corey's trisubstituted olefin synthesis [36]. The (*R*,*R*)-borolane was found to amplify the intrinsic stereochemical bias of the reaction to furnish aldol adduct **7** as an 8:1 mixture of diastereomers at C11. Using the (*S*,*S*)-borolane, a modest preference for the opposite diastereomer is observed. Transketalization followed by intramolecular oxymercuration of aldol adduct **7** forms organomercurial **8** as an equimolar mixture of diastereomers, which converge to a single stereoisomer at the stage of AB-ring fragment **1** (Scheme 5.2).

The sulfone containing C-ring fragment **2** is prepared through chelation-controlled addition of the vinyllithium **10** to aldehyde **11**, which generates the C20 stereocenter as a 6:1 mixture of diastereomers. The vinyllithium **10** is prepared in 7 steps from aldehyde **9** using a chelation-controlled propargylation to set the C23 stereocenter with 8:1 diastereoselectivity. Piers' stannylcupration method [37] is used to set the geometry of the trisubstituted olefin. Aldehyde **9** is, in turn, prepared from L-threonine, from which the C25 and C26 stereocenters derive (Scheme 5.2).

5.4.2 Total Synthesis of Bryostatin 2 (Evans 1998)

The total synthesis of bryostatin 2 by Evans employs a triply convergent route involving assembly of preformed A-, B-, and C-ring pyrans [24]. The synthesis is accomplished in 42 steps (LLS) and 72 total steps. Julia–Lythgoe olefination [34] is used to join the B-ring fragment **14** and C-ring fragment **15** through formation of C16–C17 olefin. The resulting BC-ring fragment **25** is converted to the C10 triflate **26**, which serves as an electrophile in the alkylation of the A-ring sulfone **13**. Subsequent functional group interconversions, including asymmetric Horner–Wadsworth–Emmons olefination using Fuji's chiral phosphonate [38] to install the C13 B-ring enoate, aldol dehydration with methyl glyoxylate to install the C21 enoate, and Yamaguchi macrolactonization [39], are used to complete the synthesis (Scheme 5.3).

Several different asymmetric aldol additions and hydroxy-directed carbonyl reductions are elegantly used to control and relay stereochemistry. As illustrated in the reaction of **17** and neopentyl aldehyde **16**, the construction of the A-ring fragment **13** begins with the use of Evans' signature chiral oxazolidinone auxiliary to affect the equivalent of an enantioselective acetate aldol reaction [40]. The aldehyde **18** is subjected to a substrate-directed $TiCl_2(O\text{-}i\text{-}Pr)_2$ catalyzed Mukaiyama aldol reaction with the Brassard-type diene **19** to furnish compound **20**, which is subjected to hydroxy-directed reduction of the C3 ketone using $Me_4NB(OAc)_3H$ [41], ultimately providing lactone **21**, which is converted to the A-ring fragment **13** in 5 steps (Scheme 5.3).

The synthesis of the BC-ring fragment **26** begins with the enantioselective aldol reaction of aldehyde **22** and chiral boron enolate **23** derived from (−)-DIP-Cl as described by Paterson and Brown [42]. The aldol adduct **24** is converted to the

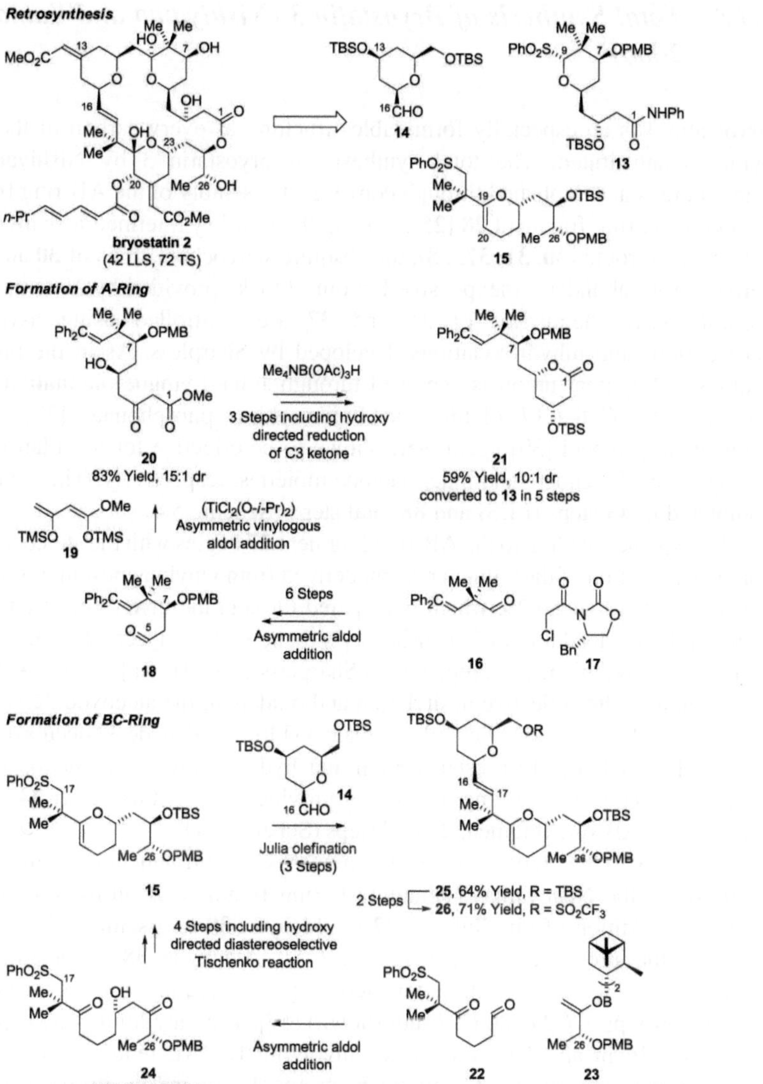

Scheme 5.3 Total synthesis of bryostatin 2 by Evans

pyran **15** in a sequence involving samarium-catalyzed Tishchenko reduction [43] to set the stereogenic center at C25, with temporary masking of the C23-hydroxy group as the *p*-nitrobenzoate. Julia–Lythgoe olefination [34] of the C-ring fragment **15** and the B-ring fragment **14** establishes the *trans*-olefin at C16–C17. The construction of the B-ring fragment **14** (not shown) is achieved through an enantioselective acetoacetate aldol reaction using a C_2-symmetric copper(II) complex [Cu(*R*,*R*)-Ph-pybox]$(SbF_6)_2$ as the chiral catalyst (Scheme 5.3) [44].

5.4.3 Total Synthesis of Bryostatin 3 (Nishiyama and Yamamura 2000)

Bryostatin 3 is an especially formidable structure, as every carbon of the C-ring pyran is substituted. The total synthesis of bryostatin 3 by Nishiyama and Yamamura is accomplished through convergent assembly of the AB-ring fragment **27** and the C-ring fragment **28** [25]. Among the four key intermediates involved in the synthetic route (**30**, **31**, **37**, **38**), the absolute stereochemistries of **30** and **38** are derived from abundant, inexpensive building blocks provided by Nature, and the absolute stereochemistries of **31** and **37** are controlled using asymmetric epoxidations and dihydroxylations developed by Sharpless. As in the preceding syntheses, fragment union is achieved through Julia–Lythgoe olefination [34] to establish the C16–C17 olefin, and Fuji's chiral phosphonate [38] and the Yamaguchi protocol [39] again were shown to be effective for installation of the B-ring exocyclic enoate and macrolactone moieties, respectively. The synthesis is completed in 43 steps (LLS) and 88 total steps (Scheme 5.4).

The sequence leading to the AB-ring fragment **27** begins with the stereocontrolled conjugate addition of the Gilman reagent derived from vinylmagnesium bromide and copper iodide to enone **29**, which is prepared through the inverse electron demand Diels–Alder cycloaddition of Danishefsky's diene with D-glyceraldehyde [45]. The indicated lithiodithiane **31**, prepared via Sharpless epoxidation [35], is alkylated by pyran **30**, and after selective desilylation and oxidation, the aldehyde **32** is formed. Vinylogous aldol addition of chiral dienolate **33** [46] to aldehyde **32** occurs diastereoselectively (24:1 dr). Transesterification and hydroxy-directed reduction ($Me_4NB(OAc)_3H$) [41] of the resulting β-ketoester provide the β-hydroxy ester **34**, which is converted to AB-ring fragment **27** in 7 steps (Scheme 5.4).

In analogy to the preparation of Masamune's C-ring fragment (Scheme 5.2), the synthesis of the Nishiyama–Yamamura C-ring fragment **28** utilizes a chelation-controlled addition of vinyllithium **37** to aldehyde **38** to assemble C17–C27 and establish the proper stereochemistry at C22. Aldehyde **38** is obtained from diacetone D-glucose through a ten-step sequence requiring deoxygenation of hydroxyl groups at C24 and C27 (not shown) [47]. The vinyllithium **37** is prepared from ester **36** in an 11-step sequence involving DIBAL reduction followed by alkyne formation using the Corey–Fuchs protocol [48] and in situ trapping of the acetylide with paraformaldehyde to provide a propargylic alcohol, which is subjected to hydroxy-directed hydroalumination-iodination to establish the C21 double bond geometry. Ester **36** is itself prepared by Sharpless asymmetric dihydroxylation of enoate **35** (Scheme 5.4) [49].

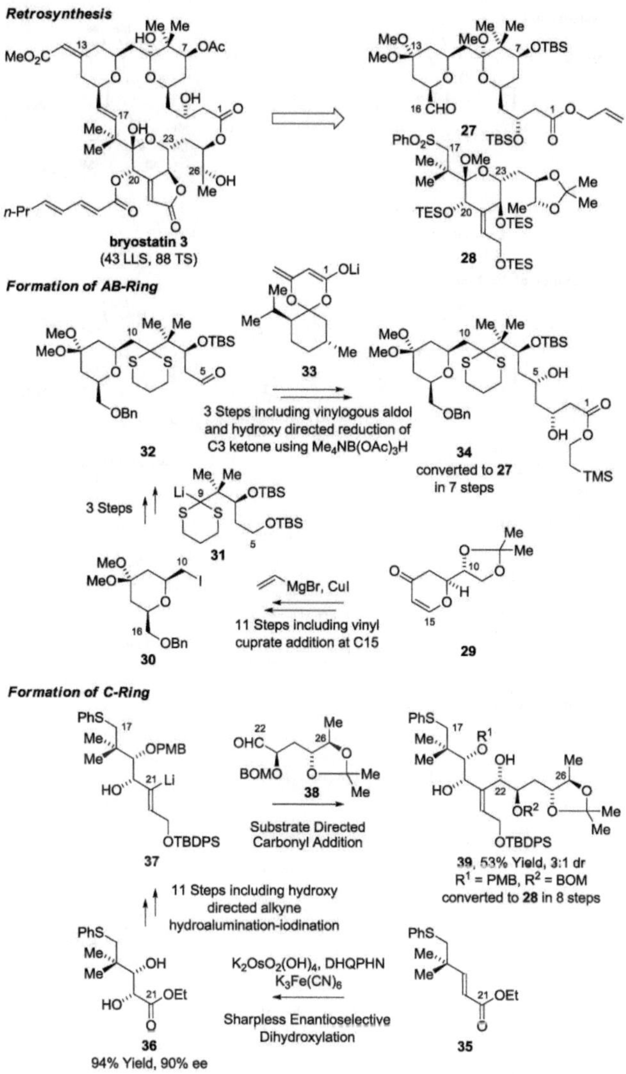

Scheme 5.4 Total synthesis of bryostatin 3 by Nishiyama and Yamamura

5.4.4 *Total Synthesis of Bryostatin 16 (Trost 2008)*

After Nishiyama and Yamamura's synthesis of bryostatin 3 [25], 9 years elapsed before the Trost group reported a stunning total synthesis of bryostatin 16, which was accomplished in only 28 steps (LLS) and 42 total steps [26]. In a significant departure from prior art, their synthesis convergently assembles the AB-ring fragment **40** and the latent C-ring fragment **41** through sequential Yamaguchi

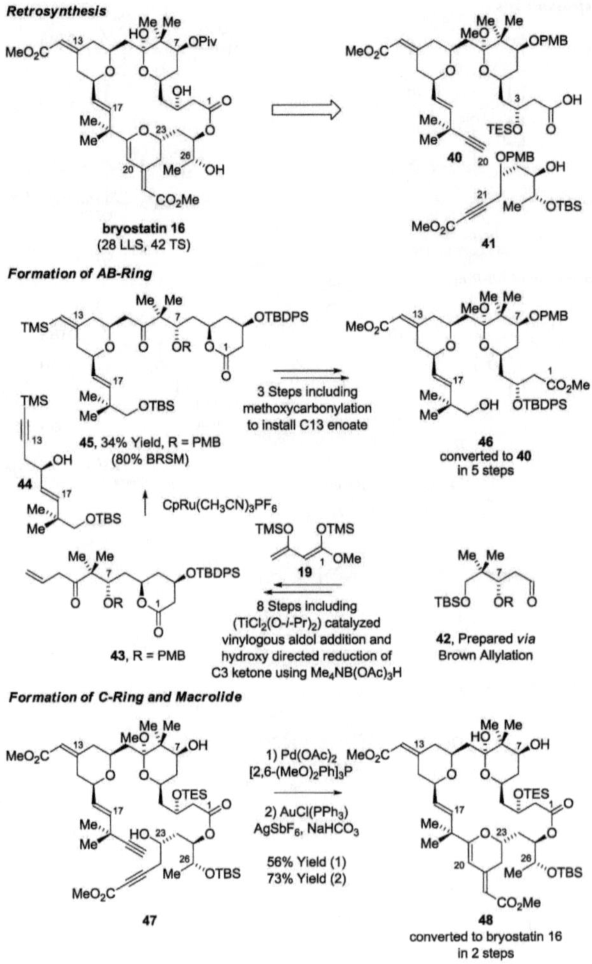

Scheme 5.5 Total synthesis of bryostatin 16 by Trost

esterification [39], followed by an ambitious intramolecular palladium-catalyzed alkyne-alkyne C–C coupling [50], and a gold-catalyzed *6-endo-dig* cyclization of the resulting acetylenic enoate to install the C-ring pyran [51]. The B-ring pyran in **40** provided yet another opportunity to employ metal catalysis, as illustrated in the ruthenium-catalyzed alkene–alkyne coupling-oxy-Michael addition reaction [52] between alkyne **44** and β,γ-enone **43** (Scheme 5.5).

As practiced in the preceding syntheses by Evans and Nishiyama and Yamamura, the A-ring fragment **43** is formed through substrate-directed vinylogous aldol reaction of the Brassard-type diene **19** and the chiral aldehyde **42**, which is prepared using Brown's protocols for enantioselective allylation [53], followed by hydroxy-directed *anti*-diastereoselective reduction of the C3 ketone ($Me_4NB(OAc)_3H$) [41].

Ruthenium-catalyzed tandem alkene–alkyne coupling-oxy-Michael addition [52] of lactone **43** and homopropargylic alcohol **44** generates the B-ring pyran in 34 % yield with control of stereochemistry at C11 and C15, as well as C13 olefin geometry. Conversion of **45** to the AB-ring fragment **46** is achieved through bromination of the exocyclic vinylsilane followed by palladium-catalyzed methoxycarbonylation of the resulting vinyl bromide, and Brønsted acid triggered transesterification–ketalization–desilylation in methanol (Scheme 5.5).

To complete the synthesis, primary alcohol **46** is transformed to diyne **47** in 7 steps through Dess–Martin oxidation of the C19 alcohol, Ohira–Bestmann alkynylation [54] of the resulting aldehyde, and Yamaguchi esterification [39] with alcohol **41** (prepared from D-galactono-1,4-lactone) to install the fragment containing the C23 homopropargylic alcohol. With diyne **47** in hand, regioselective palladium-catalyzed alkyne-alkyne C–C coupling delivers the macrocycle [50], which upon exposure to a cationic gold catalyst forms the C-ring pyran **48** [51]. Compound **48** is transformed to bryostatin 16 in two steps through pivaloylation of the C7 hydroxyl followed by removal of the silyl ether protecting groups (Scheme 5.5).

5.4.5 Synthesis of Bryostatin 1 (Keck 2011)

In 2011, following several years of study on the synthesis and evaluation of bryostatin analogues [21], Keck and coworkers reported the total synthesis of bryostatin 1, the most naturally abundant member of the bryostatin family [27]. A distinctive and powerful feature of their approach involves use of an intermolecular Prins pyran annulation developed in their laboratory [55], which convergently assembles A-ring fragment **50** and C-ring fragment **51** with concomitant B-ring formation. Additionally, a series of aldehyde allylstannations are used to establish the carbinol stereocenters at C5, C7, C11, C23, and C25. To complete the synthesis, Fuji's chiral phosphonate [38] and the Yamaguchi protocol [39] again were used to install the B-ring exocyclic enoate and macrolactone moieties, respectively. Keck's total synthesis of bryostatin 1 is accomplished in 31 steps (LLS) and 58 total steps (Scheme 5.6).

The key step in the synthesis of A-ring fragment **50** [56] is the chelation-controlled addition of allylstannane **53** to aldehyde **52**, which sets the C7 stereocenter and introduces the C8 *gem*-dimethyl moiety. Aldehyde **52** is itself prepared from 1,3-propanediol using the author's protocol for titanium-catalyzed enantioselective allylstannation [57], which sets the C5 stereocenter, followed by chelation-controlled Mukaiyama aldol addition [58] to establish the C3 stereocenter (Scheme 5.6).

A unique approach to the requisite C-ring fragment **51** is achieved through reductive cyclization of olefinic ester **55** by way of the titanium alkylidene, as described by Rainer and Nicolaou [59]. The olefinic ester **55** is prepared in ten steps from (*R*)-isobutyl lactate using consecutive chelation-controlled

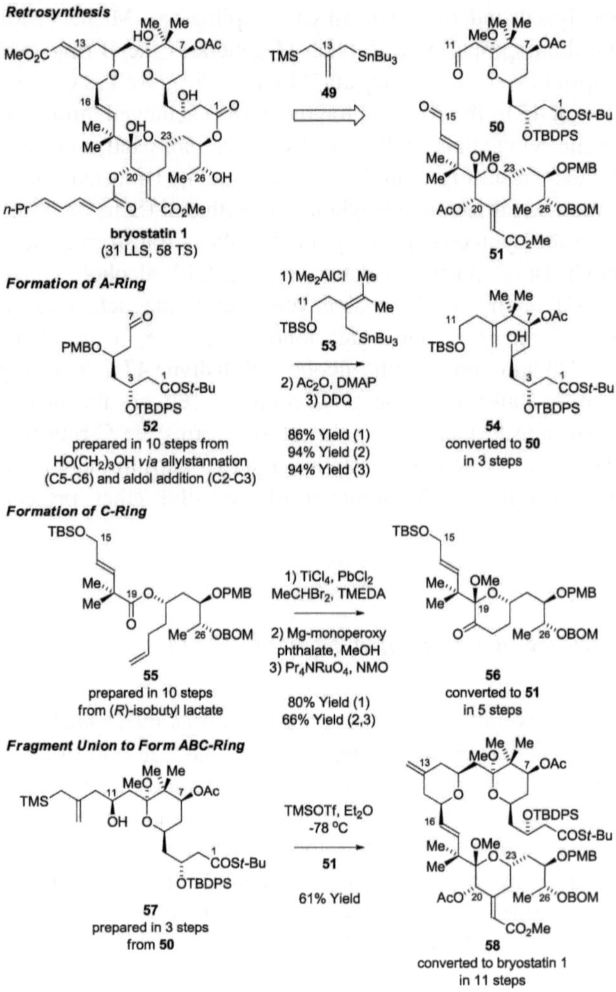

Scheme 5.6 Total synthesis of bryostatin 1 by Keck

allylstannations to set the C23 and C25 stereocenters with high levels of diastereoselectivity in each case. The C19–C20 dihydropyran obtained upon reductive cyclization of **55** is treated with magnesium monoperoxyphthalate and tetrapropylammonium perruthenate to furnish the C20 ketone **56**. Similar to Evans' approach (Scheme 5.3), the C21 enoate is introduced through aldol dehydration with methyl glyoxylate (Scheme 5.6).

Conversion of the A-ring aldehyde **50** to hydroxyallylsilane **57** is accomplished in a three-step sequence involving addition of allylstannane **49** followed by oxidation-reduction for diastereomeric enrichment at C11. With both hydroxyallylsilane **57** and aldehyde **51** in hand, intermolecular Prins pyran annulation proceeds smoothly to furnish the tricyclic compound **58** in 61 % isolated yield.

This material is converted to bryostatin 1 in 11 steps, including Yamaguchi macrolactonization, installation of the B-ring enoate using Fuji's chiral phosphonate [38] and, remarkably, selective hydrolysis of the C20 acetate in the presence of C1-macrolide, C7 acetate, as well as the C13 and C21 enoate moieties (Scheme 5.6).

5.4.6 Synthesis of Bryostatin 9 (Wender 2011)

In 1988, the Wender group, in collaboration with the groups of Pettit and Blumberg, engaged in computer-based structure–activity relationship (SAR) studies on the bryostatins [19], which led to the synthesis of several structurally simplified bryostatin analogues with promising biological activities [20]. In 2011, the Wender group reported the total synthesis of bryostatin 9 [28]. Similar to Keck's approach, A-ring fragment **59** and C-ring fragment **60** are united through Yamaguchi esterification followed by a challenging intramolecular Prins macrocyclization. As in the prior syntheses of Evans [24], Nishiyama and Yamamura [25], and Keck [27], Fuji's chiral phosphonate [38] is used to install the C13 enoate. Wender's total synthesis of bryostatin 1 is accomplished in 25 steps (LLS) and 43 total steps (Scheme 5.7).

Preparation of the A-ring fragment **59** begins with the 8 step conversion of acrolein to lactone **61** through Keck allylation [57] and Paterson–Brown asymmetric aldol [42] addition to establish the carbinol stereocenters at C3 and C5, respectively, followed by *anti*-diastereoselective hydroxy-directed reduction using $Me_4NB(OAc)_3H$ to define the C7 stereocenter [41]. Lactone **61** is converted to the *bis*-(trimethylsilyl)methyl adduct **62** through a five-step sequence involving Claisen condensation with the dienolate of ethyl acetoacetate, treatment of the resulting C9-lactol with acidic methanol to form the methyl ketal, diastereoselective reduction and TES protection of the C11 ketone, and, finally, double nucleophilic addition of $TMSCH_2MgCl$ to the C13 ester (Scheme 5.7).

A key step in the synthesis of the C-ring fragment **60** involves acid-catalyzed cyclodehydration of hydroxy ketone **63** to provide a C19–C20 dihydropyran. As in Evans' and Keck's syntheses (Schemes 5.3 and 5.6), functionalization of the C-ring dihydropyran is achieved through olefin epoxidation/ring opening, oxidation of the resulting C20 alcohol, and aldol dehydration with methyl glyoxylate to install the C21 enoate. After installing the C27 methyl group through ozonolysis of the C25–C26 olefin followed by Takai olefination [60], the C25 and C26 stereocenters are created through Sharpless asymmetric dihydroxylation (Scheme 5.7) [49].

Yamaguchi esterification of A-ring fragment **59** and C-ring fragment **60** delivers compound **65**, the substrate for silyl-Prins macrocyclization, a transformation that proceeds in 65 % isolated yield under remarkably mild conditions. Notably, after the point of convergence, only five additional steps are required to access bryostatin 9 (Scheme 5.7).

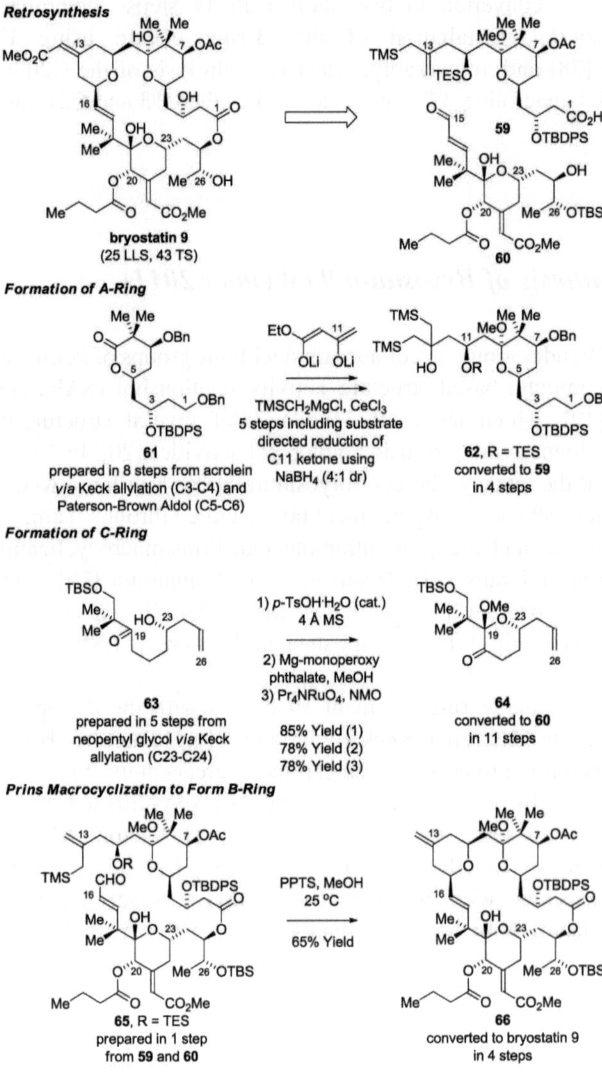

Scheme 5.7 Total synthesis of bryostatin 9 by Wender

5.5 Strategy and Retrosynthesis

In 2011, after a decade-long effort focused on the development of C–C bond forming hydrogenations and transfer hydrogenations [61], the Krische group described the total synthesis of bryostatin 7 in 20 steps (LLS) and 36 total steps, constituting the most concise synthesis of any bryostatin reported, to date [29]. Retrosynthetically, the Keck–Yu pyran annulation [55] and Yamaguchi macrolactonization [39] were viewed as the most convergent means of assembling the

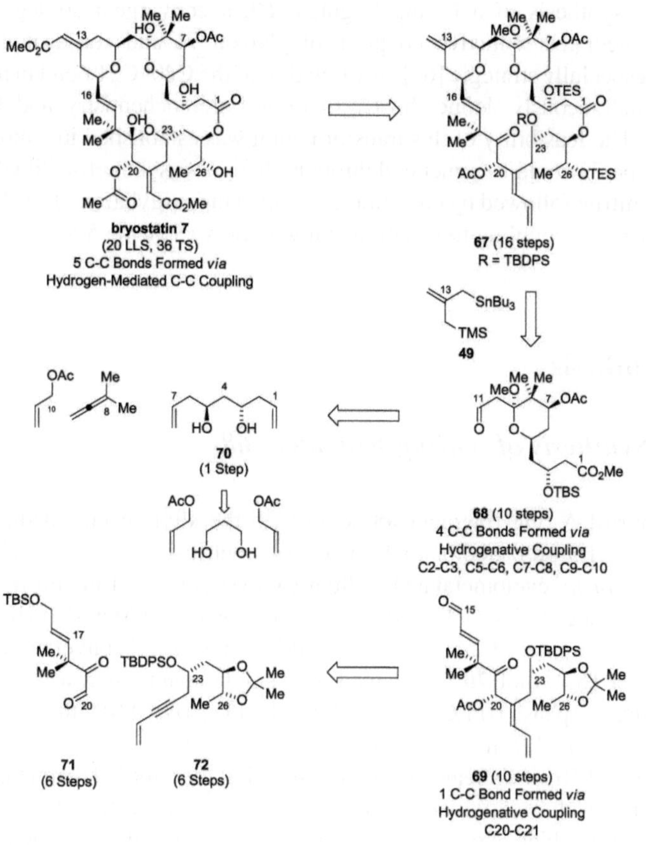

Scheme 5.8 Retrosynthetic analysis of bryostatin 7 via C–C bond forming hydrogenation and transfer hydrogenation

bryostatin skeleton. Hence, the greatest opportunity to enhance step economy appeared to reside in the identification of methods and strategies for rapid construction of appropriately functionalized A-ring and C-ring precursors, which ultimately took the form of fragments **68** and **69**, respectively (Scheme 5.8).

For the synthesis of A-ring fragment **68**, a dramatic simplification is availed through enantioselective double allylation of 1,3-propanediol to form the C_2-symmetric diol **70**, concurrently setting the C3–C5 carbinol stereocenters [62]. The power of this method for two-directional chain elongation was established in the construction of an A-ring fragment used in Evans' synthesis of bryostatin 2, which was accomplished in less than half the number of manipulations previously required [63b]. Transfer hydrogenative *tert*-prenylation [64] of a C7 aldehyde employing 1,1-dimethylallene could then establish the C7 carbinol stereochemistry and install the C8 *gem*-dimethyl moiety. Finally, transfer hydrogenative allylation [65] at C9 could be used to introduce the C11 aldehyde (Scheme 5.8).

For the synthesis of a C-ring fragment **69**, a convergent strategy involving hydrogen-mediated reductive coupling of glyoxal **71** and conjugated enyne **72** appeared especially strategic [66], as formation of the C20–C21 bond in this manner would simultaneously define the C20 carbinol stereochemistry and C21 olefin geometry. The feasibility of this transformation was established in a model system [63a]. Enyne **72** could be generated through Sharpless asymmetric dihydroxylation of crotononitrile followed by chelation-controlled propargylation of a C23 aldehyde and Sonogashira coupling to install the vinyl moiety (Scheme 5.8).

5.6 Synthesis

5.6.1 Synthesis of A-Ring Fragment 68

Efforts toward A-ring fragment **68** begin with the allyl acetate-mediated double allylation of 1,3-propanediol to form C_2-symmetric diol **70** [62]. This process employs an *ortho*-cyclometalated iridium catalyst generated in situ from $[Ir(cod)Cl]_2$, allyl acetate, 4-chloro-3-nitrobenzoic acid, and (*S*)-Cl,MeO-BIPHEP. Because the minor enantiomer of the *mono*-allylated intermediate is converted to the *meso*-diastereomer [67], diol **70** is obtained as a single enantiomer, as determined by chiral stationary phase HPLC analysis. Notably, the *mono*-TBS ether of diol **70** was previously prepared from 1,3-propanediol through a seven-step protocol involving iterative use of Brown's reagent for carbonyl allylation [68]. Treatment of diol **70** with ozone results in oxidative cleavage of the olefinic termini to deliver an unstable lactol, which is protected in situ as the *bis*-TBS ether to provide aldehyde **73** as a single isomer (Scheme 5.9).

Iridium-catalyzed transfer hydrogenation of aldehyde **73** in the presence of 1,1-dimethylallene promotes *tert*-prenylation [64] to form the secondary neopentyl alcohol **74**. In this process, isopropanol serves as the hydrogen donor, and the isolated iridium complex prepared from $[Ir(cod)Cl]_2$, allyl acetate, *m*-nitrobenzoic acid, and (*S*)-SEGPHOS is used as catalyst. Complete levels of catalyst-directed diastereoselectivity are observed. Exposure of neopentyl alcohol **74** to acetic anhydride followed by ozonolysis provides β-acetoxy aldehyde **75**. Reductive coupling of aldehyde **75** with allyl acetate under transfer hydrogenation conditions results in the formation of homoallylic alcohol **76**. As the stereochemistry of this addition is irrelevant, an achiral iridium complex derived from $[Ir(cod)Cl]_2$, allyl acetate, *m*-nitrobenzoic acid, and BIPHEP was employed as catalyst (Scheme 5.9).

Selective removal of the TBS ether of the C1 lactol in the presence of the secondary C3 TBS ether is achieved using a buffered TBAF solution in THF. Concomitant oxidation of the resulting diol using the Dess–Martin reagent provides the β,γ-enone **77**. Remarkably, treatment of a methanolic solution of β,γ-enone **77** with pyridinium *p*-toluenesulfonate triggers sequential lactone ring opening followed by formation of the cyclic ketal **78a**, along with a small quantity of the

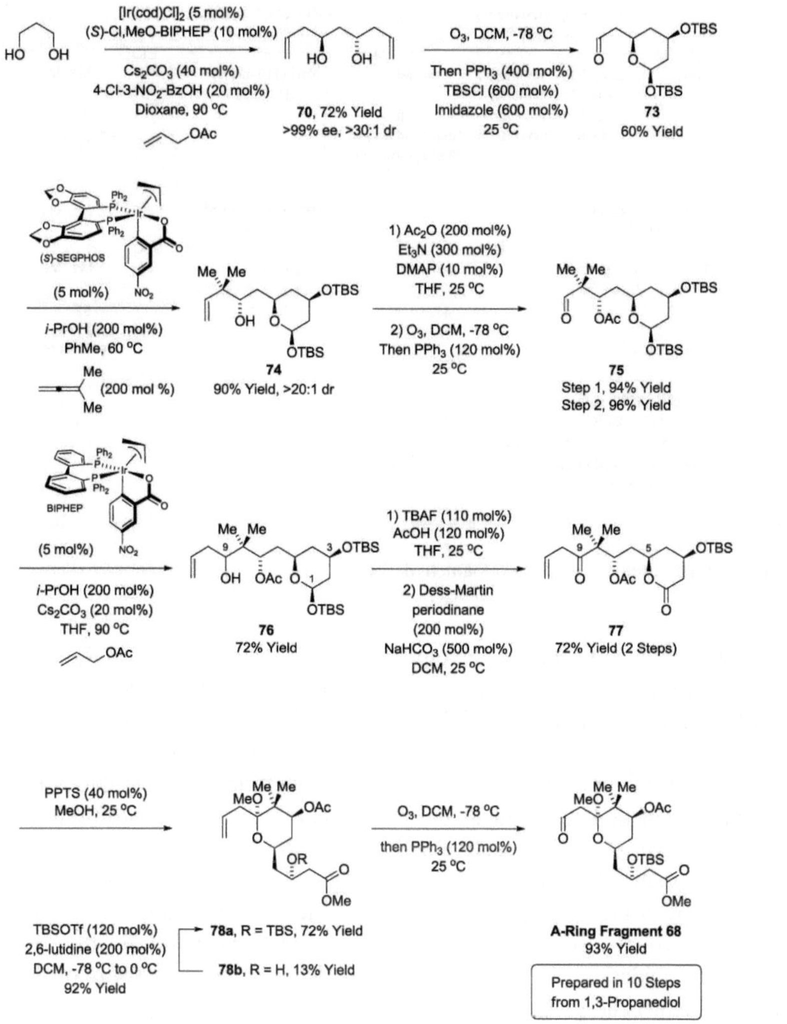

Scheme 5.9 Synthesis of A-ring fragment **68** via C–C bond forming transfer hydrogenation

C3-desilylated material **78b**, which is easily converted back to **78a**. Ozonolysis of **78a** provides A-ring fragment **68** in a total of 10 steps from 1,3-propanediol (Scheme 5.9).

5.6.2 Synthesis of C-Ring Fragment 69

The C-ring fragment **69** is prepared through the convergent assembly of glyoxal **71** and conjugated enyne **72** via hydrogen-mediated reductive coupling. The preparation of glyoxal **71** begins with acid-catalyzed aldol reaction of 3-methyl-2-butanone

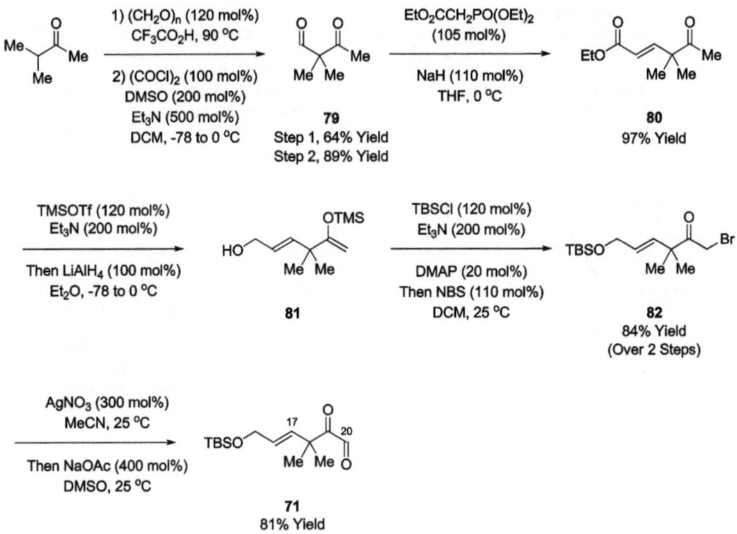

Scheme 5.10 Preparation of α-ketoaldehyde **71**

and formaldehyde to furnish the product of hydroxymethylation [69], which upon Moffatt–Swern oxidation provides the β-ketoaldehyde **79**. Horner–Wadsworth–Emmons olefination of β-ketoaldehyde **79** forms the α,β-unsaturated ester **80** as single geometrical isomer. All compounds up to this point are isolated by vacuum distillation, expediting access to large quantities of material (Scheme 5.10).

Conversion of ketone **80** to the enol silane followed by addition of lithium aluminum hydride to the reaction mixture directly provides the allylic alcohol **81** [70]. Treatment of crude allylic alcohol **81** with *tert*-butyldimethylsilyl chloride followed by *N*-bromosuccinimide furnishes the α-bromoketone **82** in 84 % yield over the two-step sequence from α,β-unsaturated ester **80**. Finally, a one-pot Kornblum oxidation [71] of α-bromoketone **82** is achieved by way of the nitrate ester to deliver the glyoxal **71**. It is worth noting that the sequence to glyoxal **71** requires only a single chromatographic purification at the second to last step (Scheme 5.10).

The synthesis of 1,3-enyne **72** begins with Sharpless asymmetric dihydroxylation of crotononitrile using "super AD-mix-β" [49b], which contains higher percentage of active catalyst and chiral ligand. The enantiomerically enriched diol **83** is converted to the acetonide and treated with diisobutylaluminum hydride to provide aldehyde **84**, which was previously prepared in six steps by an alternate route [72]. Whereas initial attempts at chelation-controlled propargylation of aldehyde **84** using the propargylmagnesium reagent provided a 2:1 mixture of diastereomers, the propargylzinc reagent provided homopropargylic alcohol as a 5:1 mixture of diastereomers. In any case, the minor stereoisomer could be converted to the desired epimer through Mitsunobu inversion. Finally, the homopropargylic alcohol is

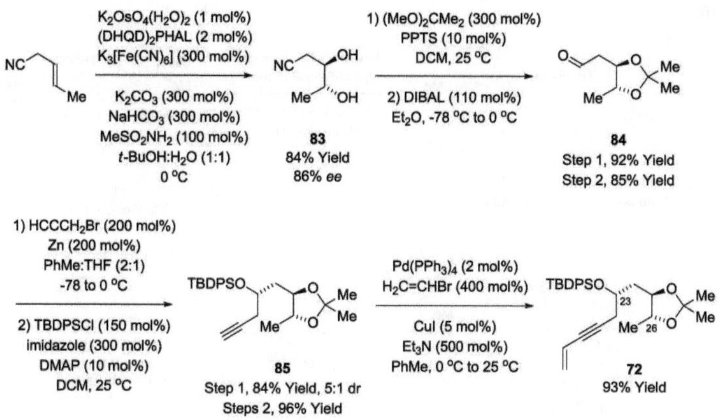

Scheme 5.11 Preparation of conjugated enyne **72**

converted to the TBDPS ether **85** and subjected to Sonogashira coupling with vinyl bromide to deliver the 1,3-enyne **72** (Scheme 5.11).

At this stage, the hydrogen-mediated reductive coupling of 1,3-enyne **72** to glyoxal **71** was explored. Gratifyingly, through application of conditions previously developed for a model system [63a], the desired α-hydroxyketone **86** is obtained in 77 % yield and 7:1 dr. Notably, although the product **86** incorporates multiple points of unsaturation, it does not suffer over-reduction under the conditions of hydrogen-mediated reductive coupling. Exposure of α-hydroxyketone **86** to acetic anhydride provides the acetate, which upon treatment with HF-pyridine participates in selective deprotection of the allylic TBS ether in the presence of the TBDPS-ether. Dess–Martin oxidation of the resulting allylic alcohol provides the C-ring fragment **69** in a total of 10 steps from 3-methyl-2-butanone or crotononitrile (Scheme 5.12).

5.6.3 *Fragment Union and Total Synthesis of Bryostatin 7*

To affect the crucial fragment union, C-ring fragment **69** is treated with allylstannane **49** in the presence of a chiral titanium Lewis acid catalyst modified by (*R*)-BINOL. Exposure of the resulting hydroxy allylsilane, which is formed as a single stereoisomer, to A-ring fragment **68** in the presence of trimethylsilyl triflate enables silyl-Prins pyran annulation to form the desired B-ring pyran **87a**. A small quantity of the elimination product **87b** is also formed; however, both compounds participate in acidic methanolysis to form triol **88b**. Carefully controlled reaction time proved crucial in this step, as short reaction times led to an increased proportion of **88a** and prolonged exposure to acidic methanol results in deprotection of the C23 TBDPS ether (Scheme 5.13).

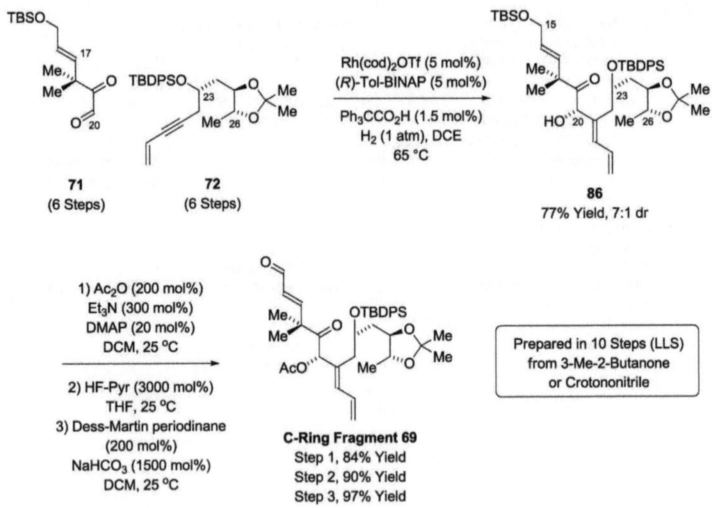

Scheme 5.12 Synthesis of C-ring fragment **69** via hydrogen-mediated reductive coupling

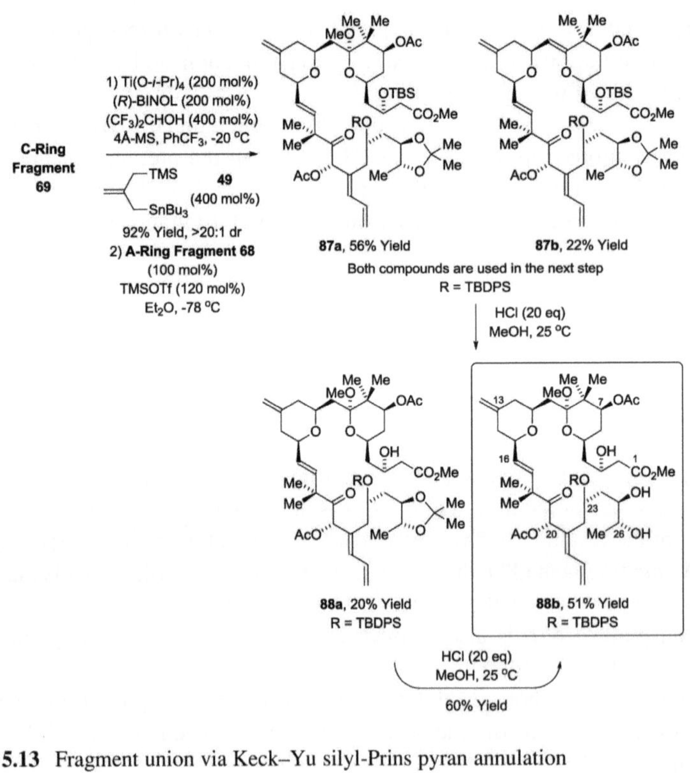

Scheme 5.13 Fragment union via Keck–Yu silyl-Prins pyran annulation

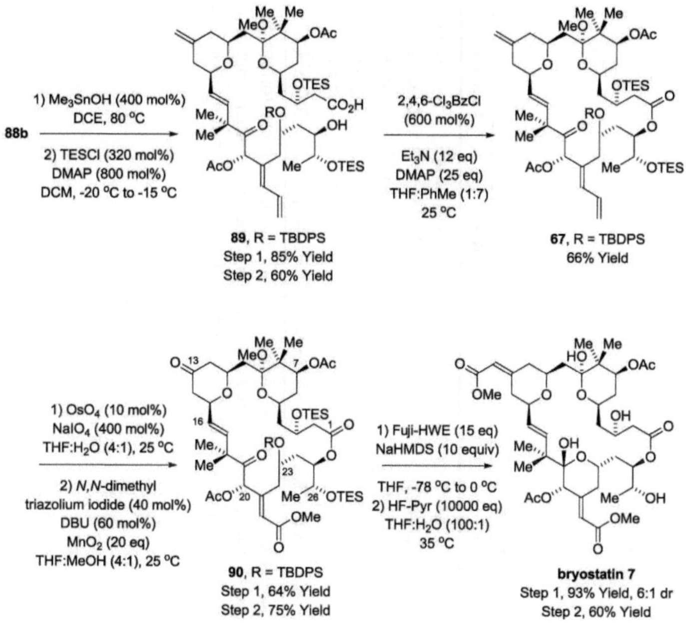

Scheme 5.14 Elaboration of triol **88b** to bryostatin 7

Elaboration of triol **88b** to bryostatin 7 requires chemoselective hydrolysis of the C1 methyl ester in the presence of the C7 and C20 acetates, macrolide formation, installation of the C13 and C21 methyl enoates, and, finally, global deprotection. The sequencing of these transformations is critical, as attempts to introduce the C21 methyl enoate to form the fully functionalized C-ring pyran in advance of macrolide formation resulted in lactonization onto the C23 hydroxyl. In the event, trimethyltin hydroxide promoted hydrolysis [73] of the C1 carboxylate of triol **88b**, and subsequent triethylsilylation of the C3 and C26 hydroxyls each occurs in a selective fashion, thus providing the *seco*-acid **89**. Yamaguchi macrolactonization [39] proceeds uneventfully to provide the macrolide **67** in 66 % yield (Scheme 5.14).

With the macrolide intact, installation of the C13 and C21 methyl enoates requires chemoselective oxidative cleavage of tetraene **67** at the terminal olefin moieties. Remarkably, under the conditions of Johnson–Lemieux oxidation [74], concomitant oxidative cleavage occurs to form the B-ring ketone and the C-ring enal. Attempted conversion of the C-ring enal to the methyl enoate using the Corey–Gilman protocol [75] failed due to cyanation of the B-ring ketone. However, the corresponding *N*-heterocyclic carbene-promoted process [76] provides the desired methyl ester **90** in good isolated yield. As practiced in prior bryostatin total syntheses, asymmetric olefination of the B-ring ketone using Fuji's chiral phosphonate [38] stereoselectively generates the B-ring enoate. Finally, treatment with HF-pyridine promotes hydrolysis of A-ring methyl ketal, desilylation of the

Table 5.1 Analysis correlating reaction type with overall efficiency in total syntheses of the bryostatins

	Masamune	Evans	Nishiyama–Yamamura	Trost	Keck	Wender	Krische
Target	Bryostatin 7	Bryostatin 2	Bryostatin 3	Bryostatin 16	Bryostatin 1	Bryostatin 9	Bryostatin 7
LLS (TS)	**41 (79)**	**42 (72)**	**43 (88)**	**28 (42)**	**31 (58)**	**25 (43)**	**20 (36)**
Skeleton constructions	11 (27 %)	10 (24 %)	10 (23 %)	11 (39 %)	9 (29 %)	9 (36 %)	8 (40 %)
Redox reactions	12 (29 %)	8 (19 %)	6 (14 %)	5 (18 %)	5 (16 %)	7 (28 %)	4 (20 %)
Protection–deprotection	13 (32 %)	12 (29 %)	21 (49 %)	9 (32 %)	10 (32 %)	7 (28 %)	5 (25 %)
Other functional group manipulations	5 (12 %)	12 (29 %)	6 (14 %)	3 (11 %)	7 (23 %)	2 (8 %)	3 (15 %)

Only transformations in the longest linear sequence (LLS) are considered. The term "skeleton constructions" refers to C–C and C–O bond formations (notwithstanding redox reactions) that directly introduce native structural features of the bryostatins without further modification. The term "other functional group manipulations" refers to steps that indirectly introduce native structural elements, the interconversion of functional groups (e.g., the introduction and removal of auxiliaries) and miscellaneous transformations that do not involve skeleton construction

C3 and C26 TES ethers and the C23 TBDPS ether, and spontaneous C-ring closure to furnish bryostatin 7 in 60 % yield (Scheme 5.14).

5.7 Conclusion

Because organic molecules are compounds of carbon and hydrogen, the formation of C–C bonds under hydrogenation conditions represents one natural endpoint in the evolution of strategies for organic synthesis. The present synthesis of bryostatin 7 serves to illustrate that even with a limited repertoire of such hydrogenative couplings, a significant, simplifying shift in retrosynthetic paradigm is availed. Indeed, analysis of the total syntheses of the bryostatins reveals how merged redox-construction events [77] enhance synthetic efficiency by circumventing discrete redox manipulations and additional steps associated with the introduction and removal of nonnative functional groups, for example, chiral auxiliaries and protecting groups (Table 5.1). These features are in line with longstanding ideals of synthetic efficiency [78].

References

1. Pettit GR, Herald CL, Doubek DL, Herald DL, Arnold E, Clardy J (1982) J Am Chem Soc 104:6846
2. Schaufelberger DE, Koleck MP, Beutler JA, Vatakis AM, Alvarado AB, Andrews P, Marzo LV, Muschik GM, Roach J, Ross JT, Lebherz WB, Reeves MP, Eberwein RM, Rodgers LL, Testerman RP, Snader KM, Forenza S (1991) J Nat Prod 54:1265
3. Hildebrand M, Waggoner LE, Liu H, Sudek S, Allen S, Anderson C, Sherman DH, Haygood M (2004) Chem Biol 11:1543
4. For selected reviews encompassing the biological properties of the bryostatins, see: (a) Hale KJ, Hummersone MG, Manaviazar S, Frigerio M (2002) Nat Prod Rep 19:413; (b) Kortmansky J, Schwartz GK (2003) Cancer Invest 21:924; (c) Wender PA, Baryza JL, Hilinski MK, Horan JC, Kan C, Verma VA (2007) Beyond natural products: synthetic analogs of bryostatin 1. In: Huang Z (ed) Drug discovery research: new frontiers in the post-genomic era, Wiley, Hoboken, NJ, pp 127–162; (d) Hale KJ, Manaviazar S (2010) Chem Asian J 5:704
5. Wall NR, Mohammad RM, Al-Katib AM (1999) Leuk Res 23:881
6. Elgie AW, Sargent JM, Alton P, Peters GJ, Noordhuis P, Williamson CJ, Taylor CG (1998) Leuk Res 22:373
7. Oz HS, Hughes WT, Rehg JE, Thomas EK (2000) Microb Pathog 29:187
8. Gescher A, Stanwell C, Dale I (1994) Cancer Top 9:3
9. (a) Nishizuka Y (1992) Science 258:607; (b) Caponigro F, French RC, Kaye SB (1997) Anti-Cancer Drugs 8:26
10. Castagna M, Takai Y, Kaibuchi K, Sano K, Kikkawa U, Nishizuka YJ (1982) Biol Chem 257:7874
11. Blumberg PM, Pettit GR (1992) In: Krosgaard-Larsen P, Christensen CB, Kodof H (eds) New leads and targets in drug research, Munksgaard, Copenhagen, pp 273–285
12. (a) Dowlati A, Lazarus HM, Hartman P, Jacobberger JW, Whitacre C, Gerson SL, Ksenich P, Cooper BW, Frisa PS, Gottlieb M, Murgo AJ, Remick SC (2003) Clin Cancer Res 9:5929;

(b) Barr PM, Lazarus HM, Cooper BW, Schluchter MD, Panneerselvam A, Jacobberger JW, Hsu JW, Janakiraman N, Simic A, Dowlati A, Remick SC (2009) Am J Hematol 84:484; (c) Ajani JA, Jiang Y, Faust J, Chang BB, Ho L, Yao JC, Rousey S, Dakhil S, Cherny RC, Craig C, Bleyer A (2006) Invest New Drugs 24:353

13. Hongpaisan J, Alkon DL (2007) Proc Natl Acad Sci USA 104:19571
14. Sun M-K, Alkon DL (2005) Eur J Pharmacol 512:43
15. (a) Etcheberrigaray R, Tan M, Dewachter I, Kuiperi C, Van der Auwera I, Wera S, Qiao L, Bank B, Nelson TJ, Kozikowski AP, Van Leuven F, Alkon DL (2004) Proc Natl Acad Sci USA 101:11141; (b) Alkon DL, Sun M-K, Nelson TJ (2007) Trends Pharmacol Sci 28:51
16. Sun M-K, Alkon DL (2006) CNS Drug Rev 12:1
17. (a) Sun M-K, Hongpaisan J, Nelson TJ, Alkon DL (2008) Proc Natl Acad Sci USA 105:13620; (b) Sun M-K, Hongpaisan J, Alkon DL (2009) Proc Natl Acad Sci USA 106:14676
18. (a) Perez M, de Vinuesa AG, Sanchez-Duffhues G, Marquez N, Bellido M, Munoz-Fernandez MA, Moreno S, Castor TP, Calzado MA, Munoz E (2010) Curr HIV Res 8:418; (b) Mehla R, Bivalkar-Mehla S, Zhang R, Handy I, Albrecht H, Giri S, Nagarkatti P, Nagarkatti M, Chauhan A (2010) PLos One 5:e11160
19. Wender PA, Cribbs CM, Koehler KF, Sharkey NA, Herald CL, Kamano Y, Pettit GR, Blumberg PM (1988) Proc Natl Acad Sci USA 85:7197
20. For selected examples of Wender's bryostatin analogues, see: (a) Wender PA, De Brabander J, Harran PG, Jiminez J-M, Koehler MFT, Lippa B, Park C-M, Shiozaki M (1998) J Am Chem Soc 120:4534; (b) Wender PA, De Brabander J, Harran PG, Jiminez J-M, Koehler MFT, Lippa B, Park C-M, Siedenbiedel C, Pettit GR (1998) Proc Natl Acad Sci USA 95:6624; (c) Wender PA, Baryza JL, Bennett CE, Bi FC, Brenner SE, Clarke MO, Horan JC, Kan C, Lacôte E, Lippa BS, Nell PG, Turner TM (2002) J Am Chem Soc 124:13648; (d) Wender PA, Baryza JL, Hilinski MK, Horan JC, Kan C, Verma VA (2007) In: Huang Z (ed) Drug discovery research: new frontiers in the post-genomic era, Wiley, Hoboken, NJ, pp 127–162; (e) Wender PA, DeChristopher BA, Schrier AJ (2008) J Am Chem Soc 130:6658; (f) Wender PA, Verma VA (2006) Org Lett 8:1893; (g) Wender PA, Baryza JL, Brenner SE, DeChristopher BA, Loy BA, Schrier AJ, Verma VA (2011) Proc Natl Acad Sci USA 108:6721
21. For selected examples of Keck's bryostatin analogues, see: (a) Keck GE, Truong AP (2005) Org Lett 7:2153; (b) Keck GE, Kraft MB, Truong AP, Li W, Sanchez CC, Kedei N, Lewin N, Blumberg PM (2008) J Am Chem Soc 130:6660; (c) Keck GE, Poudel YB, Welch DS, Kraft MB, Truong AP, Stephens JC, Kedei N, Lewin NE, Blumberg PM (2009) Org Lett 11:593; (d) Keck GE, Li W, Kraft MB, Kedei N, Lewin NE, Blumberg PM (2009) Org Lett 11:2277; (e) Keck GE, Poudel YB, Rudra A, Stephens JC, Kedei N, Lewin NE, Peach ML, Blumberg PM (2010) Angew Chem Int Ed 49:4580
22. (a) Sudek S, Lopanik NB, Waggoner L E, Hildebrand M, Anderson C, Liu H, Patel A, Sherman DH, Haygood MG (2007) J Nat Prod 70:67; (b) Lopanik NB, Shields J, Buchholz TJ, Rath CM, Hothersall J, Haygood MG, Hakansson K, Thomas CM, Sherman DH (2008) Chem Biol 15:1175; (c) Trindade-Silva AE, Lim-Fong GE, Sharp KH, Haygood MG (2010) Curr Opin Biotechnol 21:834
23. (a) Kageyama M, Tamura T, Nantz M, Roberts JC, Somfai P, Whritenour DC, Masamune S (1990) J Am Chem Soc 112:7407; (b) Masamune S (1988) Pure Appl Chem 60:1587
24. (a) Evans DA, Carter PH, Carreira EM, Prunet JA, Charette AB, Lautens M (1998) Angew Chem Int Ed 37:2354; (b) Evans DA, Carter PH, Carreira EM, Charette AB, Prunet JA, Lautens M (1999) J Am Chem Soc 121:7540
25. (a) Ohmori K, Ogawa Y, Obitsu T, Ishikawa Y, Nishiyama S, Yamamura S (2000) Angew Chem Int Ed 39:2290; (b) Ohmori K (2004) Bull Chem Soc Jpn 77:875
26. Trost BM, Dong G (2008) Nature 456:485
27. Keck GE, Poudel YB, Cummins TJ, Rudra A, Covel JA (2011) J Am Chem Soc 133:744
28. Wender PA, Schrier AJ (2011) J Am Chem Soc 133:9228
29. Lu Y, Woo SK, Krische MJ (2011) J Am Chem Soc 133:13876

30. Manaviazar S, Frigerio M, Bhatia GS, Hummersone MG, Aliev AE, Hale KJ (2006) Org Lett 8:4477
31. Trost BM, Dong G (2010) J Am Chem Soc 132:16403
32. Green AP, Lee ATL, Thomas EJ (2011) Chem Commun 47:7200
33. Masamune S, Sato T, Kim B-M, Wollmann TA (1986) J Am Chem Soc 108:8279
34. (a) Julia M, Paris JM (1973) Tetrahedron Lett 14:4833; (b) Kocienski PJ, Lythgoe B, Ruston S (1978) J Chem Soc Perk T 1 829
35. (a) Katsuki T, Sharpless KB (1980) J Am Chem Soc 102:5974; (b) Rossiter BE, Katsuki T, Sharpless KB (1981) J Am Chem Soc 103:464; (c) Gao Y, Hanson RM, Klunder JM, Ko SY, Masamune H, Sharpless KB (1987) J Am Chem Soc 109:5765
36. Corey EJ, Katzenellenbogen JA, Posner GH (1967) J Am Chem Soc 89:4245
37. Piers E, Morton HE (1980) J Org Chem 45:4263
38. (a) Tanaka K, Ohta Y, Fuji K, Taga T (1993) Tetrahedron Lett 34:4071; (b) Tanaka K, Otsubo K, Fuji K (1996) Tetrahedron Lett 37:3735
39. Inanaga J, Hirata K, Saeki H, Katsuki T, Yamaguchi M (1979) Bull Soc Chem Jpn 52:1989
40. Evans DA, Sjogren EB, Weber AE, Conn RE (1987) Tetrahedron Lett 28:39
41. Evans DA, Chapman KT, Carreira E (1988) J Am Chem Soc 110:3560
42. (a) Ramachandran PV, Xu W-C, Brown HC (1996) Tetrahedron Lett 37:4911; (b) Paterson I, Goodman JM, Lister MA, Schumann RC, McClure CK, Norcross RD (1990) Tetrahedron 46:4663
43. Evans DA, Hoveyda AH (1990) J Am Chem Soc 112:6447
44. Evans DA, Murry JA, Kozlowski MC (1996) J Am Chem Soc 118:5814
45. Danishefsky S, Kobayashi S, Kerwin Jr JF (1982) J Org Chem 47:1981
46. Demuth M, Palomer A, Sluma H-D, Dey AK, Krüger C, Tsay Y-H (1986) Angew Chem Int Ed 25:1117
47. (a) Zinner H, Wulf G, Heinatz R (1964) Chem Ber 97:3536; (b) Copeland C, Stick RV (1977) Aust J Chem 30:1269; (c) Trost BM, Klun TP (1981) J Am Chem Soc 103:1864
48. Corey EJ, Fuchs PL (1972) Tetrahedron Lett 13:3769
49. (a) Ohmori K, Nishiyama S, Yamamura S (1995) Tetrahedron Lett 36:6519; (b) Kolb HC, VanNieuwenhze MS, Sharpless KB (1994) Chem Rev 94:2483
50. Trost BM, Matsubara S, Caringi JJ (1989) J Am Chem Soc 111:8745
51. Liu Y, Song F, Song Z, Liu M, Yan B (2005) Org Lett 5409
52. Trost BM, Yang H, Wuitschik G (2005) Org Lett 7:4761
53. Brown HC, Jadhav PK (1983) J Am Chem Soc 105:2092
54. Roth GJ, Liepold B, Müller SG, Bestmann HJ (2004) Synthesis 59
55. (a) Keck GE, Covel JA, Schiff T, Yu T (2002) Org Lett 4:1189; (b) Yu C-M, Lee J-Y, So B, Hong J (2002) Angew Chem Int Ed 41:161
56. Keck GE, Welch DS, Poudel YB (2006) Tetrahedron Lett 47:8267
57. (a) Keck GE, Tarbet KH, Geraci LS (1993) J Am Chem Soc 115:8467; (b) Costa AL, Piazza MG, Tagliavini E, Trombini C, Umani-Ronchi A (1993) J Am Chem Soc 115:7001
58. (a) Mukaiyama T, Narasaka K, Banno K (1973) Chem Lett 2:1011; (b) Mukaiyama T, Izawa T, Saigo K (1974) Chem Lett 3:323; (c) Mukaiyama T, Banno K, Narasaka K (1974) J Am Chem Soc 96:7503
59. (a) Iyer K, Rainier JD (2007) J Am Chem Soc 129:12604; (b) Nicolaou KC, Postema MHD, Claiborne CF (1996) J Am Chem Soc 118:1565; (c) Nicolaou KC, Postema MHD, Yue EW, Nadin A (1996) J Am Chem Soc 118:10335
60. Okazoe T, Takai K, Utimoto K (1987) J Am Chem Soc 109:951
61. For selected reviews, see: (a) Patman RL, Bower JF, Kim IS, Krische MJ (2008) Aldrichim Acta 41:95; (b) Bower JF, Kim IS, Patman RL, Krische MJ (2009) Angew Chem Int Ed 48:34; (c) Bower JF, Krische MJ (2011) Top Organomet Chem 43:107; (d) Hassan A, Krische MJ (2011) Org Process Res Dev 15:1236
62. (a) Lu Y, Kim IS, Hassan A, Del Valle DJ, Krische MJ (2009) Angew Chem Int Ed 48:5018; (b) Han SB, Hassan A, Kim I-S, Krische MJ (2010) J Am Chem Soc 132:15559;

(c) Kumpulainen ETT, Kang B, Krische MJ (2011) Org Lett 13:2484; (d) Gao X, Han H, Krische MJ (2011) J Am Chem Soc 133:12795
63. (a) Cho C-W, Krische MJ (2006) Org Lett 8:891; (b) Lu Y, Krische MJ (2009) Org Lett 11:3108
64. (a) Han SB, Kim IS, Han H, Krische MJ (2009) J Am Chem Soc 131:6916; (b) Itoh J, Han SB, Krische MJ (2009) Angew Chem Int Ed 48:6313; (c) Bechem B, Patman RL, Hashmi ASK, Krische MJ (2010) J Org Chem 75:1795
65. (a) Kim IS, Ngai M-Y, Krische MJ (2008) J Am Chem Soc 130:6340; (b) Kim IS, Ngai M-Y, Krische MJ (2008) J Am Chem Soc 130:14891; (c) Hassan A, Lu Y, Krische MJ (2009) Org Lett 11:3112
66. (a) Hong Y-T, Cho C-W, Skucas E, Krische MJ (2007) Org Lett 9:3745; (b) Komanduri V, Krische MJ (2006) J Am Chem Soc 128:16448
67. For amplification of enantiomeric enrichment in the generation of *C2*-symmetric compounds, see: (a) Kogure T, Eliel EL (1984) J Org Chem 49:576; (b) Midland MM, Gabriel J (1985) J Org Chem 50:1143
68. (a) Smith AB III, Minbiole KP, Verhoest PR, Schelhaas M (2001) J Am Chem Soc 123:10942; (b) Rychnovsky SD, Griesgraber G, Powers JP (2000) Org Synth 77:1
69. Trost BM, Yang H, Thiel OR, Frontier AJ, Brindle CS (2007) J Am Chem Soc 129:2206
70. For a similar enolization-reduction sequence, see: Lovchik MA, Goeke A, Frater G (2007) J Org Chem 72:2427
71. Kornblum N, Frazier HW (1966) J Am Chem Soc 88:865
72. Almendros P, Rae A, Thomas EJ (2000) Tetrahedron Lett 41:9565
73. Nicolaou KC, Estrada AA, Zak M, Lee SH, Safina BS (2005) Angew Chem Int Ed 44:1378
74. For chemoselective Johnson–Lemieux oxidation of terminal olefins, see: (a) White JD, Kuntiyong P, Lee TH (2006) Org Lett 8:6039; (b) BouzBouz S, Cossy J (2003) Org Lett 5:3029
75. (a) Corey EJ, Gilman NW, Ganem BE (1968) J Am Chem Soc 90:5616; (b) Corey EJ, Katzenellenbogen JA, Gilman NW, Roman SA, Erickson BW (1968) J Am Chem Soc 90: 5618; (c) Gilman NW (1971) Chem Commun 733
76. Maki BE, Scheidt KA (2008) Org Lett 10:4331
77. For an overview of "redox economy" in organic synthesis, see: Baran PS, Hoffmann RW, Burns NZ (2009) Angew Chem Int Ed 48:2854
78. "The ideal synthesis creates a complex skeleton... in a sequence only of successive construction reactions involving no intermediary refunctionalizations, and leading directly to the structure of the target, not only its skeleton but also its correctly placed functionality." Hendrickson JB (1975) J Am Chem Soc 97:5784

Chapter 6
Serratezomine A

Julie A. Pigza and Jeffrey N. Johnston

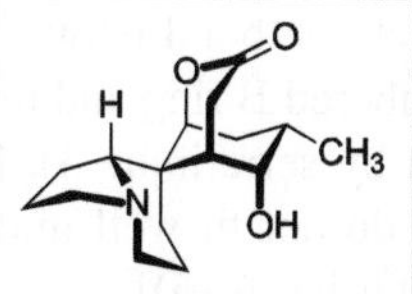

6.1 Introduction and Classification

Serratezomine A (**1**) [1] belongs to an architecturally diverse class of compounds known as the *Lycopodium* alkaloids [2]. The *Lycopodium* alkaloids have fascinated and challenged synthetic chemists since the discovery of lycopodine in 1881 [3]. They are isolated from a variety of club moss species characterized by low, mossy evergreen branches. They tend to belong to one of three groups, $C_{16}N$, $C_{16}N_2$, and $C_{27}N_3$, although the number of carbons and nitrogens may vary. The *Lycopodium* alkaloids fall into one of four main structural classes: lycopodine, lycodine, fawcettimine, and the "miscellaneous" class (Chart 6.1) [2c, 4]. The lycopodine class is characterized by four connected rings, three of which (A, B, and C) are constant and contain a quinolizidine ring (A/C), while the D ring varies in structure (see annotinine **2**). The lycodine class is similar to the lycopodine class except that

J.A. Pigza
Department of Chemistry, Queensborough Community College, Bayside, NY 11364, USA

J.N. Johnston (✉)
Department of Chemistry, Institute of Chemical Biology, Vanderbilt University, Nashville, TN 37235, USA
e-mail: jeffrey.n.johnston@vanderbilt.edu

J.J. Li and E.J. Corey (eds.), *Total Synthesis of Natural Products*,
DOI 10.1007/978-3-642-34065-9_6, © Springer-Verlag Berlin Heidelberg 2012

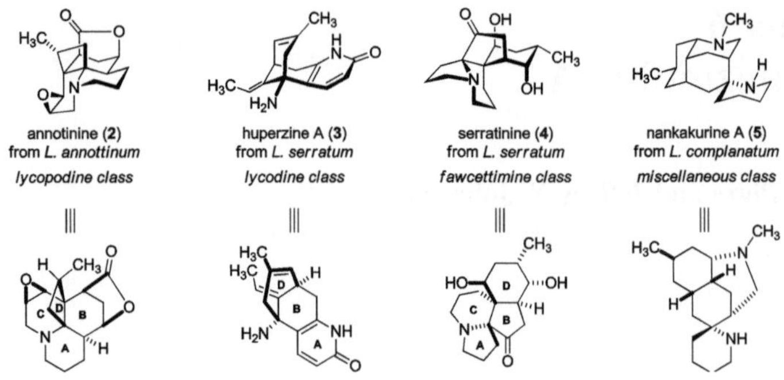

Chart 6.1 Representative *Lycopodium* alkaloids from each structural class

the A ring tends to be oxidized to a pyridine or pyridone and the C ring may or may not be intact (see huperzine A **3**). The fawcettimine class tends to contain a structural rearrangement in which the bond between the A and B rings undergoes a 1,2-shift providing a five-membered B ring and resulting in an abundance of new structural varieties (represented by serratinine **4**). Lastly, the miscellaneous class contains diverse structures that do not fit well into one of the previously defined classes [demonstrated by nankakurine A (**5**)].

6.2 Pharmacology

Of the *Lycopodium* alkaloids, huperzine A has received the most attention based on its biological activity [2c]. It has been shown to improve animal learning and memory in maze experiments and is in clinical trials for the treatment of Alzheimer's disease and myasthenia gravis in China—both related to the neurotransmitter acetylcholine [5]. Myasthenia gravis is an autoimmune disease in which the body develops antibodies that block acetylcholine receptors to prevent binding. A general muscle weakness is observed, including the muscles that control breathing, swallowing, and eye movement which can lead to a feeling of suffocation. Alzheimer's disease is a neurodegenerative disorder in which neurons in the brain stem are affected. Neuronal death results from plaque formation in the brain due to insoluble aggregrates produced by peptide breakdown (normal peptide breakdown results in soluble monomers that can be effectively handled by the body). Higher acetylcholine levels in the brain, in both diseases, show beneficial effects in patients, including memory enhancement. Huperzine A increases the amount of acetylcholine by inhibiting acetylcholinesterase (AChE) [6], an enzyme which normally breaks down acetylcholine [7]. Huperzine A reversibly binds to AChE allowing more acetylcholine to be available in the synaptic clefts between neurons in the brain. Huperzine A ($IC_{50} = 0.07\ \mu M$) is comparable to that of Aricept® (Donepezil, $IC_{50} = 0.03\ \mu M$), one of the currently marketed drugs for Alzheimer's treatment. Huperzine A is also available as an over-the-counter dietary supplement taken to enhance memory.

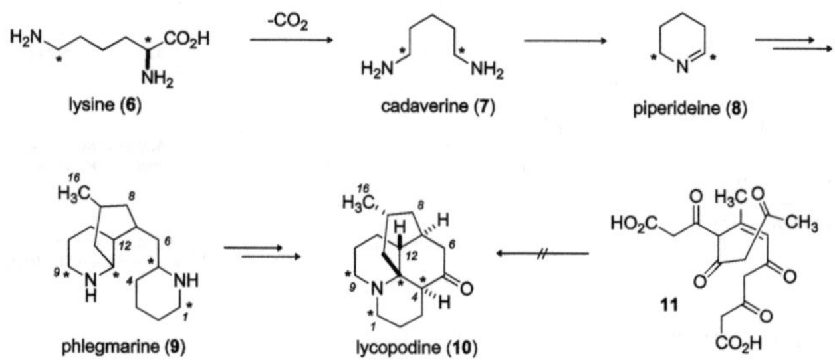

Scheme 6.1 Lysine as a biosynthetic precursor to lycopodine

6.3 Biosynthesis

The biosynthesis of the *Lycopodium* alkaloids has been of ongoing interest since the characterization of annotinine (**4**) in 1957 [4a]. It was initially proposed in the early 1960s that the alkaloids resulted from a polyacetate building block (such as **11**) derived from polyketide biosynthesis to account for the 16 carbons in lycopodine **10** (Scheme 6.1) [8]. However, there was no experimental evidence at the time to support the theory. Spenser and coworkers determined in 1968 that lysine (**6**) was incorporated into lycopodine (**10**) by a series of ^{14}C-labeling experiments [9]. In two separate experiments, lysine (**6**) was labeled at either C2 or C6 (marked with an asterisk in Scheme 6.1) and was incorporated into a *Lycopodium* species via feeding experiments. It was hypothesized that lysine undergoes a decarboxylation in the presence of a decarboxylase enzyme to form a symmetrical intermediate called cadaverine (**7**). Oxidation of **7** to the imine and cyclization results in formation of piperideine (**8**) with loss of NH_3. It was then hypothesized that two units of **8** were incorporated into lycopodine **10** via the compound phlegmarine (**9**). In support of this proposed pathway, lycopodine **10** was isolated after the feeding experiments and degraded and the byproducts were studied for radioactivity and location of the ^{14}C labels. The results of the studies showed that ^{14}C labels were found at C1, C4, C9, and C13. This showed that two units of lysine (**6**) were indeed incorporated into lycopodine, likely via piperideine (**8**).

The study above does not account for the extra six carbons acquired in the conversion of piperideine (**8**, 10 carbons) to phlegmarine (**9**, 16 carbons). It was initially proposed that the carbons were incorporated via pelletierine (**12**), which was incorporated twice into lycopodine resulting in two symmetrical "halves" of the alkaloid (Scheme 6.2). However, when ^{14}C-labeled pelletierine (**12**, label at C2) was fed to the plant, degradation studies of lycopodine revealed that only one "half" consisted of the ^{14}C label from pelletierine (the half containing C9–C16) [10]. The other half does not result from pelletierine **12** but must be something similar in structure since it does contain the piperideine unit (**8**) resulting from lysine. It was of interest then to determine the exact source of the three-carbon propionate unit in pelletierine (**12**).

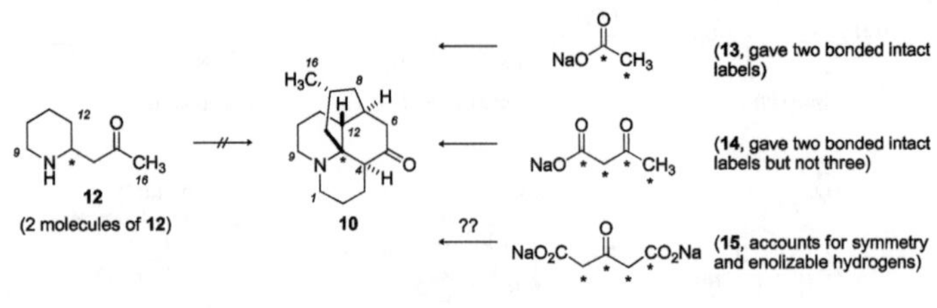

Scheme 6.2 Pelletierine as a possible precursor to lycopodine

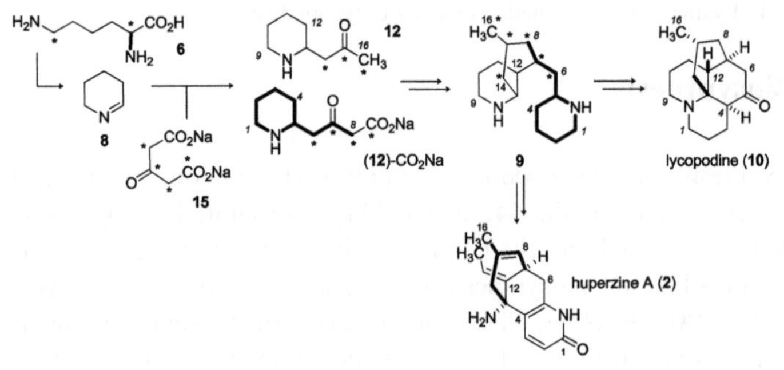

Scheme 6.3 Proof of the biosynthetic pathway to lycopodine (**10**)

Further studies by Spenser demonstrated that 1,2-^{13}C-labeled acetate (**13**) was incorporated into lycopodine but gave a distribution of the labels that did not account for the pelletierine-route that was hypothesized (Scheme 6.2) [11]. An intact 3-carbon unit was desired for testing, but labeled acetoacetate (1,2,3,4-^{13}C-acetoacetate (**14**), which could undergo decarboxylation to provide an intact 3-carbon unit) was found to give the same incorporation pattern as acetate (and therefore must have been cleaved to acetate prior to uptake). In addition, feeding studies using deuterated, ^{13}C-labeled acetate provided a loss or washout of deuterium at the C16 methyl group. This could only occur if an intermediate had formed that would provide for facile enolization. Both the equal distribution of the ^{13}C labels and loss of the deuteriums led the researchers to propose that the intermediate was symmetric, such as acetone dicarboxylic acid (**15**).

As predicted, 1,2,3,4-^{13}C-labeled acetone dicarboxylate (**15**) provided an intact three-carbon chain into lycopodine. It also helped to explain why two molecules of pelletierine (**12**) were not incorporated (Scheme 6.3) [12]. As before, lysine (**6**) is converted to piperideine (**8**) via a decarboxylation. Then a Mannich reaction of labeled **15** with **8** provides pelletierine **12**. The other "half" of the molecule to be incorporated must be pelletierine-like (**12**-CO_2Na), still containing one of the carboxylates. An aldol reaction of the two pelletierine fragments and a series of transformations leads to phlegmarine **9**. Oxidation of **9** involving imine formation between N–C5, isomerization to the enamine and then cyclization onto an imine (at N–C13), provides lycopodine **10**. Phlegmarine **9** and lycopodine **10** are proposed as

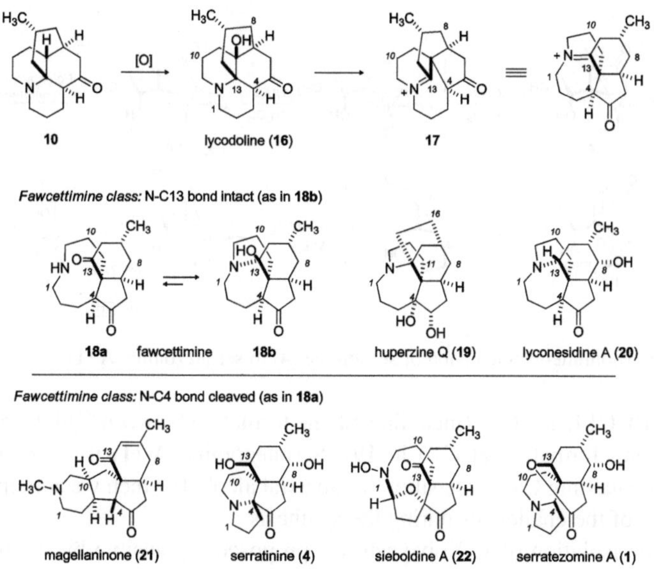

Scheme 6.4 Biosynthetic pathway to the fawcettimine class

main intermediates in the biosynthesis of other *Lycopodium* alkaloids, including huperzine A (**2**) [13].

From lycopodine **10**, the entire fawcettimine structural class of *Lycopodium* alkaloids can be established, including the target molecule serratezomine A (**1**). First, an oxidation of **10** occurs to install a tertiary hydroxyl group at C12, forming lycodoline (**16**, Scheme 6.4). Lycodoline undergoes a 1,2-carbon bond shift, initiated by the lone pair on nitrogen which displaces the hydroxyl group and forms the five-membered B ring. The resulting iminium ion (**17**) has a variety of fates providing for an array of new compounds.

Hydrolysis and cleavage of the iminium in **17** to the ketone results in a group of molecules where the N–C13 bond is no longer intact, while incomplete hydrolysis or other steps allow the N–C13 bond to remain intact. Some representative compounds from each subclass are shown to represent the structural diversity and different oxidation states that exist (Scheme 6.4) [14]. The structures are drawn in their 2D orientation to more easily highlight the differences among them. Fawcettimine **18a/b** contains an interesting keto-carbinol amine equilibration which favors the side of the carbinol amine (**18b**, N–C13 bond intact). Molecules in which the N–C13 bond is also intact are represented by huperzine Q (**19**) and lyconesidine A (**20**). When the N–C13 bond is broken by hydrolysis, this leads to structures including serratezomine A (**1**), serratinine (**4**), magellaninone (**21**), and sieboldine A (**22**)—each including a second ring formation that is not present in **18a**. The structures can be accounted for by transformations involving amine oxidation or enamine and/or enolate formation, along with subsequent bond transformations. Another interesting feature to note is the differential oxidation states along the cyclohexane ring at C8 (it is usually not

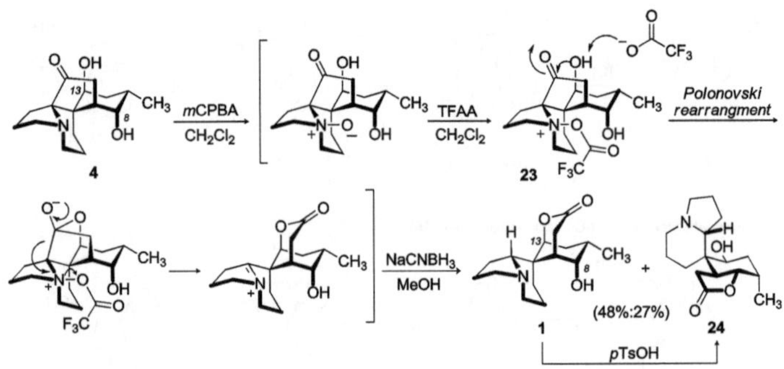

Scheme 6.5 Biomimetic conversion of serratinine (**4**) to serratezomine A (**1**)

oxygenated), C14, and C15 (including alkene formation between C14–C15 in **21** and a unique ring formation at C16 in **19**). Serratezomine A (**1**) is an example of a molecule containing C8 oxygenation (as an axial alcohol) which we anticipated to be among one of the challenges during the synthesis.

After the isolation of **1**, Kobayashi [1] proposed a plausible biosynthetic pathway to the target from the closely related serratinine (**4**) [15] by attack of the C13-OH onto the carbonyl of the cyclopentanone ring. Furthermore, he demonstrated the feasibility of this reaction via a one-pot transformation involving a Polonovski-Potier rearrangement (Scheme 6.5) [16]. Treatment of serratinine (**4**) with MCPBA resulted in formation of the *N*-oxide which was followed by trifluoroacetic anhydride to form **23a**. The free alcohol then attacks the ketone resulting in a lactone ring, breaking of a C–C bond, formation of an iminium ion, and loss of the trifluoroacetate anion. The resulting iminium ion (**23b**) can then be reduced by $NaCNBH_3$ to give **1** in 48 % yield. By-product **24** was also observed during the reaction, resulting from a *trans*-lactonization (27 %), which could be made exclusively from **1** upon treatment with *p*-TsOH.

6.4 Previous Synthetic Work

Based on their unique structures and complicated ring skeletons, the *Lycopodium* alkaloids have attracted sustained attention from the synthetic community [2e, 17]. Our synthesis of serratezomine A (**1**) [18] currently stands alone, but the structural similarity between serratezomine A and serratinine (**4**) compels a discussion of prior work towards the latter. This includes syntheses of the framework [19] of serratinine (**4**) or similar molecules and one total synthesis [20].

6.4.1 Total Synthesis of Serratinine

Inubushi's synthesis of racemic serratinine commences with a Diels–Alder reaction of butadiene with substituted quinone derivative **26** (available in four steps from **25**,

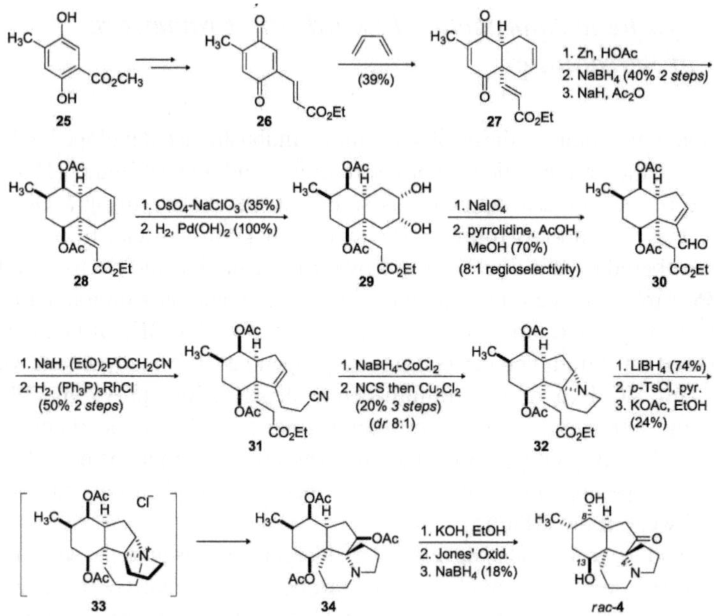

Scheme 6.6 Inubushi synthesis of *rac*-serratinine (**4**)

86 % yield) to provide dienone **27** (Scheme 6.6) [20]. Zinc reduction of the doubly unsaturated alkene and then $NaBH_4$ ketone reduction followed by acetylation resulted in unsaturated ester **28**. Dihydroxylation of the electron-rich alkene followed by hydrogenation of the conjugated alkene led to diol **29**. Periodate cleavage of the diol and Mannich cyclization of the dialdehyde in the presence of pyrrolidine resulted in regioselective formation of aldehyde **30**, containing two of the four rings of serratinine (8:1 ratio of regioisomers). Use of piperidine or Al_2O_3 instead of pyrrolidine resulted in selective formation of the undesired regioisomer [21].

Subsequent Wittig reaction of the aldehyde incorporates the nitrogen and reduction of the newly installed conjugated alkene using Wilkinson's catalyst furnished nitrile **31**. The nitrile was selectively reduced in the presence of the ester using $NaBH_4/CoCl_2$. Chlorination of the amine with NCS and in situ aziridination via a nitrene intermediate gave the desired aziridine diastereomer (**32**) resulting from addition to the less hindered, convex face of the cyclopentene. Reduction of the ester, tosylate formation, and displacement with the aziridine nitrogen formed an aziridinium ion (**33**) in which the aziridine nitrogen is part of three different ring systems. Displacement of the strained aziridinium ring at the less hindered carbon with an acetate anion provided **34**. It was then necessary to invert the configuration of the methyl and adjacent acetate on the six-membered ring. Therefore, global acetate deprotection was accomplished with KOH and then the three alcohols were oxidized with H_2CrO_4, which fortuitously caused epimerization of the methyl group to the correct configuration. $NaBII_4$ provided reduction of the two six-membered ring ketones in preference to the five-membered ring ketone, providing *rac*-serratinine **4**.

6.4.2 Synthetic Approaches Towards the Framework of Serratinine

In addition to the total synthesis of serratinine, Inubushi also developed a divergent synthesis to the framework of both serratinine and fawcettimine (**18a/b**). The beginning of the Inubushi approach is similar to before, requiring a Diels–Alder reaction, but this time to form **35** (Scheme 6.7) [19a]. Instead of aziridination, a nine-membered lactam ring is formed by nitrile reduction and cyclization to form amide **36**. Amide and ketone reduction, *N*-protection, and reoxidation to the ketone was followed by epoxidation to form **37**. When exposed to $BF_3{\cdot}OEt_2$, the nitrogen in **37** forced an intramolecular epoxide ring opening. The intermediate was then treated with KOH to effect trifluoroacetate deprotection, providing **39**, which contains the tetracyclic core of serratinine. Interestingly, if the α-epoxide (**38**) was exposed to $BF_3{\cdot}OEt_2$, no cyclization was observed, but instead elimination and epoxide opening transpired. This intermediate was then carried forward to produce fawcettimine (**18a/b**).

Mehta and coworkers also developed a divergent approach to the serratinine and fawcettimine cores via epoxidation but through an alternative strategy to gain access to the nine-membered ring in **42** [19b]. They used an ozonolysis of a cyclooctene to break the ring and allow introduction of the nitrogen. To begin, conjugate addition of a Grignard onto enone **40** (available from cyclooctadiene) provided an adduct that could be readily modified via a Wacker oxidation and Michael reaction to tricyclic **41**. A chemoselective ketone protection as the acetal, reduction of the remaining ketone, and protection with a TBS group provided **42**. A reductive ozonolysis to the diol and conversion to a leaving group allowed for the amine to be introduced with a tosylate protecting group. Lewis acid catalyzed elimination of the OTBS provided **43**, which was epoxidized in a similar approach to the Inubushi work.

Zard's strategy to the serratinine core implemented a radical cascade to install the indolizidine core [19c]. A Pauson–Khand reaction of acyclic alkyne **44** provided the eastern portion of the molecule with high diastereoselectivity (**45**, Scheme 6.7). Modification of the side chain to the benzoyl hydroxamide radical precursor occurred in short order. Exposure of **45** to $^{n}Bu_3SnH$ and a radical initiator (ACCN) formed a nitrogen-centered radical (**46**) that added in a 5-*exo*-trig fashion to the enone. The resulting tertiary radical added to the terminal alkene through a 6-*endo*-trig transition state to provide the tetracyclic core (**47**). If the chlorine was absent, the tertiary carbon radical instead underwent a 5-*exo*-trig process providing the pyrrolizidine core rather than an indolizidine. Tetracyclic **51** was then converted to 13-deoxyserratine in three steps.

Lastly, Livinghouse's approach to the tricyclic core of **4** is highlighted by an acylnitrilium spirocyclization [19d]. Exposure of isonitrile **48** to thiophenyl acetyl chloride provides an intermediate chloramine (**49**) that, in the presence of a silver salt, eliminates the chloride to form a nitrilium ion ($R{-}N \equiv C^{+}{-}R$). The nitrilium is attacked by the nucleophilic aromatic ring to furnish spirocyclic imine **50** in high

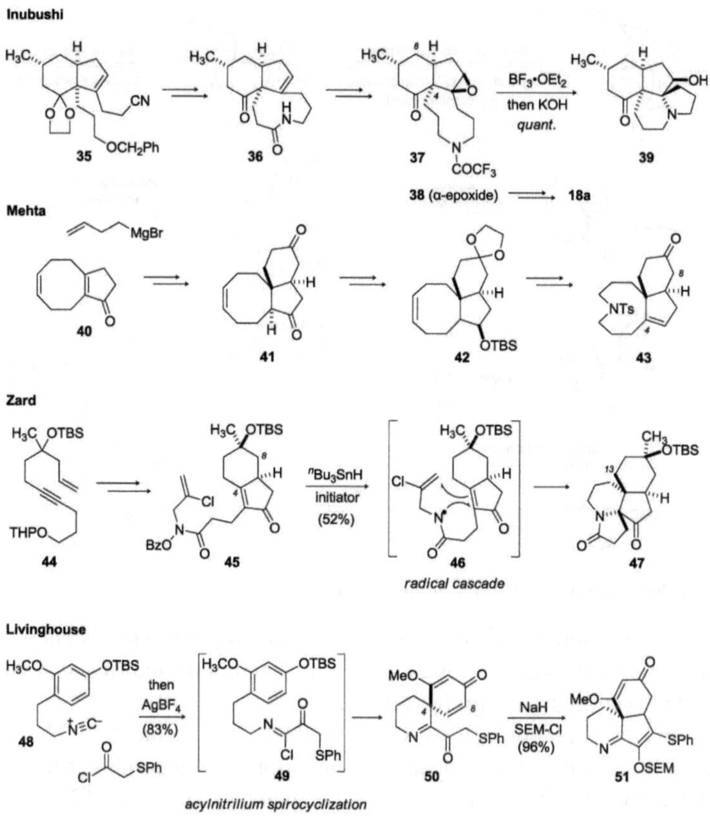

Scheme 6.7 Approaches to the serratinine framework

yield. Desymmetrization of the dienone occurs via enolate formation of the α-thioketone and regioselective, intramolecular Michael addition to yield tricyclic **51** containing a variety of functional groups for further elaboration.

6.5 Strategy and Retrosynthesis

The structure of serratezomine A (**1**) was determined by ^{1}H and ^{13}C NMR as well as 2D NMR techniques including COSY, HOHAHA, HMQC, and HMBC in CD_3OD [1]. In addition, **1** exhibited moderate cytotoxicity against murine lymphoma L1210 cells ($IC_{50} = 9.7$ μg/mL) and human epidermoid carcinoma KB cells ($IC_{50} > 10$ μg/mL) and demonstrated strong anticholinesterase activity. Our interest was stimulated by the structural complexity and backbone architecture of **1**. Serratezomine A (**1**) was a particularly inspiring synthetic target

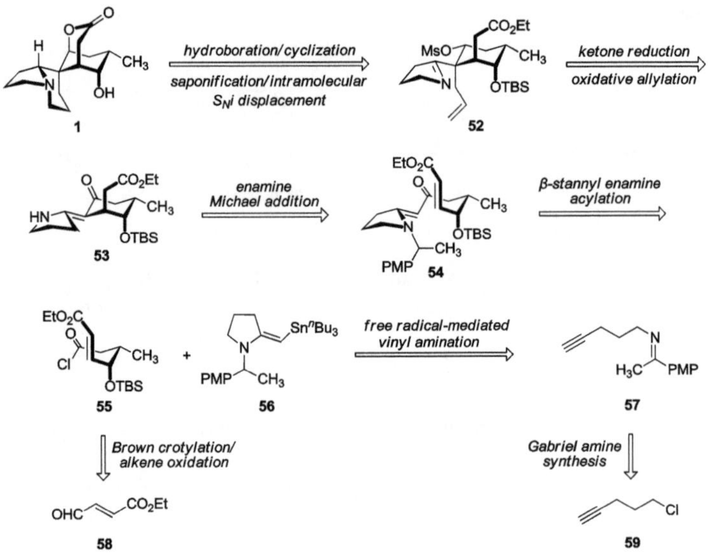

Scheme 6.8 Retrosynthesis of serratezomine A (**1**)

containing a *seco*-serratinine-type skeleton, six contiguous stereocenters, and an all-carbon spirocyclic center embedded within four connected ring systems plus a challenging, axial hydroxyl group at C8.

Highlights of our approach to **1** are outlined in Scheme 6.8. The first two disconnects involve the piperidine and lactone rings. The piperidine ring formation is brought about by functionalization of the pendant alkene in **52** to provide a leaving group for imine cyclization and then reduction to the bicyclic amine. The lactone ring is formed after saponification of the ester and intramolecular (S_Ni) displacement of the mesylate. Target **52** provides an intermediate that contains all of the required carbons in the natural product. Installation of the allyl group in a diastereoselective manner to form the quaternary stereocenter in **52** was expected to be one of the most challenging points in the synthesis. This was planned via an allylation of vinylogous amide **53**. The cyclohexanone ring in **53** was anticipated to result from an intramolecular Michael addition onto the α,β-unsaturated ester in **54** brought about by *N*-deprotection. Selectivity in this step was anticipated to be driven by minimization of $A^{1,3}$-strain [22] and to favor an axial ethyl acetate substituent as depicted in **53**. The *N*-protected vinylogous amide (**54**) represents the convergent point in our synthesis. It was desired via acylation of β-stannylenamine **56** with acid chloride **55**. The β-stannylenamine was synthesized by methodology developed earlier by us, involving a nonconventional radical-mediated 5-*exo*-trig cyclization of an imine [23, 24]. The required imine (**57**) is accessible in three steps from commercially available chloride **59** by a Gabriel amine synthesis. The acid chloride (**55**) could be synthesized from α,β-unsaturated aldehyde **58** using a Brown crotylation and functionalization of the terminal alkene. A key aspect in the synthesis would be to use the two stereocenters in **55** to direct the formation of all remaining stereocenters in **1**.

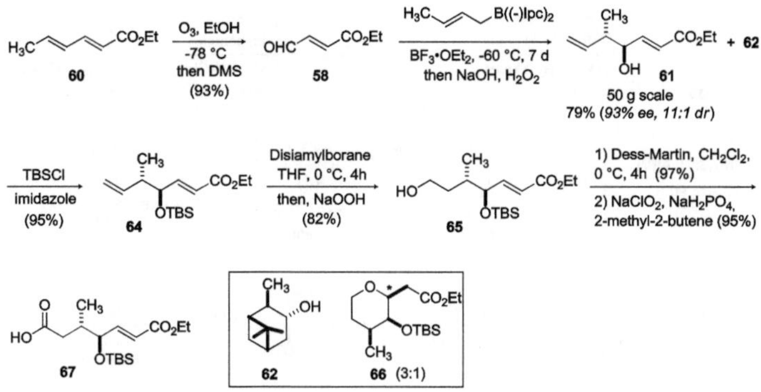

Scheme 6.9 Synthesis of carboxylic acid (**67**)

6.6 Synthesis

The synthesis began with selective ozonolysis of the terminal alkene in commercially available ethyl sorbate (**60**, Scheme 6.9). The α,β-unsaturated aldehyde (**58**) was isolated after vacuum distillation in 93 % yield [25]. Next, the homoallylic alcohol was installed in the *anti*-configuration using a Brown crotylation. This reaction is especially difficult with unsaturated aldehydes and tends to provide lower yields and selectivities [26]. Fortunately, using a modified large dry ice bath that could maintain cold temperatures for an extended period of time, the reaction could be performed on a 50-gram scale providing the desired product with good enantioselectivity and yield (**61**, 93 % ee, 79 % yield). The structure of the main diastereomer was confirmed by formation of both (*R*)- and (*S*)-Mosher esters in which NMR analysis led to the determination of relative stereochemistry as depicted for alcohol **61** [27]. The majority of the terpene alcohol (**62**, a by-product from the oxidation of the chiral Ipc auxiliary) could be fractionally distilled, and the crude reaction mixture containing the two alcohols (**61** and any remaining **62**) was subjected to TBS protection to afford a mixture of TBS ethers. The TBS group provided greater separation of the two products by column chromatography and allowed for the isolation of pure **64**.

After successful installation of the first two stereocenters, our attention was focused on elaboration of the terminal alkene in **64** (Scheme 6.9). Treatment with disiamylborane followed by oxidative workup afforded primary alcohol **65** in good yield (70–85 %). A side product containing a mixture of two diastereomers (**66**) was also observed and resulted from conjugate addition of the alkoxide formed during basic workup onto the unsaturated ester. Maintaining the temperature at 0 °C by a slow, dropwise quench during the oxidative workup was necessary to minimize the amount of the undesired cyclization product (**66**). Subsequent oxidation of the primary alcohol **65** using Dess–Martin periodinane [28] and a Pinnick oxidation afforded carboxylic acid **67** [29].

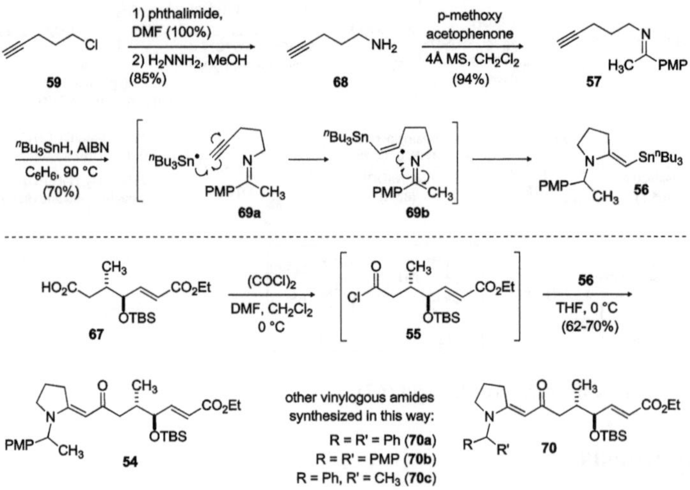

Scheme 6.10 β-Stannylenamine formation and coupling reaction

The stage was now set for the convergent coupling with the β-stannyl enamine linchpin (**56**). Starting from pentynyl chloride **59**, a Gabriel amine synthesis was utilized to form primary amine **68** in 85 % yield (Scheme 6.10). Condensation with *p*-methoxyacetophenone generated imine **57** (94 %). Free radical-mediated aminostannation was carried out using slow addition of $^nBu_3SnH$ and AIBN to a refluxing, degassed benzene solution of imine **57** [23, 24]. The stannane radical adds to the terminal position of the alkyne (as in **69a**) to form vinyl radical **69b**. The 5-*exo-trig* cyclization of the vinyl radical onto the azomethine nitrogen provides a stabilized tertiary carbon radical.[1] This radical is quenched by propagative hydrogen atom transfer from $^nBu_3SnH$ to form β-stannylenamine **56**. The enamine was used in unpurified form for the subsequent coupling reaction with acid chloride **55** (formed from acid **67** using oxalyl chloride). The resulting oil was chromatographed to provide the coupled product (**54**) in yields ranging from 62–70 %. Vinylogous amides (**70a–c**) [27b], with different *N*-protecting groups, were synthesized utilizing a similar protocol to find the optimal substrate for the next deprotection/cyclization step.

Various methods were investigated to achieve the deprotection/cyclization using vinylogous amides **54** and **70a–c**. These included complexation with Lewis acids (to activate the vinylogous amide) with or without added hydride reducing agents, metal-catalyzed reduction, and oxidation. Many of the conditions were unable to stimulate any reaction, and those that did either caused an unwanted TBS deprotection (upon exposure to Lewis acids) or oxidative cleavage of the vinylogous amide alkene forming the *N*-protected pyrrolidinone (using DDQ or

[1] At least one radical-stabilizing group is required, either an aromatic ring or a trifluoromethyl group, see [23]b.

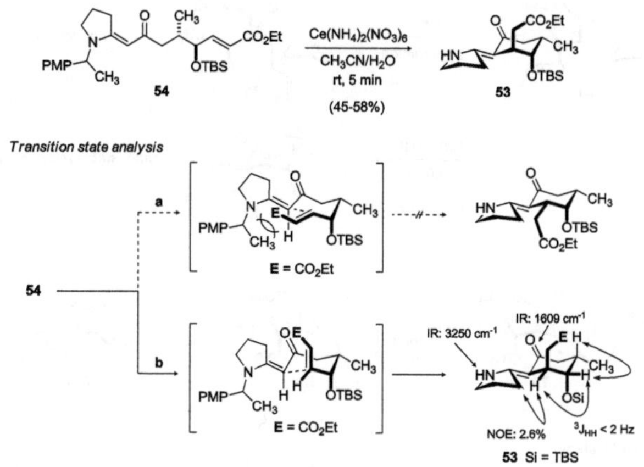

Scheme 6.11 Oxidative deprotection and Michael reaction

CAN). Fortunately, using CAN [30] with vinylogous amide **54** under short reaction times provided the desired, cyclized product **53** (in yields ranging from 45 to 58 %, Scheme 6.11).

Only one diastereomer of the product was observed, and this was expected based on examination of the developing $A^{1,3}$-strain in the transition state. An interaction between the large ethyl acetate substituent and the nitrogen protecting group would disfavor the ethyl acetate substituent in the equatorial position (Scheme 6.11, path *a*). Alternatively, an axial ester would minimize strain with the pyrrolidine ring (path *b*). Data obtained for **53** confirmed our prediction. Using 1D NOE studies and coupling constants, the stereochemistry of the ester side chain as well as those created by the Brown crotylation was confirmed. In addition, the vinylogous amide was established as *Z* based on an NOE between the equatorial hydrogen and a methylene hydrogen on the pyrrolidine ring.

With the cyclohexanone ring in place, our efforts were then focused on addition of the allyl group (which would complete all of the carbons necessary in **1**). We anticipated that this step would prove to be challenging but were surprised to see that standard conditions to install an allyl group onto the vinylogous amide in **53** were not successful (**71**, Scheme 6.12) [31]. Instead, a new ring system was formed containing a vinylogous imide (**72**). The mechanism of this transformation would involve deprotonation of the vinylogous amide hydrogen in **53** and enolization to form **73a**. A σ-bond rotation (**73b**) and cyclohexane ring flip (**73c**) places the imine in proximity to the ester and allows cyclization to form **72**. Imide **72** was observed even using Et_3N-treated silica gel. A way to prevent imide formation would be to reduce the ester, but at the time, our reduction attempts also led to **72** due to the basicity of standard reducing agents (even less basic $NaBH_4/CeCl_3$ still caused the cyclization).

Due to the instability of **53** and the ready conversion to **72**, we reassessed our initial strategy for installing the allyl group (Scheme 6.12). Since **72** still contains a

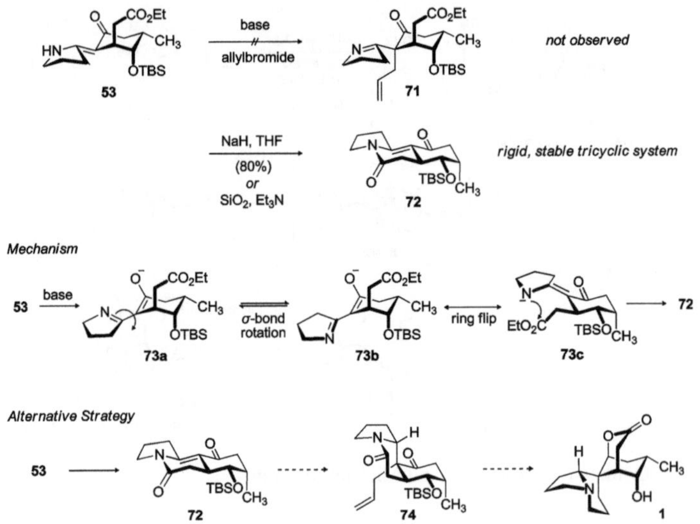

Scheme 6.12 Formation of a new tricyclic system (**72**) and an alternative route to **1**

vinylogous amide, we could potentially install an allyl group (as in **74**) and carry forward with the synthesis, leaving the additional step of amide bond cleavage to form **1**.

Our efforts to progress imide **72** are summarized in Scheme 6.13.[2] To install an allyl group, an alkene reduction was desired. Hydride reagents such as $LiBH_4$ caused over-reduction to alcohol **75**. Selective alkene reduction was achieved using metal-catalyzed hydrogenation over Pd/C (56 % yield) or using hydroboration (84 %) [32] to provide ketone **76** (while PtO_2 provided over-reduction to alcohol **75**, 80 % yield) [33]. Gratifyingly, the pyrrolidine stereocenter in both **75** and **76** was set in the correct orientation as required in **1**, indicating that substrate control provided the needed configuration. Interestingly, using a bulky hydride reagent (*L*-Selectride) provided the opposite stereocenter at C4 (as in **78**) [34].

With ketone **76** in hand, we attempted deprotonation and allylation. Disappointingly, no allylation was observed. However, starting material was epimerized to the *trans*-decalone and could be optimized using NaH (**77**, 87 % yield). Ketone epimer **77** could be formed in one step using a dissolving metal reduction [35] (**72** → **77**, 63 % yield) but again introduction of an allyl halide into the reaction did not allow for allylation [36]. Small amounts of *O*-allylated product **79** were observed during these reactions and could be enhanced using KO^tBu and HMPA (61 % yield). It was then anticipated that a Claisen rearrangement could convert **79** to a *C*-allylation product. Heating **79** in a sealed tube was successful but

[2] All chair conformations depicted in the schemes throughout this chapter were elucidated using 2D NMR techniques and coupling constants.

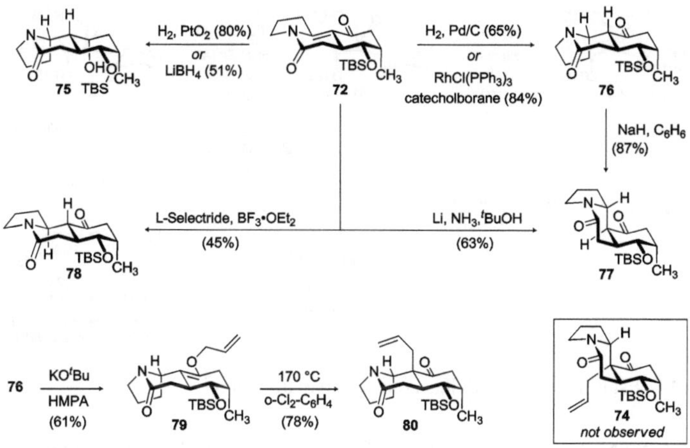

Scheme 6.13 Summary of attempts using imide **72**

unfortunately gave the undesired *C*-allyl product (**80**).[3] While this new route was at first promising, the inability to form *C*-allyl **74** by a variety of means encouraged us to further pursue our original route, which must now involve reduction of the ester in **53** to prevent the cyclization to imide **72**.

After much experimentation, it was found that Red-Al selectively reduced the ester in the presence of the vinylogous amide (**53**) to provide alcohol **81** (81 % yield), which was subsequently protected as the pivalate (**82**, 93 % yield, Scheme 6.14). Different allylation conditions were attempted next but again to no avail (to form **85**). Interestingly, if **82** and an allyl halide were exposed to Ag(I) salts and heated, small amounts of allylated products were observed (including *N*- and *C*-alkylation). This indicated that the vinylogous amide was inherently nucleophilic, so an attempt to capitalize on this behavior was tested via exposure of **82** to two oxidants, NCS and MCPBA, providing the α-chloroketoimine (**83**) and α-hydroxyketoimine (**84**), respectively [37, 38]. Both were reasonable substrates to allow introduction of an allyl group via reductive enolate formation; however, exposure of either compound to SmI_2 provided intractable mixtures [39]. The use of either $Fe(acac)_3$ [40] or HMPA [41] provided a cleaner reaction, but no allylated product (**85**) was observed, and only reduced vinylogous amide **82** was isolated.

The insight gained from exposure of **36** to oxidants led us to the successful work from the Flowers group utilizing a cerium-mediated oxidative allylation of vinylogous amides [42]. Their investigations used a more soluble source of Ce(IV)

[3] Using Pcmodel showed that while the energy difference between non-allylated *cis*-decalone **77** and *trans*-decalone **76** was relatively small (0.13 kcal/mol), the difference between allylated *cis*-decalone **74** and *trans*-decalone **80** is much greater (2.4 kcal/mol favoring *trans*-decalone **80**). Inspection of the bond angles in each six-membered ring showed that the cyclohexanone ring suffered the most torsional strain in **74** due to the allyl group. Pcmodel v.8 was developed by Dr. Kevin Gilbert, Serena Software, Bloomington, IN. We are grateful to Dr. Gilbert and Dr. Gajewski (Indiana University) for their assistance.

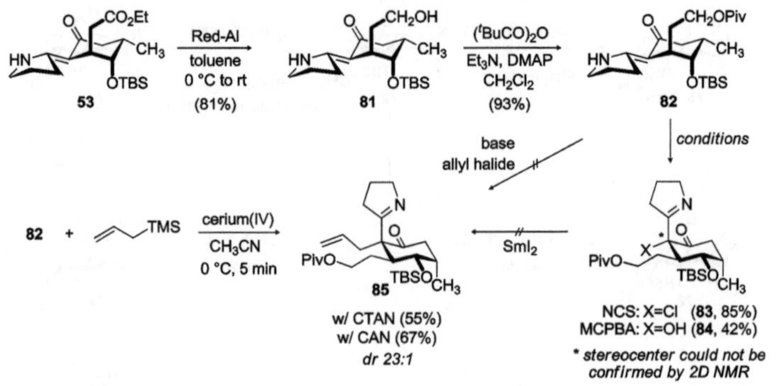

Scheme 6.14 Ester reduction and diastereoselective allylation

known as CTAN ($Ce(NBu_4)_2(NO_3)_6$) [43] along with allyl silane as the source of the three-carbon chain. Initial reactions provided low yields (20 %), but optimization of the reaction conditions included increasing the amount of allyl silane to two equivalents (45 % yield), degassing the CH_3CN solvent (55 % yield), and switching from CTAN to CAN (58–67 %). Most significant is that the allylation product was found by 2D NOESY to be the correct diastereomer and is formed almost exclusively during the reaction (**85**, dr 23:1, measured by isolation of the minor diastereomer on large scale).[4] The success of the reaction demonstrates an *umpolung* reactivity where the vinylogous amide is first oxidized to a radical cation and then attacked by a nucleophilic allyl equivalent.

With the challenging quaternary stereocenter set correctly and all carbons in place, our attention was then turned towards ketone reduction to form the β-alcohol as required in the natural product. Reduction with $NaBH_4$ in THF at room temperature provided more of the undesired α-alcohol, likely due to an imine-directed borane delivery to the top face. To direct hydride addition to the opposite face, chelating conditions with a Lewis acid were used to effectively block the top face (Et_2AlCl along with $^n Bu_3SnH$ as a reducing agent) to afford the β-alcohol in good yield and moderate diastereoselectivity (**86**, Scheme 6.15) [44]. We decided to try lactone formation as it was only one step to the diol (**87**) after a simple pivalate deprotection (Na, MeOH, 86 %). Subjecting diol **87** to a variety of oxidation conditions including Ag_2CO_3/C_6H_6 (Fetizon oxidation) [45], Dess–Martin, PCC, and TEMPO proved fruitless, as no lactone **88** was observed. This was surprising since diol **87** is in the correct chair conformation to undergo lactonization. Our suspicion was that the nucleophilic imine nitrogen might be involved in the formation of side products that result once the primary alcohol in **87** is oxidized [46].[5]

---

[4] The diastereoselection observed is likely due to attack by allyl silane when the chair conformation of the cyclohexane ring is as shown in **85**, which places the bulky OTBS out of the way in the equatorial position.

[5] Cyclization of the nucleophilic imine was observed any time the primary alcohol was converted to either a leaving group or a carbonyl-containing functional group.

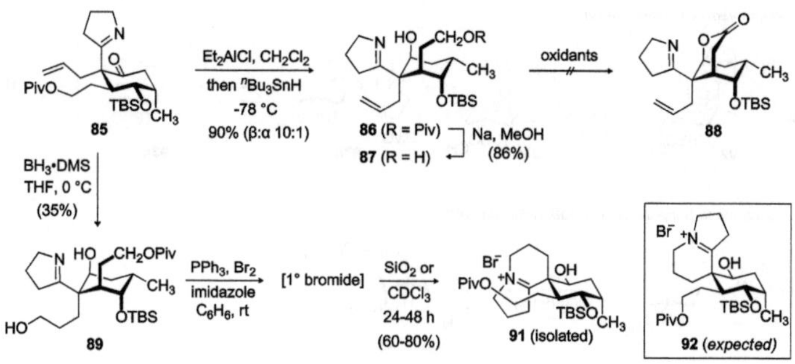

Scheme 6.15 Reactions of β-alcohol **86**

Since our attempts at lactonization were thwarted, we decided our best strategy was to focus on alkene functionalization to allow piperidine ring formation. Treatment of the alkene in **86** with either BH_3·THF or BH_3·DMS followed by oxidative workup provided the desired primary alcohol **89** but in low yields (35 % with up to 25 % recovered starting material, Scheme 6.15).[6] To progress forward, alcohol **89** was then exposed to PPh_3 and Br_2 to convert to the primary bromide, which was not isolated but was observed by ^{1}H NMR, and found that upon sitting or during chromatography underwent cyclization to form the iminium salt. 2D NMR data on the new product, in particular NOESY and HMBC, pointed towards an unexpected epimerization of the spirocyclic center forming the unexpected iminium **91** (rather than the expected product of cyclization, **92**).

The epimerization likely occurs through an enamine *retro*-aldol reaction after formation of the initial cyclized product (**92**) (Scheme 6.16) [47]. First, a ring opening of **92** forms the enamine-aldehyde (**93a**). Rotation about the C–C σ-bond in **93a** provides intermediate **93b** in which enamine addition to the aldehyde to reclose the ring would give **93c**. After protonation of the enolate, **91** would result with an overall epimerization of the spirocyclic carbon. In addition to the 2D NMR data, we also planned a complement of experiments to support the epimerization assignment.

First, **91** was subjected to a PtO_2-catalyzed iminium reduction to provide the amine as a single diastereomer (**94**, 83 %, Scheme 6.16). At this point 2D NMR techniques more clearly highlighted the correlations with the new methine adjacent

[6] A survey of boron reagents, including 9-BBN, disiamyl borane, and catechol borane with Wilkinson's catalyst [$RhCl(PPh_3)_3$], did not provide **89**. Increasing the borane amount, reaction time, or temperature also proved to be detrimental to the yield. We thought that the imine may be binding irreversibly to the borane, so Lewis acids were added along with the borane but again with no improvement in yield. We were eventually able to increase the yield later in the synthesis by using a modified workup with DMAP (*vide infra*).

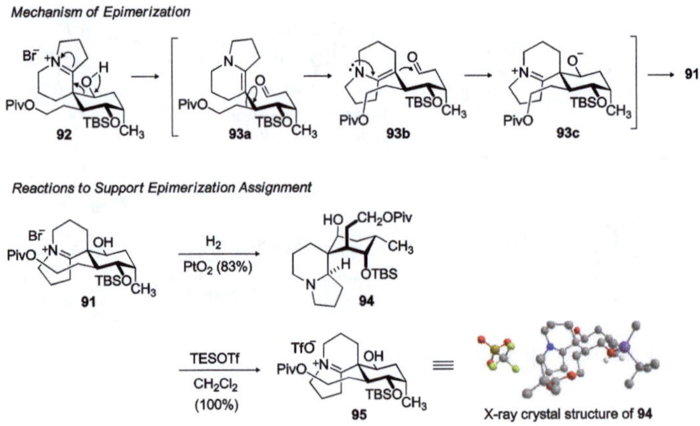

Scheme 6.16 Late stage epimerization of the spirocyclic stereocenter

to the nitrogen.[7] Second, conversion of the bromide counterion in **91** to a triflate (using TESOTf) gave a solid in which a crystal structure could be obtained allowing for unambiguous proof of the epimerization (**95**). The secondary alcohol remained in the β-orientation during the epimerization, likely due to the fact that this position is now equatorial within the new chair conformation of **91**.

Our next reactions focused on protection of the β-alcohol in **86** by a variety of means but were largely unsuccessful.[8] At this point, we took a step back to review our initial strategy using β-alcohol [49]. While it is the stereocenter required in the lactone ring of serratezomine A, we decided to approach lactonization in a different way (other than by oxidation of a diol, as we attempted in **87** → **88**, Scheme 6.15). This new strategy uses the α-alcohol which is protected to form a good leaving group (as in **96**, Scheme 6.17). Based on our prior experience, we felt that lactonization should be the last step so piperidine ring formation would be carried out next, as before (the protecting group on the α-alcohol will prevent epimerization). Then after pivalate deprotection and oxidation to a carboxylate, the resulting carboxylate (**97**) could undergo a nucleophilic displacement of the leaving group providing the lactone ring and leaving only a TBS deprotection to furnish **1**.

Reduction of ketone **85** with $NaBH_4$ in THF formed the α-alcohol selectively (**98**, >23:1 dr, 55 % yield) which was then protected as the mesylate

[7] The difficult part in assigning **91** was in distinguishing the 3 methylene carbons in the five-membered ring versus those in the six-membered ring of the indolizidine ring system. When reduction of the iminium was carried out to form **94**, the new methine could now be correlated via COSY to the five-membered ring. Now that the five- and six-membered rings were able to be distinguished and again along with HMBC and NOESY were able to establish the epimerization and the assignment of the new stereocenter.

[8] This is likely due to four groups on the ring being in the axial position. We were successful by protecting the alcohol as the mesylate, but further reactions did not prove fruitful in moving forward in the total synthesis. These reactions, as well as optimization of the hydroboration described, are discussed in detail in [48].

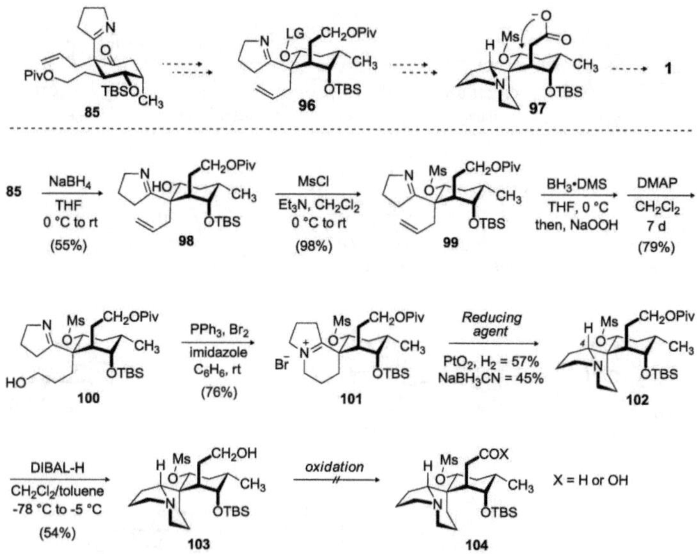

Scheme 6.17 New strategy using α-alcohol **98**

(**99**, Scheme 6.17). Hydroboration conditions were carried out and were optimized to provide a 79 % yield of the alcohol (**100**). The increased yield resulted from new workup conditions that were found after analysis of very polar fractions from column chromatography. These fractions showed broad peaks by ^{1}H NMR suggesting borane complexation [50]. To displace the borane from the imine nitrogen, DMAP was added to selectively bind the borane [51]. Treating the crude reaction mixture with DMAP after every reaction resulted in a doubling of the yield (from 35 to 79 %). Alcohol **100** was then treated with bromine, PPh_3 and imidazole to afford the primary bromide, which again cyclized slowly upon standing to afford the iminium salt (**101**). The subsequent reduction utilizing either hydride reducing agents or metal-catalyzed hydrogenation provided the amine (**102**). The ^{1}H NMR peaks of **102** were uniformly broad, suggesting two interconverting chair conformations as shown. As a result, 2D NMR analysis to determine the facial selectivity of the reduction step could not be conducted. However, the presence of only one set of peaks in the NMR strongly suggested the reduction was highly stereoselective to provide the desired isomer, as a *syn*-pentane interaction between the pyrrolidine ring and ester side chain would highly disfavor the undesired isomer.[9]

Pivalate deprotection of **102** provided primary alcohol **103**. Oxidation attempts to form the aldehyde or carboxylic acid (**104**) were again not successful, even with the nitrogen as part of the indolizidine ring (rather than before as the imine, for **87** → **88**). The nitrogen still affected our ability to oxidize the remote alcohol

[9] The one-step Polonovski rearrangement by Kobayashi, which contains the reduction of a similar iminium ion, also supports this prediction. See [16].

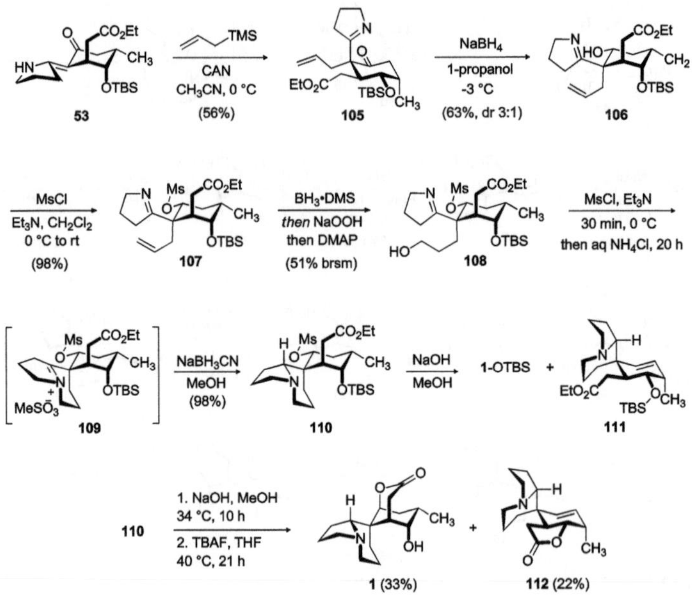

Scheme 6.18 α-Alcohol strategy with the ethyl ester in place

properly [46]. Based on our strategy, oxidation of the primary alcohol (**103**) was required as a result of our early reduction of the ester (Scheme 6.12). At the time, it was needed to prevent the undesired cyclization to imide **72** which occurred during allylation attempts. Now that we have discovered nonbasic conditions to install the allyl group, we decided to go back and try it on vinylogous amide **53** (with the intact ester). This final strategy was the key to completion of the synthesis of **1** (Scheme 6.18).

CAN-mediated oxidative allylation of vinylogous amide **53** was carried out. Fortunately, no cyclization of the imine onto the pendant ester was observed in the product (**105**), and the high diastereoselectivity for the allylation was maintained. Ketone reduction in the following step had to be modified from the original conditions ($NaBH_4$ in THF) due to low yields. Experimentation with a variety of solvents and other reducing agents (KBH_4, $LiBH_4$, and *L*-Selectride) led to the conditions that ultimately provided the best combination of diastereoselectivity (3:1) and yield (63 %) to provide α-alcohol **106** (using 1-propanol at −3 °C). The undesired β-alcohol diastereomer that was produced was reoxidized using Dess Martin Periodinane and recycled to provide higher material throughput. Mesylate protection (**107**, 98 %) and hydroboration followed by treatment of the crude reaction mixture with DMAP provided alcohol **108** (51 %). After some experimentation, the desired iminium salt could be formed in quantitative yield by mesylation of **108**, rather than formation of the corresponding bromide, followed by treatment with aqueous NH_4Cl (**109**). When ^1H NMR showed complete conversion of the crude reaction mixture to **109**, it was then treated with $NaCNBH_3$ to afford the

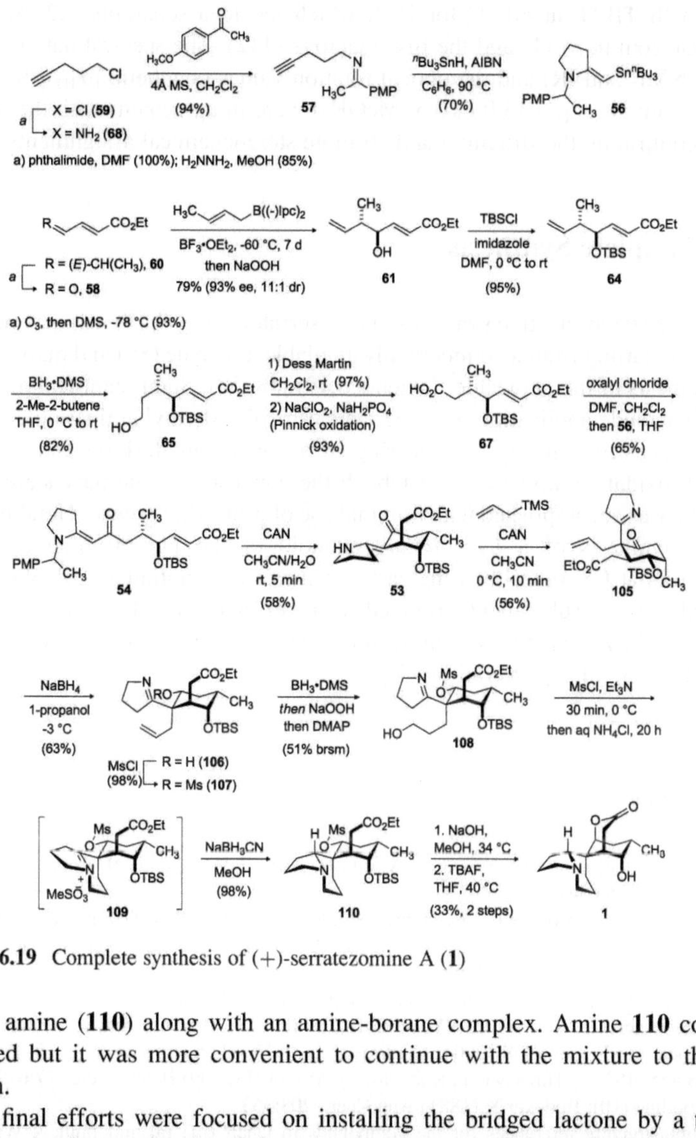

Scheme 6.19 Complete synthesis of (+)-serratezomine A (**1**)

tertiary amine (**110**) along with an amine-borane complex. Amine **110** could be separated but it was more convenient to continue with the mixture to the next reaction.

The final efforts were focused on installing the bridged lactone by a tandem saponification/intramolecular S_Ni cyclization strategy (Scheme 6.18). Employing different reagents for the saponification including NaOTMS, K_2CO_3, and LiOH led to either decomposition or formation of an alkene by facile elimination of the mesylate (**111**, formed from a chair flip of **110** and β-elimination) [52]. Treatment with warm, aqueous NaOH for 10 h afforded a 1.1:1 mixture of the lactone with the OTBS group attached (**1-OTBS**) and alkene **111**, as indicated by NMR analysis. Purification attempts of the crude mixture were unsuccessful due to co-elution of the products. Therefore, TBS deprotection was carried out directly after the lactonization. After exposure of **110** to NaOH, the crude reaction mixture was treated with TBAF at 40 °C for 20 h which led to a separable 3:2 mixture of (+)-serratezomine A (**1**) and the fused lactone (**112**). The spectral data (^{1}H NMR and ^{13}C NMR and IR) and the optical rotation values (synthetic $[\alpha]_D$ +9.5 (*c* 0.3, MeOH), natural $[\alpha]_D$ +13.0 (*c* 0.5, MeOH)) were in agreement with the reported values, confirming the structural and absolute stereochemical assignments.

6.7 Complete Synthesis

Overall, the stereoselective synthesis of (+)-serratezomine A (**1**) was accomplished in 15 steps starting from a commercially available aldehyde (**58**) and marks the first total synthesis of this molecule (Scheme 6.19) [18, 53]. Significant features of this synthesis include application of a free radical-mediated vinyl amination developed by us to construct the pyrrolidine ring in a convergent fashion and a cerium-mediated oxidative allylation to establish the congested, quaternary stereocenter. All of this was accomplished with minimal use of protecting groups. Notably, this is the first synthesis of a *Lycopodium* alkaloid of the fawcettimine class with oxygenation at C8, other than the early synthesis of serratinine [20]. We believe the axial oxygen substituent provided a powerful driving force for quaternary carbon epimerization but was ultimately addressed by reassessing our strategy and without sacrificing overall brevity.

References

1. Isolation and structure determination: Morita H, Arisaka M, Yoshida N, Kobayashi J (2000) J Org Chem 65:6241
2. For a general review of the *Lycopodium* alkaloids, see: (a) Ayer WA, Trifonov LS (1994) Ch. 3 Lycopodium alkaloids. In: Cordell GA, Brossi A (eds) The alkaloids: chemistry and pharmacology, vol 45. Academic, New York, NY, pp 233–266; (b) Maclean DB (1985) Ch. 5 Lycopodium alkaloids. In: Brossi A (ed) The alkaloids: chemistry and pharmacology, vol 26. Academic, New York, NY, pp 241–296; (c) for a thorough and informative overall reference: Ma X, Gang DR (2004) Nat Prod Rep 21:752–772; (d) Ayer WA (1991) Nat Prod Rep 8:455–463; (e) Hirasawa Y, Kobayashi J, Morita H (2009) Heterocycles 77:679–729
3. Lycopodine (**10**): Bödeker K (1881) Ann Chem 208:363
4. Isolation/structure references for the compounds in Chart 6.1: (a) annotinine - Wiesner K, Valenta Z, Ayer WA, Fowler LR, Francis JE (1958) Tetrahedron 4:87; Wiesner K, Ayer WA, Fowler LR, Valenta Z (1957) Chem Ind (London):564; (b) huperzine A—Liu J-S, Zhu Y-L, Yu C-M, Zhou Y-Z, Han Y-Y, Wu F-W, Qi B-F (1986) Can J Chem 64:837; full NMR assignments: Zhou B-N, Zhu DY, Huang M-F, Lin L-J, Lin L-Z, Han X-Y, Cordell GA (1993) Phytochemistry 34:1425; (c) serratinine - Inubushi Y, Ishii H, Yasui B, Hashimoto M, Harayama T (1966) Tetrahedron Lett 8:1537 and later structural revision: Nishio K, Ishii H, Inubushi Y, Harayama T (1969) Tetrahedron Lett 11:861; (d) nankakurine A—Hirasawa Y, Morita H, Kobayashi J (2004) Org Lett 6:3389 and later structural revision at the spirocyclic center: Hirasawa Y, Kobayashi J, Obara Y, Nakahata N, Kawahara N, Goda Y, Morita H (2006) Heterocycles 68:2357
5. (a) Kozikowski AP, Tückmantel W (1999) Acc Chem Res 32:641; (b) Liu JS, Zhu YL, Yu CM, Zhou YZ, Han YY, Wu FW, Qi BF (1986) Can J Chem 64:837

6. An ^{1}H NMR test for AChE inhibition has been described: Li Y, Yin G, Wei W, Wang H, Jiang S, Zhu D, Du W (2007) Biophys Chem 129:212
7. (a) Jiang H, Luo X, Bai D (2003) Curr Med Chem 10:2231; (b) Desilets AR, Gickas JJ, Dunican KC (2009) Ann Pharmacother 43:514
8. Conroy H (1960) Tetrahedron Lett 1/31:34
9. Gupta RN, Castillo M, MacLean DB, Spenser ID, Wrobel JT (1968) J Am Chem Soc 90:1360
10. (a) Castillo M, Gupta RN, Ho YK, MacLean DB, Spenser ID (1970) J Am Chem Soc 92:1074; (b) This same finding of incorporation of only one half of pelletierine was also demonstrated in the alkaloid cernuine: Gupta RN, Ho YK, MacLean DB, Spenser ID (1970) J Chem Soc D Chem Comm:409
11. The ^{13}C labels allow for use of ^{13}C NMR in which the peaks containing the ^{13}C label enrichment show satellites corresponding to a $^{1}J_{CC}$ coupling (~30 Hz). The distribution pattern of ^{13}C labels found was more complicated than one single pathway and pointed toward both pelletierine and cocaine-derived alkaloid pathways, both which incorporate two acetate units: Hemscheidt T, Spenser ID (1993) J Am Chem Soc 115:3020
12. Hemscheidt T, Spenser ID (1996) J Am Chem Soc 118:1799
13. For more background information and proposals of the biosynthesis of lycopodine see: Blumenkopf TA, Heathcock CH (1985) In: Pelletier SW (ed) Alkaloids: chemical and biological perspectives, vol 3. Wiley, New York, NY, pp 185–240 and also those in [2]
14. Isolation/structure references for each of the compounds in Scheme 4: (a) fawcettimine—Burnell RH (1959) J Chem Soc:3091; (b) huperzine Q—Tan C-H, Ma X-Q, Chen G-F, Zhu D-Y (2002) Helv Chim Acta 85:1058; (c) lyconesidine A—Hirasawa Y, Morita H, Kobayashi J (2002) Tetrahedron 58:5483; (d) magellaninone—Loyola LA, Morales G, Castillo M (1979) Phytochemistry 18:1721; (e) serratinine—see [4c]; (f) Hirasawa Y, Morita H, Shiro M, Kobayashi J (2003) Org Lett 5:3991; (g) serratezomine A—see [1]
15. The biosynthetic pathway from a lycopodine derivative to serratinine **4** was proposed previously: (a) Inubushi Y, Ishii H, Yasui B, Harayama T (1966) Tetrahedron Lett 7:1551; (b) Inubushi Y, Ishii H, Yasui B, Harayama T (1968) Chem Pharm Bull 16: 101; (c) the references in (a) and (b) both have the incorrect stereocenter at C4. This was not established unambiguously until a crystal structure of a derivative was obtained: see [4c]
16. Morita H, Kobayashi J (2002) J Org Chem 67:5378
17. Representative examples of recent *Lycopodium* alkaloid total syntheses (2009-current): (a) (+)-lucidulene: Barbe G, Fiset D, Charette AB (2011) J Org Chem 76:5354; (b) (+)-fawcettimine and (+)-lycoflexine: Yang Y-R, Shen L, Huang J-Z, Xu T, Wei K (2011) J Org Chem 76:3684; (c) (+)-fastigiatine: Liau BB, Shair MD (2010) J Am Chem Soc 132:9594; (d) (+)-clavolonine: Laemmerhold KM, Breit B (2010) Angew Chem Int Ed 49:2367; (e) *rac*-lycopladine A: De Lorbe JE, Lotz MD, Martin SF (2010) Org Lett 12:1576; (f) *rac*-lycodine: Tsukano C, Zhao L, Takemoto Y, Hirama M (2010) Eur J Org Chem:4198; (+)-complanadine A: (g) Yuan C, Chang C-T, Axelrod A, Siegel D (2010) J Am Chem Soc 132: 5924; (h) Fischer DF, Sarpong R (2010) J Am Chem Soc 132:5926; (i) *rac*-nankakurines A and B and luciduline: Cheng X, Waters SP (2010) Org Lett 12:205; (j) (+)-nankakurines A and B: Altman RA, Nilsson BL, Overman LE, de Alaniz JR, Rohde JM, Taupin V (2010) J Org Chem 75:7519; (k) *rac*-lycoposerramine R: Bisai V, Sarpong R (2010) Org Lett 12:2551; (l) lycopodine: Yang H, Carter RG (2010) J Org Chem 75:4929; (m) (+)-lycoflexine: Ramharter J, Weinstabl H, Mulzer J (2010) J Am Chem Soc 132:14338; (n) lycoposerramine C and phlegmariurine A: Nakayama A, Kogure N, Kitajima M, Takayama H (2009) Org Lett 11:5554; (o) (−)-huperzine: Koshiba T, Yokoshima S, Fukuyama T (2009) Org Lett 11:5354; (p) lycoposerramines X and Y: Tanaka T, Kogure N, Kitajima M, Takayama H (2009) J Org Chem 74:8675; (q) (+)-lyconadin A: West SP, Bisai A, Lim AD, Narayan RR, Sarpong R (2009) J Am Chem Soc 131:11187
18. Chandra A, Pigza JA, Han J-S, Mutnick D, Johnston JN (2009) J Am Chem Soc 131:3470
19. Syntheses of the core of serratinine: (a) via synthesis of 8-deoxyserratinine and fawcettimine (**18a/b**): Harayama T, Takatani M, Inubushi Y (1979) Tetrahedron Lett 20:4307; (b) via a

different approach: Mehta G, Reddy MS, Radhakrishnan R, Manjula MV, Viswamitra MA (1991) Tetrahedron Lett 32:6219; (c) via synthesis of 13-deoxyserratine (not shown): Cassayre J, Gagosz F, Zard SZ (2002) Angew Chem Int Ed 41:1783; (d) via a route to serratine (constitutional isomer of **4** at C8 and C9): Luedtke G, Livinghouse T (1995) J Chem Soc Perk T I:2369
20. (a) Harayama T, Ohtani M, Oki M, Inubushi Y (1974) J Chem Soc Chem Commun:827; (b) Harayama T, Ohtani M, Oki M, Inubushi Y (1975) Chem Pharm Bull 23: 1511; (c) reviewed in: Inubushi Y, Harayama T (1981) Heterocycles 15:611
21. (a) Harayama T, Takatani M, Inubushi Y (1989) Chem Pharm Bull 28:1276; (b) Woodward RB, Sondheimer F, Taub D, Heusler K, McLamore WM (1952) J Am Chem Soc 74:4223
22. Hoffmann RW (1989) Chem Rev 89:1841
23. (a) Johnston JN, Plotkin MA, Viswanathan R, Prabhakaran EN (2001) Org Lett 3:1009; (b) Nugent BM, Williams AL, Prabhakaran EN, Johnston JN (2003) Tetrahedron 59:8877; (c) Prabhakaran EN, Nugent BM, Williams AL, Nailor KE, Johnston JN (2002) Org Lett 4:4197
24. (a) Viswanathan R, Mutnick D, Johnston JN (2003) J Am Chem Soc 125:7266; (b) Viswanathan R, Prabhakaran EN, Plotkin MA, Johnston JN (2003) J Am Chem Soc 125:163; (c) Srinivasan JM, Burks HE, Smith CR, Viswanathan R, Johnston JN (2005) Synthesis:330
25. Veysoglu T, Mitscher LA, Swayze JK (1980) Synthesis:807
26. (a) Toshima K et al (2001) J Org Chem 66:1708; For a general crotylation reference, see (b) Brown HC, Bhat KS (1986) J Am Chem Soc 108:293
27. (a) Using Kakisawa analysis. Ohtani I, Kusumi T, Kashman Y, Kakisawa H (1991) J Am Chem Soc 113:4092; Full details of the analysis are given in: (b) Mutnick Daniel M (2003) M.S. Thesis, Indiana University, Bloomington, IN
28. (a) IBX was prepared according to: Frigerio M, Santagostino M, Sputore S (1999) J Org Chem 64:4537; (b) Dess–Martin was prepared from IBX using (step B only): Boeckman RK, Shao P, Mullins JJ (2000) Org Syn Coll 10:696; Boeckman RK, Shao P, Mullins JJ (2000) Org Syn Coll 77:141 (2000).
29. Bal BS, Childers WE Jr, Pinnick HW (1981) Tetrahedron 37:2091
30. (a) Maulide N, Vanherck J, Gautier A, Markó IE (2007) Acc Chem Res 40:381; (b) Nair V, Balagopal L, Rajan R, Mathew (2004) J Acc Chem Res 37:21; (c) Wright JA, Jinquan Y, Spencer JB (2001) Tetrahedron Lett 42:4033; (d) Molander GA (1992) Chem Rev 92:29 (see p 30)
31. (a) Greenhill JV, Moten MA (1983) Tetrahedron 39:3405; (b) Chen YL, Mariano PS, Little GM, O'Brien D, Huesmann PL (1981) J Org Chem 46:4643
32. Mannig D, Noth H (1985) Angew Chem Int Ed 24:878
33. (a) Pu X, Ma D (2004) Angew Chem Int Ed 43:4222; (b) Ma D, Pu X, Wang J (2002) Tetrahedron Asymmetry 13:2257; (c) Fujimoto R, Kishi Y; Blount JF (1980) J Am Chem Soc 102:7154
34. (a) Alcaide B, Almendros P, Alonso JM, Redondo MC (2003) J Org Chem 68:1426; (b) Han GH, LaPorte MG, Folmer JJ, Werner KM, Weinreb SM (2000) J Org Chem 65:6293; (c) Comins DL, Libby AH, Al-awar RS, Foti CJ (1999) J Org Chem 64:2184
35. Brennan JP, Saxton JE (1985) Tetrahedron Lett 26:1769
36. For examples of conjugate reduction and C-alkylation with alkyl/allyl bromides see: (a) Saha AK, Das S, Mukherjee D (1993) Synth Commun 23:1821; (b) Stork G, Logusch EW (1980) J Am Chem Soc 102:1218 and 1219; (c) Stahl P, Kissau L, Mazitschek R, Huwe A, Furet P, Giannis A, Waldmann H (2001) J Am Chem Soc 123:11586
37. For chlorination/bromination of vinylogous amides in a similar fashion see: (a) Pierron C, Garnier J, Levy J, LeMen J (1971) Tetrahedron Lett 12:1007; (b) Jirkovsky I (1974) Can J Chem 52:55
38. The α-chloroketone is proposed as an intermediate in the NaOCl oxidation of a vinylogous amide: Staskun B (1988) J Org Chem 53:5287
39. For examples, see: (a) Molander GA (1992) Chem Rev 92:29; (b) Molander GA, Hahn G (1986) J Org Chem 51:1135; (c) Enholm EJ, Schreier JA (1995) J Org Chem 60:1110

40. The reaction proceeded cleanly with $Fe(acac)_3$ present but resulted in many products without it, see: Molander GA, McKie JA (1991) J Org Chem 56:4112
41. Inanaga J, Ishikawa M, Yamaguchi M (1987) Chem Lett:1485
42. (a) Zhang Y, Raines AJ, Flowers RA II (2004) J Org Chem 69: 6267; (b) For pioneering work in this area using β-diketones: Hwu JR, Chen CN, Shiao SS (1995) J Org Chem 60:856
43. (a) Preparation of CTAN: Muathen HA, Indian J Chem *30B*:522; (b) oxidation potential of CTAN vs. CAN: Zhang Y, Flowers RA II (2003) J Org Chem 68:4560
44. Evans DA, Allison BD, Yang MG, Masse CE (2001) J Am Chem Soc 123:10840
45. For examples see: (a) "Silver(I) Carbonate on Celite," Fetizon M (2005) In: Crich D, Fuchs PL, Molander G, Paquette LA (eds) Electronic encyclopedia of reagents for organic synthesis, Wiley, Sussex; (b) Lafontaine JA, Provencal DP, Gardelli C, Leahy JW (2003) J Org Chem 68:4215
46. In their synthesis of (−)-sarain A, Overman et al. reported a similar difficulty in the oxidation of an alcohol proximal to a tertiary amine in the molecule: Garg NK, Hiebert S, Overman LE (2006) Angew Chem Int Ed 45:2912
47. (a) Atarashi S, Choi JK, Ha DC, Hart DJ, Kuzmich D, Lee CS, Ramesh S, Wu SC (1997) J Am Chem Soc 119:6226; (b) Hamelin O, Deprés J-P, Greene AE, Tinant B, Declercq J-P (1996) J Am Chem Soc 118:9992
48. For more details and the completion of the synthesis see: Chandra A (2010) Ph.D. Thesis, Vanderbilt University, Nashville, TN
49. Workup until this point has been a collaboration, as referenced in: Pigza JA (2008) Ph.D. Thesis, Indiana University, Bloomington, IN. and a postdoc in our group at the time, Dr. Jeong-Seok Han
50. (a) Dewar MJS, Gleicher GJ, Robinson BP (1964) J Am Chem Soc 86:5698; (b) Dewar MJS, Rona P (1967) J Am Chem Soc 89:6294; (c) Polívka Z, Ferles M (1969) Collect Czechoslov Chem Commun 34:3009; (d) Polívka Z, Kubelka V, Holubová N, Ferles M (1970) Collect Czechoslov Chem Commun 35:1131; (e) Wille H, Goubeau J (1972) Chem Ber 105:2156; (f) Le Toumelin, J.-B.; Baboulène, M. *Tetrahedron: Asymmetry* (1997) 8:1259 (g) Midland MM, Kazubski A (1992) J Org Chem 57:2953
51. For intramolecular hydroboration of homoallylic and bishomoallylic amine boranes by activation of I_2 at elevated temperatures, see: (a) Scheideman M, Wang G, Vedejs E (2008) J Am Chem Soc 130:8669; (b) Scheideman M, Shapland P, Vedejs E (2003) J Am Chem Soc 125:10502; (c) Wang G, Vedejs E (2009) Org Lett 11:1059
52. (a) with NaOTMS: Laganis ED, Chenard BL (1984) Tetrahedron Lett 25:5831; (b) with K_2CO_3: Ziegler FE, Klein SI, Pati UK, Wang TF (1985) J Am Chem Soc 107:2730
53. Pigza JA, Han J-S, Chandra A, Mutnick D, Pink M, Johnston JN (2012) J Org Chem doi: 10.1021/jo302333s

Chapter 7
Hypocrellin/Cercosporin

Carol A. Mulrooney, Erin M. O'Brien, and Marisa C. Kozlowski

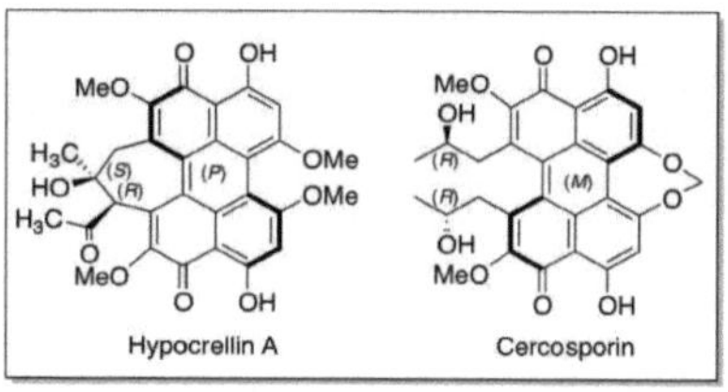

7.1 Introduction

Hypocrellin and cercosporin are perylenequinones, compounds with unique structural and biological properties that are all characterized by a pentacyclic core containing an extended quinone system (Chart 7.1) [1]. A large number of perylenequinone natural products, typically brightly colored pigments, have been isolated from molds, aphids, and plant species. These natural products have been categorized into three different classes. The simplest, belonging to class A, are C_{20} compounds without carbon substituents. Class B perylenequinones are the most prevalent and include the natural products cercosporin (**3**) and hypocrellin A (**7**). The Class B compounds have been mostly isolated from mold species and have structurally more complex substitution. Class C perylenequinones include the erythroaphins like rhodoaphin-*be* (**2**), isolated from aphids. This chapter will focus on Class B natural products, which exhibit unique photochemical and biological properties. The perylenequinones of this class are characterized by a pentacyclic core that is twisted out of planarity due to steric interactions between

C.A. Mulrooney • E.M. O'Brien • M.C. Kozlowski (✉)
Department of Chemistry, University of Pennsylvania, 231 South 34th Street, Philadelphia, PA 19104-6323, USA

J.J. Li and E.J. Corey (eds.), *Total Synthesis of Natural Products*,
DOI 10.1007/978-3-642-34065-9_7, © Springer-Verlag Berlin Heidelberg 2012

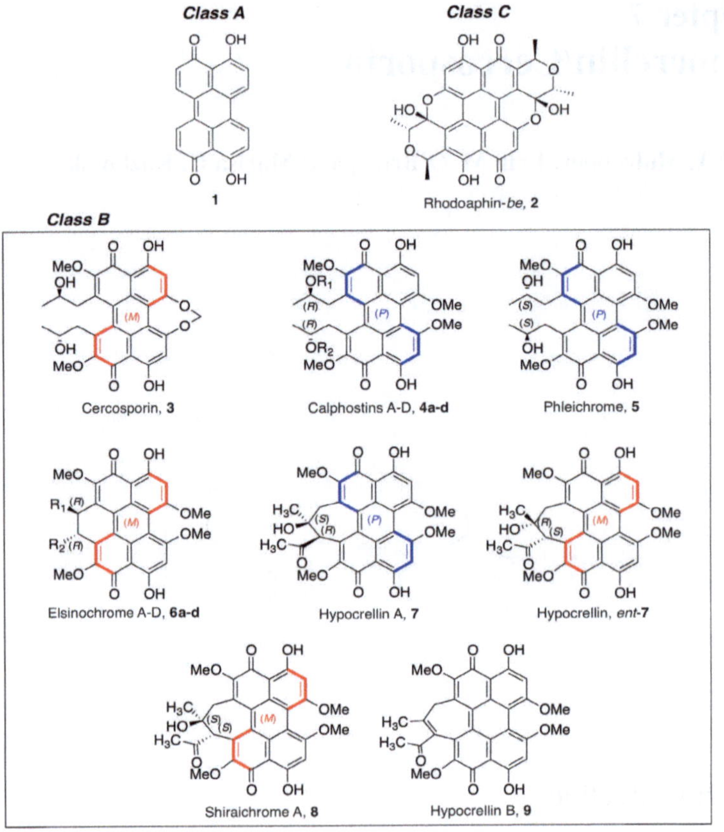

Chart 7.1 Perylenequinone natural products

the C2,C2′ methoxy and C7,C7′-carbon substituents that force a shallow helix with a twist of 20°. This element of stereochemical complexity, in addition to the stereocenters present in the C7,C7′-substituents, has added to the difficulty in the total syntheses of these natural products, and multiple research groups have targeted the synthesis of a number biologically active Class B perylenequinones [1, 2].

The "purple speck disease" of soybeans is caused by the fungus *Cercospora kikuchii*, and additional cash crops such as tobacco, corn, sugar, and coffee are damaged by fungi from the genus *Cercospora* [3]. Cercosporin (**3**, Chart 7.1), initially isolated in 1957 from *Cercospora kikuchii* and subsequently from other *Cercospora* species, was found to be the phototoxin responsible for the destructive nature of the pathogen [4]. For these reasons, the natural product was extensively studied, yet its structure was not elucidated until the 1970s [5].

The hypocrellins (**7**–**9**) are Class B perylenequinones that have also been the subject of much investigation due to their biological activity, having been used in Chinese folk medicine for the treatment of vitiligo, psoriasis, and other diseases for centuries. Hypocrellin A, shiraichrome A, and hypocrellin B have been isolated from the mold species *Shiraia bambusicola*, and hypocrellin from the species *Hypocrellin bambusae* [6–8]. The naming of these compounds in the literature differs among reports. For example, hypocrellin has been referred to as hypocrellin A in several studies, and hypocrellin A as shiraiachrome B. Here, we employ the names and structural assignments suggested by Mondelli et al. [9].

In addition to cercosorin, other photoactive members of class B include calphostins A–D (**4a–d**), isolated from *Cladosporium cladosporioides* [10]; phleichrome (**5**), isolated from *Cladosporium phlei* [11]; and the elsinochromes (**6**) [12], isolated from several species of the genus *Elsinoe*. In this chapter, background on the syntheses and biological activity of the calphostins and phleichrome is included, because of their structural similarity to hypocrellin and cercosporin.

7.2 Biological Activity

Perylenequinones are photosensitizers, that is, they exhibit light-induced biological activity. Upon irradiation, they generate singlet oxygen from atmospheric oxygen, and this property makes them candidates for photodynamic therapy (PDT). PDT is a treatment for many conditions including multiple forms of cancer, macular degeneration, and skin diseases such as acne and psoriasis [13]. Currently, the photosensitizers used in PDT are porphyrins, chlorophylls, or dyes. The possibility of introducing perylenequinone compounds or use in PDT is under extensive investigation.

The phototoxicity of cercosporin has been widely studied due to its destruction of cash crops such as tobacco, soybeans, sugar, corn, beets, bananas, and coffee [14]. Biological evaluation of other Class B perylenequinones have revealed light-dependent activity against multiple tumor cell lines [15–17]. Furthermore, they have displayed antiviral [18] and immunotherapeutic [19] properties. The hypocrellins, especially, have shown much promise for use in photodynamic therapy against cancer [20]. Hypocrellin A (**7**) has been studied in clinical trials for the treatment of skin diseases such as vitiligo [21] and is a proposed photosensitizer for PDT due to its high quantum yield of singlet oxygen and its low dark toxicity [22].

The problems of high lipophilicity and subsequent low bioavailability of the perylenequinones have been approached by altering their formulations [23]. Another limitation to the use of perylenequinones as therapeutic agents is the poor extinction coefficient in the longer wavelength of light (650–800 nm) that is needed to penetrate deep tissue. Many derivatives of hypocrellin have been

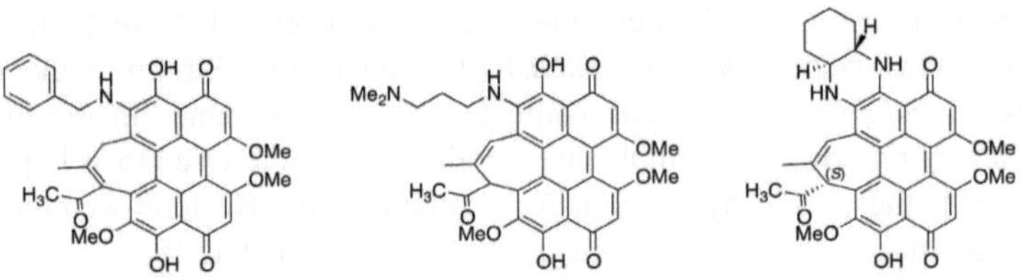

10 11 12

Chart 7.2 Derivatives of hypocrellins

prepared in an attempt to improve these properties,[1] but the lack, until recently, of a total synthesis of hypocrellin has hindered a full understanding of their photophysical and biological properties.

Of the derivatives that have been developed, some have addressed these concerns and improved activity. For example, the amine-substituted hypocrellin A (**10**) listed in Chart 7.2 has a strong absorption shoulder at 628 nm. This derivative was reported to be cytotoxic to the gastric adenocarcinoma MGC803 cell line. Its photopotentiation factor (LD_{50} dark/LD_{50} light) is more than 200-fold, compared to hypocrellin A, which is fourfold in the same assay [24]. Hypocrellin B derivative SL017 (**11**) was chosen as a clinical candidate for both PDT and SDT (sonodynamic therapy) for actinic keratosis, for acne vulgaris, and for permanent hair removal [25]. The cyclic diamine derivative SL052 (**12**) was selected as a clinical candidate against prostate cancer [25].

Further biological evaluations have discovered that the perylenequinones selectively bind the C1 regulatory domain of protein kinase C (PKC) [26, 27]. PKC has been recognized as a significant target for anticancer and antiviral drugs. The catalytic domain-acting PKC inhibitors such as staurosporine (PKC, $IC_{50} = 9$ nM) and other indolocarbazoles are highly potent but not specific for PKC, as they also inhibit other protein kinases [28]. However, the perylenequinones are both potent and specific kinase inhibitors compared to protein kinase A (PKA) and polyphosphate kinase (PPK) [Calphostin C (**4c**): PKC, $IC_{50} = 0.05$–0.46 μM; PKA, $IC_{50} = >100$ μM; PPK, $IC_{50} = >100$ μM; Cercosporin (**3**): PKC, $IC_{50} = 0.6$–1.3 μM; PKA, $IC_{50} = >500$ μM; PPK, $IC_{50} = >180$ μM]. This selectivity is presumed to arise from selective action on a regulatory domain unique to PKC [26, 27]. Due to limitations of available synthetic methods, much of the evaluation of the perylenequinones as PKC inhibitors has been confined to natural product derivatives or simple analogs. A number of calphostin analogs have been patented: [29]. However, improved total syntheses of cercosporin and hypocrellin have enabled the evaluation of complex analogs not available from the natural products (see below).

[1] For a review of perylenequinone derivatives see [1c] and references therein.

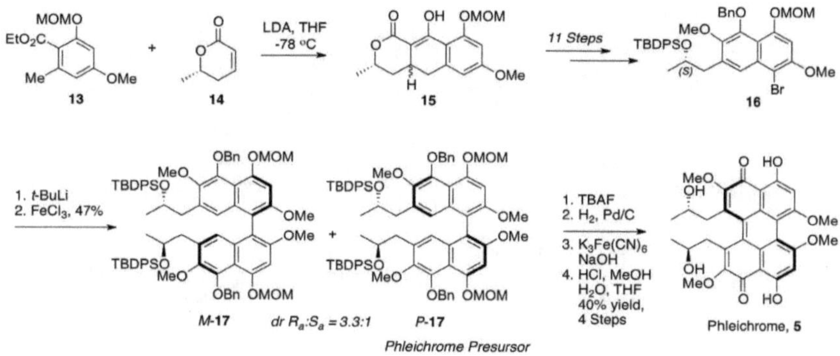

Scheme 7.1 Broka phleichrome synthesis

7.3 Previous Synthetic Work

7.3.1 Synthesis of (−)-Phleichrome and (−)-Calphostin A,D [30]

Sixteen years after the reported isolation, Chris Broka published the first total syntheses of perylenequinone natural products phleichrome and calphostins A and D [30]. The synthetic approach, originally designed for phleichrome, involved dimerization of the highly functionalized enantiopure naphthalene **16** to binaphthyl **17**, followed by a *para*-phenoxy radical coupling and subsequent oxidation to the perylenequinone (Scheme 7.1). The enantiopure secondary alcohols of **16** directed the stereochemistry of the newly formed biaryl axis. The cornerstone of the Broka synthesis involved installation of the stereogenic hydroxypropyl group via a condensation reaction between the anion derived from **13** (four steps from 2-bromo-3,5-dimethoxytoluene) and (*S*)-6-methyl-5,6-dihydropyran-2-one **14**. While condensation of **14** to **13** was a novel and high-yielding approach, further elaboration to the desired naphthalene **16** required 11 steps. Dimerization of **16** yielded biaryl **17** with 3.3:1 diastereoselectivity in favor of the undesired (*M*,*S*,*S*)-isomer *M*-**17**. The minor diastereomer from the first biaryl coupling attempt, *P*-**17**, was purified via chromatography, debenzylated, and oxidized to the perylenequinone using $K_3Fe(CN)_6$. Removal of the MOM ethers completed the first total synthesis of phleichrome.

To access the calphostins, (*S*)-**16** was subjected to Mitsunobu conditions followed by the two-step dimerization protocol (lithiation, $FeCl_3$) resulting in biaryl (*P*)-**17** as the major diastereomer (Scheme 7.2). In this case, the relative stereochemistry matched the stereochemistry of calphostins A and D. Elaboration following the same protocol as for **5** completed the first total synthesis of calphostin D (**4d**). Calphostin A (**4a**) was also synthesized via benzoate protection of the secondary alcohols.

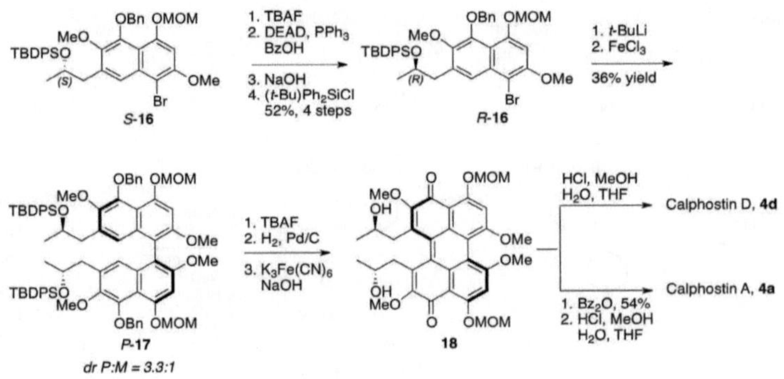

Scheme 7.2 Broka syntheses of calphostin A and D

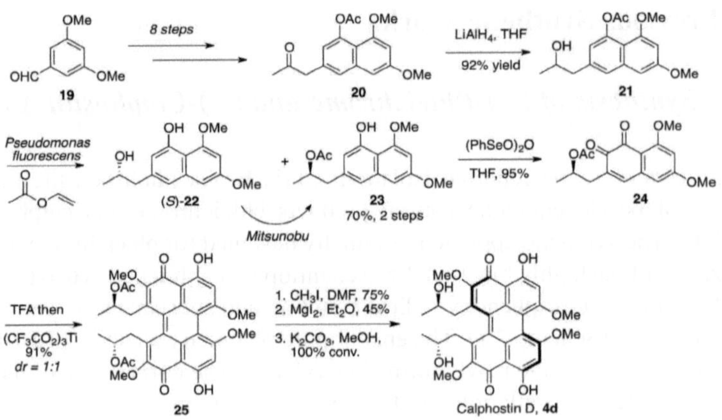

Scheme 7.3 Hauser synthesis of calphostin D

7.3.2 Synthesis of (−)-Calphostin D [31]

In 1994, the Hauser laboratory reported a total synthesis of calphostin D [31]. Akin to the earlier Broka synthesis, the Hauser approach involved the initial generation of an enantiopure naphthalene followed by biaryl formation and oxidation to the perylenequinone. The key difference between the approaches is the manner in which the secondary alcohol was generated as a single enantiomer. Commencing with aldehyde **19**, the Hauser group synthesized naphthalene **20** in eight steps, which was reduced to the alcohol (**21**, Scheme 7.3). This racemate was resolved using an enzymatic acylation protocol with lipase *Pseudomonas fluorescens*, providing **23** in 42 % yield (99 % ee). The opposite alcohol enantiomer (*S*)-**22** (58 % yield; 78 % ee) was inverted via Mitsunobu conditions and resubjected to the resolution process, supplying **23** in a combined yield of 70 %. Hauser's dimerization to the perylenequinone was also novel; *ortho*-naphthoquinone **24** was generated with $(PhSeO)_2O$, which dimerized in the presence of TFA to **25**. Although the acid

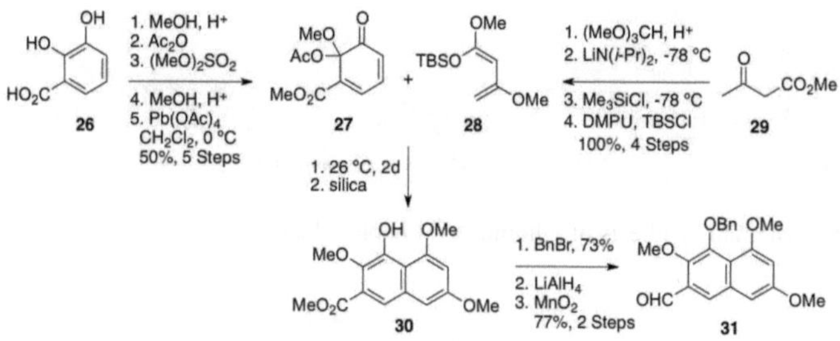

Scheme 7.4 Coleman naphthalene synthesis

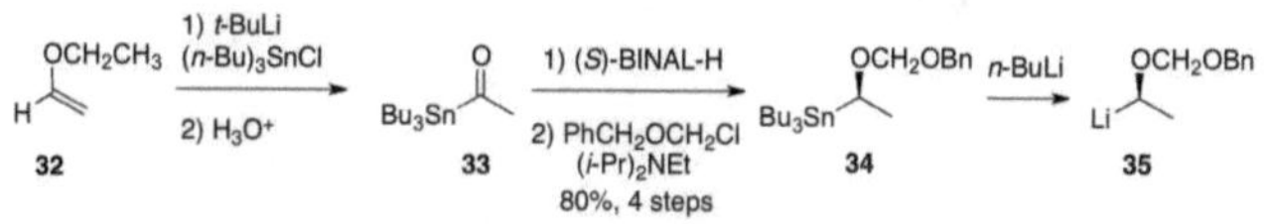

Scheme 7.5 Chiral lithium synthesis

catalyzed process was non-diastereoselective, three further steps including separation of the diastereomers by chromatography completed the total synthesis of calphostin D (15 steps).

7.3.3 Synthesis of (−)-Phleichrome and (−)-Calphostin A [32a]

During the same year as Hauser's reported synthesis of calphostin D, the Coleman laboratories published an alternative approach to this family of natural products [32]. This report included the enantioselective syntheses of both phleichrome and calphostin A and, like the previous approaches, started from an enantioenriched naphthalene.

The Coleman synthesis commenced with a [4+2] cyclization to achieve naphthalene **30** (Scheme 7.4). This convergent process required four steps to obtain diene **27** and five steps to attain dieneophile **28**. Three further functional group transformations were required to complete aldehyde **31**. This novel process realized the installation of all the carbon and oxygen substituents in a regioselective manner.

Coleman established the hydroxypropyl stereochemistry via addition of a homochiral α-alkoxyalkyl organometallic species. This reagent was prepared in high enantiomeric excess using a Noroyi BINAL-H reduction of organostannane **33**, which was transmetallated with *n*-BuLi to achieve the desired organolithium reagent **35** (Scheme 7.5). Both enantiomers of **35** could be obtained via this route.

Organolithium reagent **35** was added to aldehyde **31** (Scheme 7.6) to obtain alcohol **36** as an inconsequential 1:1 mixture of diastereomers. The benzylic alcohol was removed using a Barton two-step radical deoxygenation protocol, followed by electrophilic aromatic bromination to provide the desired coupling partner **37**.

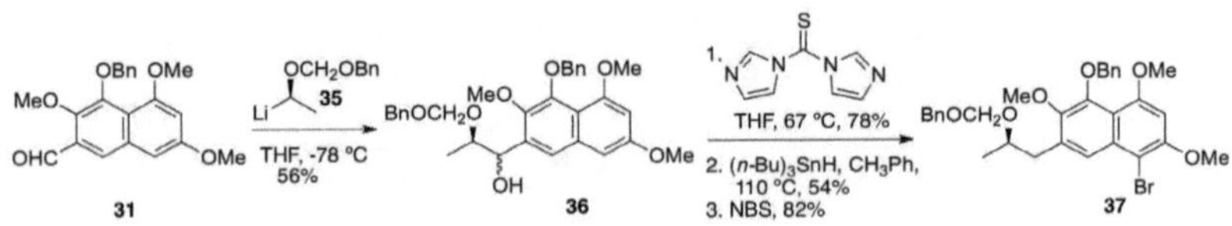

Scheme 7.6 Coleman synthesis of a homochiral naphthalene

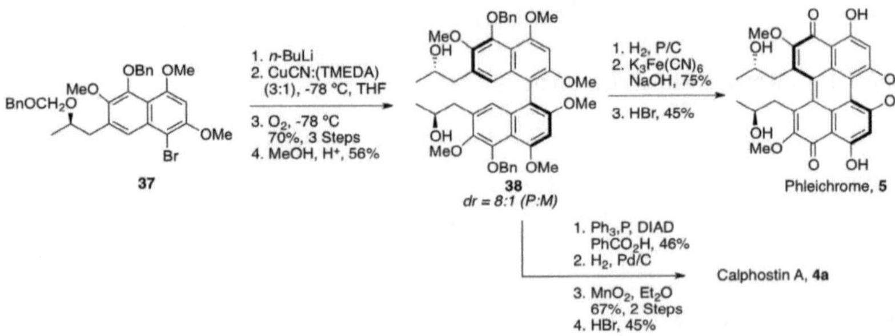

Scheme 7.7 Coleman synthesis of phleichrome and calphostin A

To attain the desired biaryl for phleichrome, Coleman adapted the copper-mediated coupling reaction pioneered by Lipshutz; (*S*)-bromide **37** was lithiated followed by transformation to the higher order cuprate upon treatment with a CuCN·TMEDA complex (Scheme 7.7). Exposure of this species to oxygen at low temperatures resulted in the desired biaryl **38** with a high degree of atropdiastereoselectivity (8:1), favoring the (*P*)-biaryl configuration. Compound **38** was further elaborated to phleichrome **5** upon protecting group removal and, mirroring the initial Broka report, oxidation to the perylenequinone with $K_3Fe(CN)_6$. Coleman's work, as Broka's, proved that the stereochemistry of the side chain could direct the atropconfiguration of the biaryl, albeit with moderate diastereoselectivity. To access the unattainable (*P*,*R*,*R*) configuration of calphostin A, Coleman subjected compound **38** to a double Mitsunobu reaction. Optimized conditions inverted both appendage stereocenters to achieve the desired benzoate groups of calphostin A in place. For this synthesis, *para*-phenoxy radical coupling and oxidation were accomplished in one step with MnO_2. Finally, acidic methyl ether removal completed the synthesis of calphostin A (**4a**).

7.3.4 Synthesis of (−)-Calphostin A–D [33a]

In 2000, the Merlic research group reported concise total syntheses of calphostins A and D and the first total syntheses of calphostins B and C [33]. This original approach to the naphthalene fragment relied on a chromium carbene intermediate to obtain the required regioselection. The synthesis began with phosphonate **39** (three steps), followed by a Horner-Wadsworth-Emmons olefination of **39** with

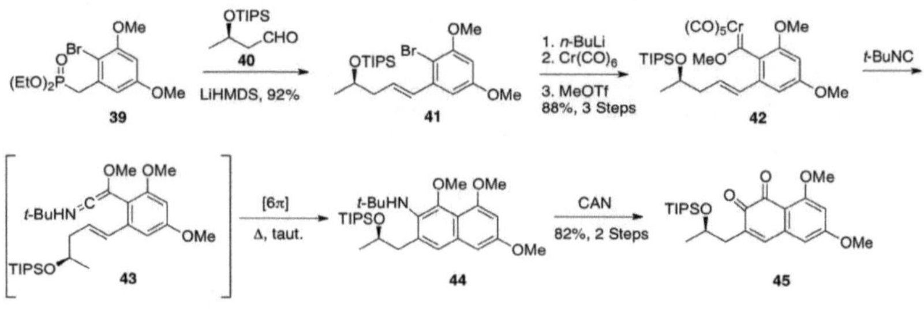

Scheme 7.8 Merlic naphthalene synthesis

aldehyde **40**, affording the *E*-alkene **41** as the only product.[2] Halogen-metal exchange with *t*-BuLi, followed by conversion to the carbene complex with Cr $(CO)_6$, and subsequent methylation yielded key intermediate **42**. Conversion to the ketenimine with *t*-BuNC, followed by electrocyclization of intermediate **43** in a 6π-fashion, yielded naphthylamine **44**. Oxidation of **44** with ceric (IV) ammonium nitrate afforded the desired *ortho*-naphthoquinone **45** (Scheme 7.8).

Merlic used a biomimetic approach to dimerize **45** to **46** with $FeCl_3$. While the reaction proceeded in good yield, the diastereomeric ratio was low [*dr* = *1.2:1* (*P*:*M*)] and resulted in the loss of both TIPS-protecting groups. With the secondary alcohols now unprotected, attempts to methylate **46** resulted in complex reaction mixtures. To solve this problem, benzoate **47** was synthesized. Analogous to the Hauser synthesis, dimerization was accomplished with TFA and oxygen. Biaryl **48** was isolated as a 1:2 ratio of atropisomers favoring the undesired (*M*)-biaryl. This mixture was thermally equilibrated in refluxing toluene to a 3:1 ratio now favoring the (*P*)-biaryl. Merlic rationalized the thermodynamic selectivity based on a stabilizing π-interaction between the benzoate side chains and the perylenequinone pentacyclic core. Such an interaction is only conformationally attainable with the (*P*,*R*,*R*) isomer. With the desired biaryl in hand, Merlic further elaborated (*P*)-**48** by methylation of the C6,C6′-hydroxy groups and selective deprotection of the C3, C3′-methoxy groups with MgI_2 to yield calphostin A (**4a**). Deprotection of the benzoate groups afforded calphostin D (**4d**) (Scheme 7.9).

With a concise route to enantiopure **49** identified, Merlic was able to effect desymmetrization by controlled methanolysis to access calphostins B (**4b**) and C (**4c**) (Scheme 7.10). The carbonate linkage of calphostin C was installed via reaction of the free alcohol with phosgene and 4-acetoxyphenol, thus completing the first synthesis of calphostin C in 16 steps.

[2] Enantiopure **40** was synthesized in two steps from commercially available methyl-3-hydroxybutyrate.

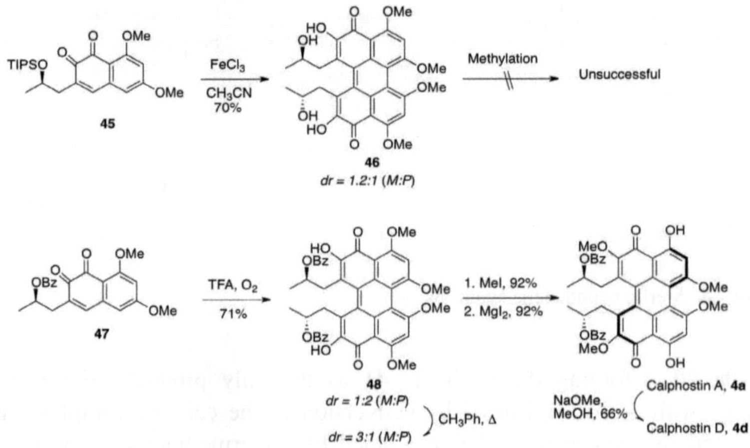

Scheme 7.9 Merlic syntheses of calphostins A and D

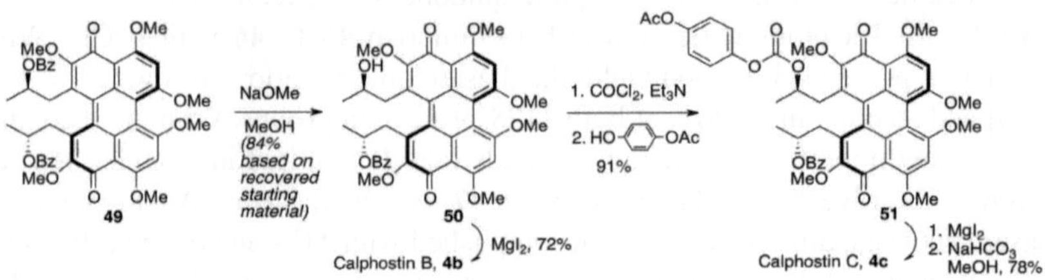

Scheme 7.10 Merlic syntheses of calphostins B and C

7.4 Conformational Properties

The atropisomerization of the helical configuration entails movement of the C2, C2′- and C7,C7′-groups past one another (Fig. 7.1), and the barrier to this isomerization varies substantially for the different perylenequinones. While the calphostins (**4**) and phleichrome (**5**) are atropisomerically stable at room temperature and require temperatures over 110 °C to isomerize, the additional seven-membered ring bridge at the C2,C2′-positions in cercosporin (**3**) lowers the barrier allowing it to atropisomerize at 37 °C [34].

The barrier to atropisomerization is also lower for the hypocrellins due to introduction of the seven-membered ring bridging the C7,C7′-positions. In fact, hypocrellin A (**7**) and hypocrellin (*ent*-**7**) both exist as rapidly atropisomerizing mixtures of diastereomers at room temperature, as revealed by NMR studies by Mondelli. Here, two sets of sharp peaks of the resultant diastereomers are observed in the NMR spectra of each natural product [35]. Figure 7.2 presents the structures of hypocrellin A (**7**) and its atropisomer, *atrop*-**7**, which exist as an equilibrium

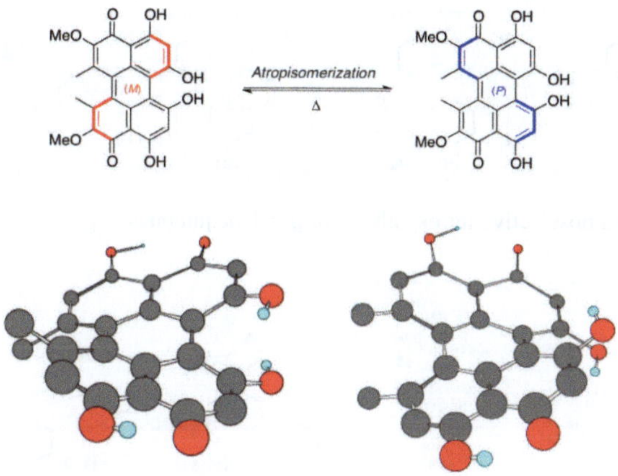

Fig. 7.1 Atropisomerization of perylenequinones

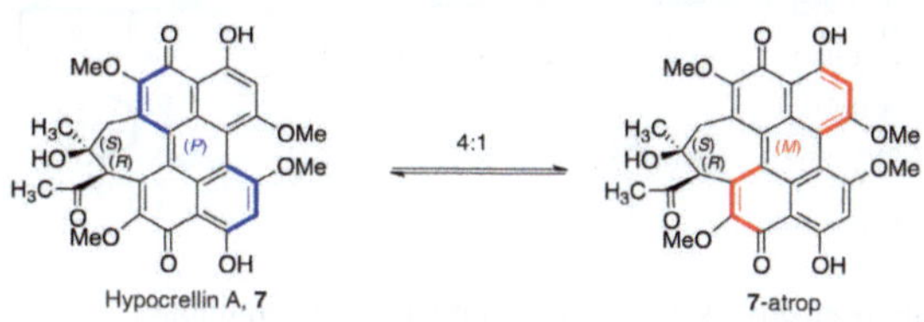

Fig. 7.2 Atropisomerization of hypocrellin A

mixture favoring the more stable (*P*)-diasteromer. Due to the predominance of one diastereomer over another, hypocrellin A (**7**), hypocrellin (*ent*-**7**), and shiraichrome A (**8**) exhibit CD spectra characteristic of helical stereochemistry.

On the other hand, hypocrellin B (**9**) exhibits no peaks in the CD spectrum [36]. Although it is possible that the presence of a double bond in the seven-membered ring of hypocrellin B could lead to a planar structure accounting for this phenomenon, calculations indicate that the ground state of **9** contains a similar stereogenic helix as the related congeners hypocrellin A (**7**), hypocrellin (*ent*-**7**), and shiraichrome A (**8**) [37]. However, for **9**, the atropisomers are enantiomers, and the CD signals of the two atropisiomers cancel each other.

7.5 Strategy and Retrosynthesis

The common denominator for all the previous approaches to the perylenequinones is a diastereoselective oxidative coupling of homochiral 2-naphthols. Our group proposed an alternative enantioselective convergent synthesis that would greatly

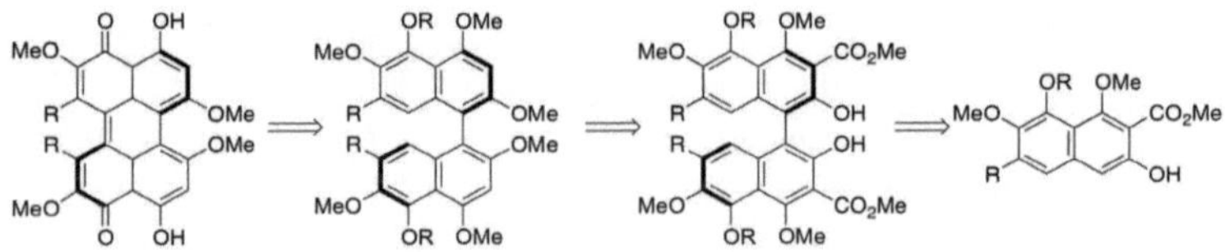

Scheme 7.11 Enantioselective retrosynthesis of perylenequinones

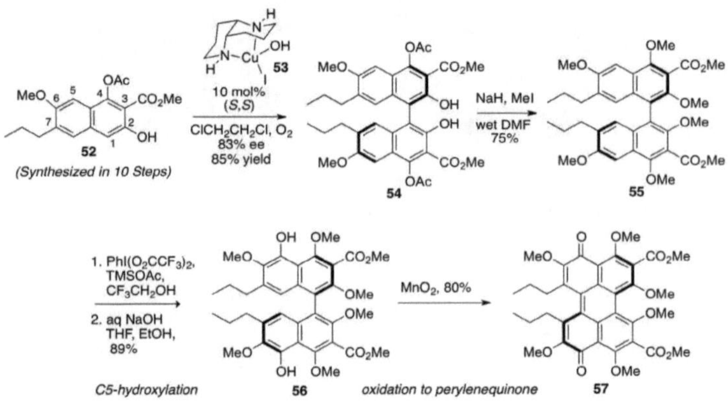

Scheme 7.12 Enantioselective synthesis of a model perylenequinone

facilitate an examination of the structural aspects responsible for the photodynamic properties. We have developed a copper-catalyzed enantioselective naphthol coupling [38] that can prepare highly substituted binaphthols in high yield and enantiomeric excess. Thus, attention turned to the synthesis of a binaphthalene that could undergo stereoretentive oxidation to a perylenequinone (Scheme 7.11).

After investigating a number of substituted naphthols with the enantioselective coupling, we discovered that those with the substitution pattern of **52** (Scheme 7.12) provided the optimum yields and enantioselectivities [39]. The C2-hydroxyl group and the C3-ester are necessary for chelation to the copper catalyst. A C4-acetoxy group provides the optimum electron density in the ring system for this reaction. Although there are multiple groups tolerated at C6 and C7, substitution at C5 proved detrimental to the yield and enantioselectivity of the biaryl coupling. Since C5,C5′-oxygenation is needed to obtain the required perylenequinone, we investigated several possible aromatic hydroxylation methods on a model system (Scheme 7.12). Success was obtained by applying an aromatic oxidation reported by Kita using PhI $(OCOCF_3)_2$ [40]. The product of this dihydroxylation was then oxidized by MnO_2 to the perylenequinone **57** with retention of axial stereochemistry.

In addition to installation of the C5,C5′-hydroxyls, removal of the C3,C3′-esters was needed to provide the substitution pattern of the natural products. However,

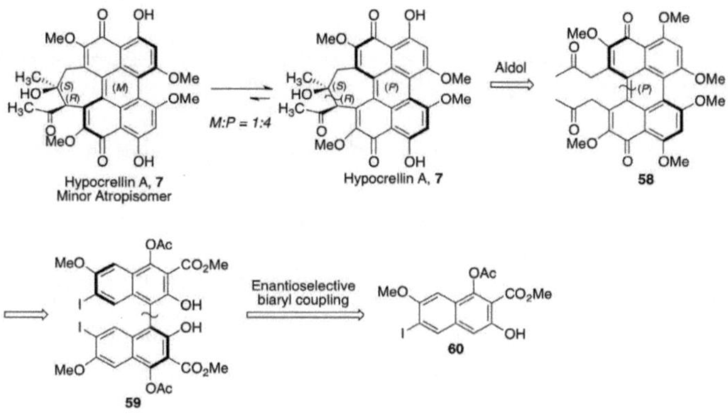

Scheme 7.13 Kozlowski retrosynthesis of hypocrellin A

conventional aromatic decarboxylation protocols [41] require temperatures that would atropisomerize the biaryl. As a consequence, we developed a palladium-catalyzed decarboxylation protocol [42], which worked well in these systems. With this chemistry in hand, we undertook the synthesis of cercosporin as well as the related calphostin and phleichrome, exploring several possible methods to introduce the C7,C7′-groups (see below).

For the synthesis of the hypocrellins, the additional complexity inherent in the stereochemistry of the 7-membered ring needs to be addressed. The previously reported diastereoselective syntheses of perylenequinones could not be readily applied to hypocrellin A (**7**), since the oxidation state required in a precursor such as **58** would result in loss of the hydroxypropyl stereochemistry, for which great effort had been expended (see Schemes 7.1, 7.3, 7.5, 7.6, and 7.8). For the synthesis of hypocrellin A, we envisioned a chiral transfer reaction in which enantiomerically pure perylenequinone **58** would be synthesized first and then would relay its stereochemical information in an intramolecular aldol cyclization, forming the 7-membered ring of hypocrellin A (Scheme 7.13). No such diastereoselective 1,8-diketone aldol reaction has been used in a natural product synthesis. The stereochemistry transfer was further complicated by the dynamic state of the helical axis of **7**. The helical stereochemistry of perylenequinone **58** was stable at room temperature and needed to direct the stereochemistry of the intramolecular aldol reaction, but following formation of the seven-membered ring, the integrity of the helical axis is lost due to the rapid atropisomerization of **7** (4:1, *P*:*M*), even at low temperature. A catalytic asymmetric oxidative 2-napthol coupling of **60** would set the biaryl axis, by means of a copper diaza-*cis*-decalin catalyst.

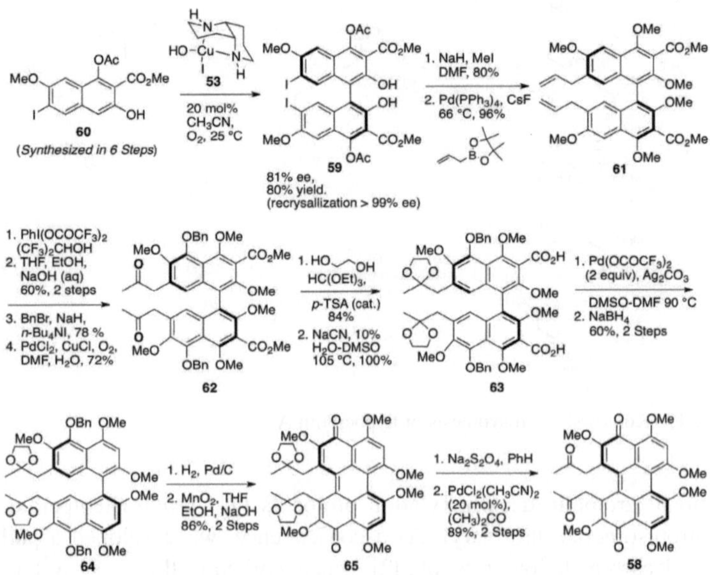

Scheme 7.14 Synthetic steps to hyprocrellin A perylenequinone retron

7.6 Synthesis

7.6.1 Synthesis of (−)-Hypocrellin A

The synthesis of the desired aldol perylenequinone precursor **58** is outlined in Scheme 7.14 [43]. Enantioselective oxidative biaryl coupling of **60** demonstrated the effectiveness of the copper 1,5-diaza-*cis*-decalin catalyst system, providing **59** in 81 % ee (>99 % ee, 1 trituration) and 80 % yield. Intermediate **61** was obtained via a one-pot deprotection/methylation of the C4,C4′-hydroxyl groups followed by a bis-Suzuki coupling to install the C7,C7′-allyl substituents. After C5,C5′-hydroxylation using $PhI(O_2CCF_3)_2$, a Wacker oxidation converted the allyl groups to the requisite ketones. The next step, palladium-mediated decarboxylation, required hydrolysis of the methyl esters. This transformation proved difficult via basic hydrolysis but was achieved in quantitative yield by an S_N2 displacement with NaCN in DMSO. The palladium-mediated decarboxylation of the resultant bis-acid provided **64**, and the final steps to provide intermediate **58** included oxidative cyclization with MnO_2 and $PdCl_2$ deprotection of the acid-sensitive ketals in **65**.

With a synthesis of **58** completed, the key intramolecular diketone aldol cyclization was investigated. Precedent for this type of 1,8-dicarbonyl aldol reaction is rare, although an aldol reaction has been proposed in the biosynthetic pathway to the hypocrellins. The only reported examples of such diketone aldol cyclizations involve multicyclic or bridged bicyclic systems, and of these no examples exist for 1,8-diketones forming 7-membered rings. MM2 calculations indicated that a

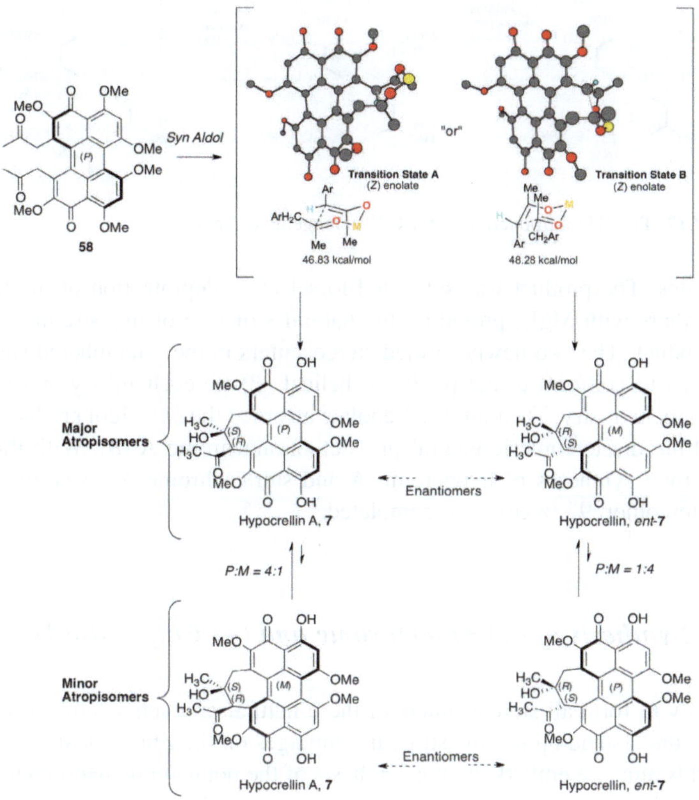

Scheme 7.15 Diketone aldol transition states to hypocrellin A

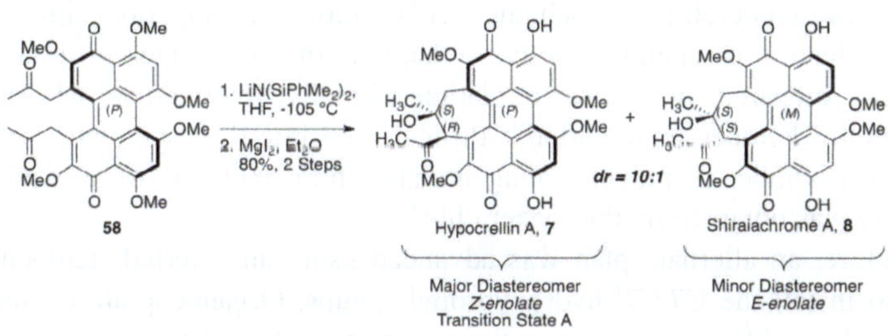

Scheme 7.16 Completion of hypocrellin A

(*Z*)-enolate of **58** would adopt a closed chair-like transition state that favored exposure of one face of the ketone by 1.5 kcal/mol, leading to the *syn*-aldol product corresponding to hypocrellin A (Scheme 7.15).

Exposure of **58** to $LiN(SiMe_2Ph)_2$ at −105 °C provided the aldol product with the desired 7-membered ring present of the hypocrellins (Scheme 7.16). This *syn*-aldol adduct was anticipated from a (*Z*)-enolate and silazide bases give predominately

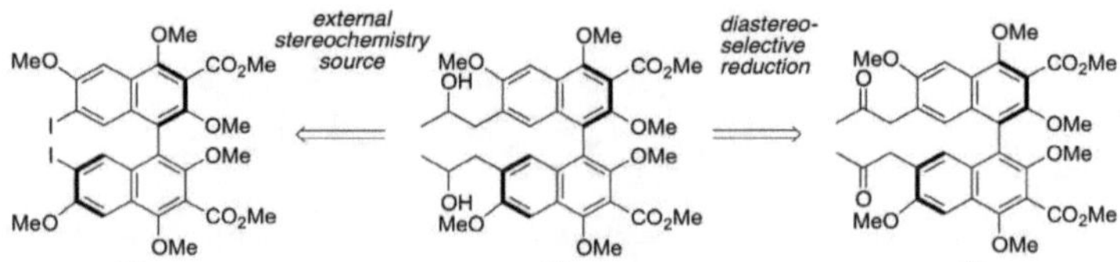

Scheme 7.17 Possible introduction of C7,C7′-stereogenic centers

(*Z*)-enolates. The product was subjected to selective deprotection of the C4,C4′-methyl ethers with MgI_2, providing the natural structure of hypocrellin A as the major product. The two newly formed stereocenters in the 7-membered ring were determined to conform to the predicted helical (*P*)-stereochemistry and the *syn*-aldol stereochemistry. The minor (*E*)-enolate afforded the *anti* aldol product, which matched the diastereomeric natural product shiraiachrome A (**8**). With this step, the first total syntheses of hypocrellin A and shiraiachrome A (*syn*:*anti* = 10:1; syn diastereomer, 92 % ee) were completed.

7.6.2 Synthesis of (+)-Phleichrome and (+)-Calphostin D

The ability to form all stereoisomers of these natural products is crucial toward a thorough understanding of how structure impinges on their biological activity. To further this aim, we embarked on a synthesis of the nonnatural enantiomers of the natural products (−)-phleichrome and (−)-calphostin D. These products contain stereogenic C7,C7′-hydroxypropyl groups, and two possible means of introducing these groups were investigated, both using intermediates readily available from the synthesis of hypocrellin A (Scheme 7.17). Initial investigations involved the diastereoselective or enantioselective reductions of the ketones in model biaryl **68**, readily prepared by Wacker oxidation of diallyl intermediate **61**. However, benzyl methyl ketones are remarkably difficult substrates for asymmetric reduction. Although a number of reducing reagents and chiral methods were explored, no selectivity was obtained on this system [44].

Therefore, an alternate plan was advanced using an external stereochemistry source to install the C7,C7′-hydroxypropyl groups. Organocuprate opening of a homochiral epoxide would also have the advantage in that all possible stereoisomers would be accessible. Due to the high reactivity of electron-rich aryl cuprates, this method was first explored in a model system. Although very sensitive, the cuprate of iodonaphthalene **69** underwent alkylation with (*R*)-propylene oxide successfully to form compound **70** in high yield (Scheme 7.18). With this precedent in hand, the more complex biscuprate and double epoxide alkylation was pursued en route to calphostin and phleichrome [44, 45].

The synthetic routes to the (+)-calphostin and (+)-phleichrome are presented in Scheme 7.19. Compound **66**, the enantiomer of the intermediate used in the

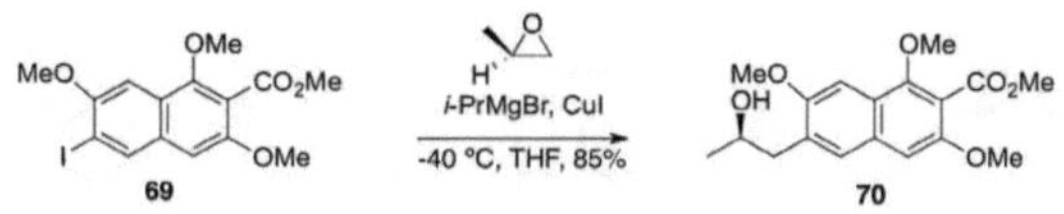

Scheme 7.18 Organocuprate-mediated epoxide opening

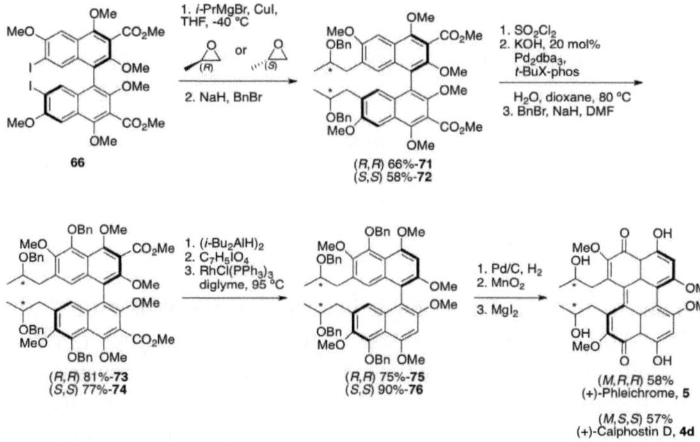

Scheme 7.19 Syntheses of (+)-phleichrome and (+)-calphostin D

hypocrellin synthesis, was synthesized following the same protocol outlined in Scheme 7.14. Formation of the biscuprate and double epoxide alkylation were undertaken following the method described above (see Scheme 7.18). Notably, both (*R*)- and (*S*)-propylene oxide were equally successful in the biscuprate alkylation, and after bis-benzylation, provided the diastereomers (*M,R,R*)-**71** and (*M,S,S*)-**72** in good yield. Next, the C5,C5′-oxidation was undertaken. Here, we discovered that our protocol using Kita's method was very sensitive to any change in the oxidation potential of the substrate, and reoptimization was needed for each new precursor. As a result, a more robust C5,C5′-oxidation utilizing a palladium-catalyzed *O*-arylation was pursued [46]. First, C5,C5′-chlorination was effected using sulfuryl chloride. Subsequent hydroxylation using KOH and a palladium catalyst [47] proceeded extremely well and was found to be effective for a range of very hindered substrates. Although the product bisphenol was unstable, immediate protection with benzyl bromide provided the stable intermediates (*M,R,R*)-**73** and (*M,S,S*)-**74**.

The next task was removal of the C3,C3′-esters. Although the palladium-catalyzed decarboxylation protocol performed well in previous systems, a competing C-H insertion reaction was discovered with the methylidene bridge needed for cercosporin (see below). Since reexamination of alternate decarboxylation methods [48] led to no success, a decarbonylation strategy was explored [49]. Formation of the requisite dialdehyde was best accomplished by overreduction using DIBAL and

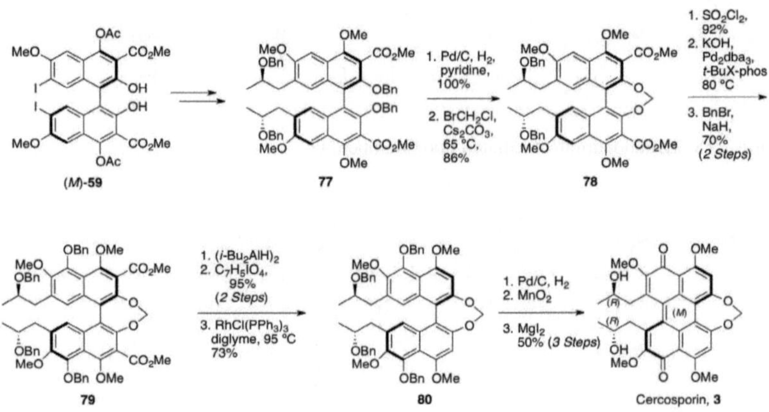

Scheme 7.20 Synthesis of cercosporin

then oxidation with IBX. Decarbonylation with excess Wilkinson's catalyst to (*M*,*R*,*R*)-**75** and (*M*,*S*,*S*)-**76** then proceeded smoothly under oxygen-free conditions.

The remaining three steps of the synthesis could all be conducted under mild conditions. Removal of the four benzyl ethers using palladium-catalyzed hydrogenolysis was followed by oxidative cyclization with MnO_2 to provide the central perylenequinone core. Deprotection of the C4,C4′-methyl ethers with MgI_2 revealed the enantiomers of the natural products phleichrome (*ent*-**5**) and calphostin D (*ent*-**4d**), respectively. Each synthesis required a total of 17 steps, *ent*-**5** with an overall yield of 5.3 % (average of 87 % per step) and *ent*-**4d** with an overall yield of 5.2 % (average of 87 % per step). With all stereoisomers of the natural products available, it was now possible to assess the impact of the helical and centrochiral centers on the biological activity.

7.6.3 *Synthesis of (+)-Cercosporin*

The studies into the syntheses of (+)-calphostin and (+)-phleichrome proved crucial toward the total synthesis of cercosporin. A key difference between these natural products is the C2,C2′-methylene bridge contained in cercosporin. As mentioned above, this bridge lowers the barrier to atropisomerization and adds an additional level of complexity to the synthesis. Our group reported a first-generation synthesis of cercosporin in 23 steps and 0.4 % overall yield [50], but turned attention toward improving this sequence. A second-generation synthesis optimizing the C5,C5′-hydroxylations and C3,C3′-decarboxylations is presented in Scheme 7.20. Starting from common intermediate (*M*)-**59**, orthogonal protection of the C2,C2′-positions was installed in the form of benzyl ethers via a Mitsunobu reaction. After introduction of the 2-hydroxypropyl groups via organocuprate formation and epoxide opening (see Scheme 7.18 above), chemoselective debenzylation of the more labile

C2,C2′-benzyl ethers of **77** was accomplished using Pd/C poisoned with pyridine. Installation of the methylene bridge was difficult, presumably due to disfavorable entropy in forming the seven-membered ring, along with the added ring strain. Ring formation was most facile with the methylidene equivalent $BrCH_2Cl$ instead of $BrCH_2I$ or $BrCH_2Br$, even though heating to 65 °C (well below the atropisomerization threshold of a binaphthalene)[3] was required.

Implementation of the C5,C5′-hydroxylation protocol as described in Scheme 7.19 above (**71**/**72** to **73**/**74**) provided further efficiencies. The C5,C5′-chlorination proceeded uneventfully, but the chloro to alkoxy interchange was difficult and required optimization of the reaction conditions. The catalyst system derived from Pd_2dba_3 and the X-phos(*t*-Bu) ligand proved to be effective in the coupling with KOH to provide the desired bisphenol. The resulting product was highly unstable and decomposed under a one-pot alkylation protocol. Isolation of the bisphenol under carefully controlled conditions followed by immediate benzylation (BnBr, NaH, DMF) furnished key intermediate **79** in 70 % yield.

As described in the transformation of **73** and **75** to **74** and **76**, respectively (see Scheme 7.19 and accompanying text), decarboxylation of **79** proved difficult. With **79**, we observed a competing C-H insertion reaction onto the methylidene in the seven-membered ring. Fortunately, the alternate decarbonylation of the dialdehyde with Wilkinson's catalyst proceeded in good yield. The temperature (85 °C in diglyme) required for this process was well tolerated, and no atropisomerization of the binaphthalene was observed, in accord with atropisomerization barriers measured for this series [44]. After benzyl deprotection of **80**, the bisphenol was oxidized by MnO_2 to afford the perylenequinone. This final cyclization step was delayed until the end of the route because formation of the perylenequinone reduces the dihedral angle between the upper and lower portions from 50° in **80** to 20° in **3**, further lowering the atropisomerization barrier [34]. Directed deprotection of the C4,C4′-methyl ethers was accomplished selectively with MgI_2, completing the first total synthesis of cercosporin (**3**) in 20 steps and 2.8 % overall yield (86 % average yield per step). Notably, in this synthesis none of the diastereomer of **3** arising from the facile atropisomerization was observed.

7.7 Synthesis of Perylenequinone Analogs

Perylenequinones generate singlet oxygen when exposed to light and also selectively bind to the C1 regulatory domain of PKC, a well-studied cancer target [26, 27]. These properties allow for two distinct benefits in cancer therapy: (1) direction of a specific wavelength of light to the region of the tumor and (2) sequestration in cells with high levels of PKC. As a consequence, derivatives were targeted to both increase the photodynamic activity and to also probe PKC binding. The perylenequinone

[3] See reference [45] for a study of atropisomerization thresholds for model biaryls and perylenequinones.

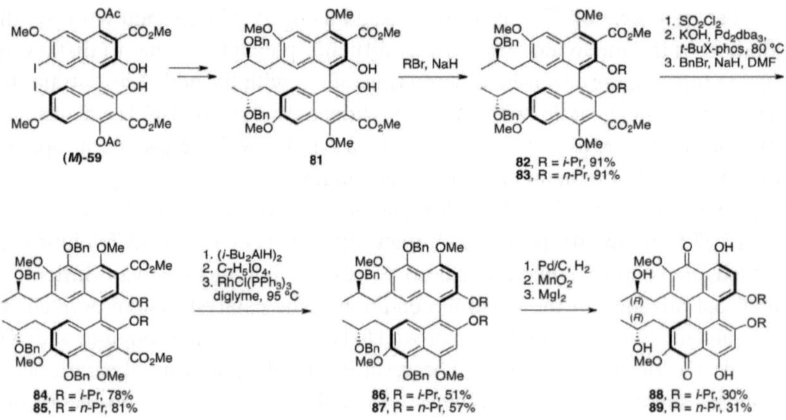

Scheme 7.21 Synthesis of C2,C2′-substituted (+)-calphostin D analog

chromophore was extended through C3,C3′-substitution to increase the photo-response in the photodynamic window (600–900 nm). Additionally, analogs of the natural products were designed and synthesized in a systematic manner to examine the effects of the helical stereochemistry, C2,C2′-substitution, C3,C3′-substitution, and C7,C7′-substitution, which have previously been hypothesized to have the greatest effect on PKC inhibition [27]. Most perylenequinone analogs examined previously were obtained from transformations on the natural products themselves; generation of the above substitution patterns from the natural products is difficult or not feasible. On the other hand, our synthetic strategy allows rapid and convergent synthesis of these analogs.

The C2,C2′-substitution was modified to include the more hydrophobic isopropoxy and *n*-propoxy groups (**88** and **89**) to take advantage of an unfilled hydrophobic pocket in the PKC C1 domain.[4] These novel perylenequinones could be synthesized by selective alkylation of an advanced intermediate (**81**, Scheme 7.21) available from the cercosporin synthesis (Scheme 7.20). The remaining steps followed the protocols for the syntheses of calphostin/phleichrome (Scheme 7.19) and cercosporin (Scheme 7.20) providing unnatural congeners **88** and **89** (Scheme 7.21). The reduced yield for the last three steps is presumably due to the larger C2,C2′-*n*-propoxy and -*iso*propoxy groups which slow the perylenequinone formation.

In addition to the possible C2,C2′-substituent interactions with a hydrophobic pocket in the regulatory domain site, the much higher affinity of calphostin A (Chart 7.1 **4a**, $R^1 = R^2 =$ COPh, $IC_{50} = 0.25\ \mu M$) compared to calphostin D (**4d**, $R^1 = R^2 =$ H, $IC_{50} = 6.4\ \mu M$) also points to interaction of the C7,C7′-portion with the binding site [51]. The C2,C2′- and C7,C7′-analogs would be invaluable in understanding the relative importance of each of these interactions. To determine

[4] See Supporting Information of reference [45].

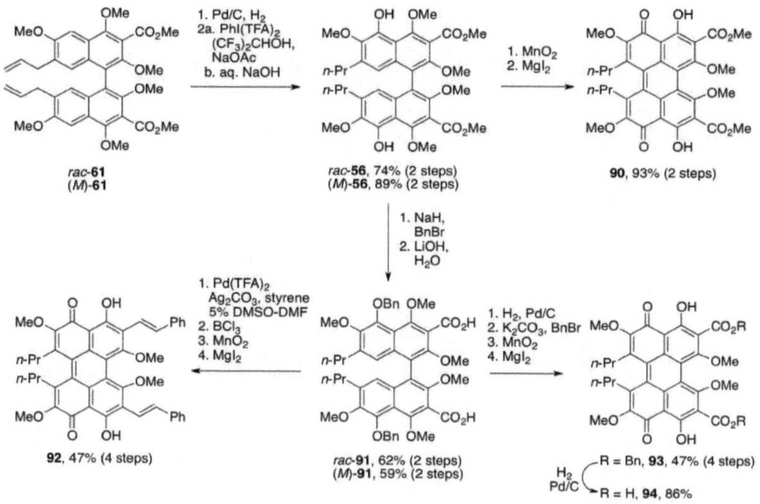

Scheme 7.22 Synthesis of C3,C3′-substituted C7,C7′-propyl perylenequinone analogs

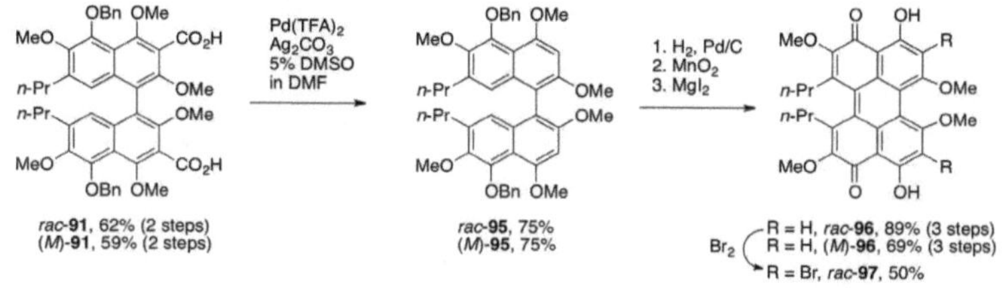

Scheme 7.23 Synthesis of C3,C3′-substituted C7,C7′-propyl perylenequinone analogs

whether the stereogenic hydroxyl groups on the C7,C7′-substituents of the natural products were critical to binding, the simpler C7,C7′-propyl substitution was incorporated (see Scheme 7.12). Based upon the improved activity of this compound (see below) relative to the parent natural products (*ent*-**4d** and *ent*-**5**), a series of derivatives (**90–97**) with the C7,C7′-propyl groups were designed with different substituents at the C3,C3′-positions (Scheme 7.22). Specifically, bromo (**97**), ester or acid (**90, 93, 94**), and vinyl (**92**) C3,C3′-substitution were synthesized with the goal of increasing the absorption wavelength of the perylenequinone chromophore. The use of the C3,C3′-methyl ester in our synthetic strategy allowed the preparation of all of these structures either directly or by decarboxylative functionalization.

We initially prepared racemic analogs to readily probe effects of differing C3, C3′ substitution. The synthesis of **90–97** (Schemes 7.22 and 7.23) commenced with racemic bisallyl **61** (see Scheme 7.14). Subsequent hydrogenation yielded the bis-*n*-propyl compound, and here the lack of functionalization of the C7,C7′-groups allowed facile C5,C5′-hydroxylation with Kita's reagent [40] to provide **56**, as we discovered during our work on the model system (Scheme 7.12).

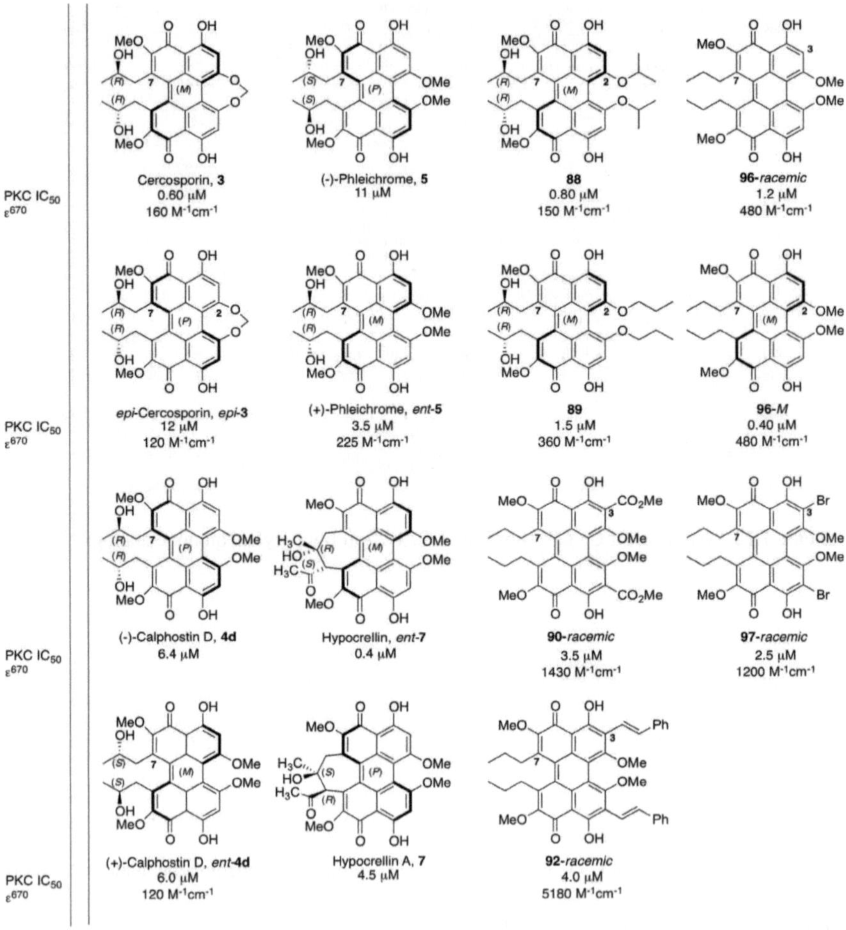

Chart 7.3 Potency (IC_{50}) of selected perylenequinones against PKC

Compound **56** is the first branch point intermediate in the analog syntheses, furnishing **90** upon oxidative cyclization with MnO_2 and deprotection with MgI_2. Intermediate **56** was also benzylated to protect the C5,C5′-naphthols in preparation for ester hydrolysis, which provided the next key branch point intermediate, bisacid **91**. Ester hydrolysis here with aqueous base was surprising facile relative to intermediate **62**, en route to (+)-calphostin D (Scheme 7.14). Presumably, the smaller C7,C7′-groups alleviate the steric gearing that hinders the reactivity of the C3,C3′-esters.

Bisacid **91** was used toward three different targets. For the first, a palladium-catalyzed decarboxylative Heck reaction followed by perylenequinone formation provided bis-styryl derivative **92** (Scheme 7.22) [52]. For the second, the C5,C5′-benzyl ethers were cleaved, and the more acidic carboxylic acids were then selectively benzylated using BnBr and K_2CO_3 (Scheme 7.22). This re-esterification was necessary because the carboxylic acids did not survive the MnO_2-mediated oxidative cyclization nor was selective hydrogenation of the bisbenzyl ester of **91** successful. Perylenequinone formation and demethylation yielded **93**, which could then be further debenzylated to the perylenequinone bisacid **94**. Third, protodecarboxylation of **91** (Scheme 7.23), using a palladium source, provided **95** which was then transformed to perylenequinone **96** following established protocols. Perylenequinone **96** was treated with Br_2 to provide bisbromide **97**. Subsequent enzyme assays with this analog series revealed high activity for **96**. We therefore synthesized (*M*)-**96** since we had established that the (*M*)-isomers are more potent (see below).

The potencies against PKC of the analogs prepared in Scheme 7.22 and the corresponding natural products and their isomers are listed in Chart 7.3, along with their absorption at 670 nm (in the therapeutic window). Several trends can be seen from this data. First, the (*M*)-isomers are consistently more active than the (*P*)-isomers, regardless of the stereochemistry around the C7,C7′-2-hydroxypropyl groups. For the (*M*)-isomers, the (*R*,*R*)-stereochemistry around the C7,C7′-groups was slightly better than the (*S*,*S*)-stereochemistry [3.5 μM for (+)-**5** vs. 6.0 μM for (+)-**4d**]. As mentioned above, the racemic perylenequinone **96** with C7,C7′-unsubstituted propyl groups had fairly strong potency of 1.2 μM, supporting that hydrophobic groups are optimal at this position. The (*M*)-isomer had excellent potency, equal to the most potent natural products cercosporin (**3**) and hypocrellin (*ent*-**7**).

As predicted, the C2,C2′-isopropoxy and propoxy groups are more potent than the corresponding methoxy (0.8 and 1.5 μM vs. 6.4 μM). In addition, C3,C3′-substitution has an effect on both the potency and photopotentiation factor. Although the absorption at 670 nm was improved with C3 substitution, the potency against PKC decreased for analogs **90**, **92**, and **97**.

The development of new synthetic methodology has successfully enabled the investigation of structure-activity relationships (SAR) of perylenequinone agents for use in photodynamic therapy. Simplified analogs, such as (*M*)-**96**, that have potency equal to the natural product hypocrellin and superior chromophores to improve photoactivation in the therapeutic window were prepared.

References

1. Reviews, see: (a) Weiss U, Merlini L, Nasini G (1987) Prog Chem Org Nat Prod 52:1–71; (b) Bringmann G, Günther C, Ochse M, Schupp O, Tasler S (2001) Prog Chem Org Nat Prod 82:1–249; (c) Mulrooney CA, O'Brien EM, Morgan BJ, Kozlowski MC (2012) Eur J Org Chem 21:3887–3904; (d) Kozlowski MC, Morgan BJ, Linton EC (2009) Chem Soc Rev 38:3193–3207
2. Lown JW (1997) Can J Chem 75:99–119
3. (a) Kikuchi R (1924) The Scientific Researches of the Alumni Association of the Utsunomiya Agricultural College (Japan) 1:7–25; (b) Tomoyasu R (1925) The Scientific Researches of the

Alumni Association of the Utsunomiya Agricultural College (Japan) 2:53–73; (c) Matsumoto J, Tomoyasu R (1925) Ann Phytopathol Soc Jpn 1:1–14
4. Kuyama S, Tamura T (1957) J Am Chem Soc 79:5725–5726
5. (a) Lousberg RJJ, Weiss U, Salemink CA, Arnone A, Merlini L, Nasini G (1971) Chem Commun 1463–1464; (b) Yamazaki S, Ogawa T (1972) Agric Biol Chem 36:1707–1718; (c) Yamazaki S, Okube A, Akiyama Y, Fuwa K (1975) Agric Biol Chem 39:287–288
6. Chen WS, Chen YT, Wang XY, Friedrichs E, Puff H, Breitmaier E (1981) Liebigs Ann Chem 1880–1885
7. Hypocrellin (*ent*-**7**): Chen WS, Chen YT, Wang XY, Friedrichs E, Puff H, Breitmaier E (1981) Liebigs Ann Chem 1880–1885
8. Hypocrellin A (**7**) and Hypocrellin B (**9**): (a) Kishi T, Tahara S, Tsuda M, Tanaka C, Takahashi S (1991) Planta Med 57:376–379 (shiraiachrome A is called hypocrellin B here); (b) Wu H, Lao XF, Wang QW, Lu RR (1989) J Nat Prod 52:948–951 (hypocrellin A is called shiraiachrome B here)
9. For the correct structural assignment of shiraiachrome A, see: (a) Mazzini S, Merlini L, Mondelli R, Scagloni L (2001) J Chem Soc Perkin Trans 2 409–416; Which corrects: (b) Smirnov A, Fulton DB, Andreotti A, Petrich JW (1999) J Am Chem Soc 121:7979–7988
10. (a) Kobayashi E, Ando K, Nakano H, Iida T, Ohno H, Morimoto M, Tamaoki T (1989) J Antibiot 42:1470–1474; (b) Iida T, Kobayashi E, Yoshida M, Sano H (1989) J Antibiot 42:1475–1481
11. Yoshihara T, Shimanuki T, Araki T, Saramura S (1975) Agric Biol Chem 39:1683–1684
12. Weiss U, Ziffer H, Batterham TJ, Blumer M, Hackeng WHL, Copier H, Salemink CA (1965) Can J Microbiol 11:57–66
13. (a) McCaughan JS Jr (1999) Drugs Aging 15:49–68; (b) Detty MR (2001) Expert Opin Ther Pat 11:1849–1860; (c) Lane N (2003) Sci Am 288(38–41):43–45; (d) Dolmans DEJGJ, Fukumura D, Jain RK (2003) Nat Rev Cancer 3:380–387; (e) Triesscheijn M, Baas P, Schellens JHM, Stewart FA (2006) Oncologist 11:1034–1044; (f) Kessel D (2008) J Porphyrins Phthalocyanines 12:877–880
14. For a review of these studies, see: Daub ME, Ehrenshaft M (2000) Ann Rev Phytopath 38:461–490, and references therein
15. Paul BT, Babu MS, Santhoshkumar TR, Karunagaran D, Selvam GS, Brown K, Woo T, Sharma S, Naicker S, Murugesan R (2009) J Photochem Photobiol B Biol 94:38–44
16. (a) Chiarini A, Whitfield JF, Pacchiana R, Armato U, Dal Pra I (2008) Biochim Biophys Acta 1783:1642–1653; (b) Olivo M, Ali-Seyed M (2007) Int J Oncol 30:537–548; (c) Guo B, Hembruff SL, Villeneuve DJ, Kirwan AF, Parissenti AM (2003) Breast Canc Res Treat 82:125–141; (d) Ma L, Tai H, Li C, Zhang Y, Wang Z-H, Ji W-Z (2003) World J Gastroenterol 9:485–490; (e) Ali SM, Olivo M, Yuen GY, Chee SK (2001) Int J Oncol 19:633–643; (f) Dubauskas Z, Beck TP, Chumura SJ, Kovar DA, Kadkhodaian MM, Shrivastav M, Chung T, Stadler WM, Rinker-Schaeffer CW (1998) Clin Cancer Res 4:2391–2398; (g) Vandenbogaerde AL, Delaey WS, Vantieghem AM, Himpens BE, Merlevede WS, De Witte PA (1998) Photochem Photobiol 67:119–125; (h) Zhang J, Cao E-H, Li J-F, Zhang T-C, Ma W-J (1998) J Photochem Photobiol B 43:106–111; (i) Pollack IF, Kawecki S (1997) J Neuro-oncol 31:255–66; (j) Diwu Z (1995) Photochem Photobiol 61:529–539
17. Wang HK, Xie JX, Chang JJ, Hwang KM, Liu SY, Ballas LM, Jiang JB, Lee KH (1992) J Med Chem 35:2717–2721
18. (a) Hudson JB, Imperial V, Haugland RP, Diwu Z (1997) Photochem Photobiol 65:352–354; (b) Hirayama J, Ikebuchi K, Abe H, Kwon K-W, Ohnishi Y, Horiuchi M, Shinagawa M, Ikuta K, Kamo N, Sekiguchi S (1997) Photochem Photobiol 66:697–700
19. Leveugle B (2001) US Patent 7,157,477, 24 Oct 2001
20. Diwu Z, Lown JW (1990) Photochem Photobiol 52:609–616
21. Hu YZ, An JY, Jiang LJ, Chen DW (1995) J Photochem Photobiol A 89:45–51
22. He YY, An JY, Jiang LJ (1999) Dyes Pigm 41:93–100

23. Zhou L, Wang W, Feng Y, Wei S, Zhou J, Yu B, Shen J (2010) Bioorg Med Chem Lett 20:6172–6174
24. (a) Zhang W-G, Weng M, Pang S-Z, Zhang M-H, Yang H-Y, Zhao H-X, Zhang Z-Y (1998) J Photochem Photobiol B Biol 44:21–28; (b) Xu S, Chen S, Zhang M, Shen T, Xhao Y, Liu Z, Wu Y (2001) Biochim Biophys Acta 1537:222–232
25. Quest Pharmatech URL: http://www.questpharmatech.com/sl017.htm
26. (a) Kobayashi E, Ando K, Nakano H, Tamaoki T (1989) J Antibiot 42:153–155; (b) Kobayashi E, Nakano H, Morimoto M, Tamaoki T (1989) Biochem Biophys Res Commun 159:548–553; (c) Bruns RF, Miller FD, Merriman RL, Howbert JJ, Heath WF, Kobayashi E, Takahashi I, Tamaoki T, Nakano H (1991) Biochem Biophys Res Commun 176:288–293
27. Diwu Z, Zimmermann J, Meyer T, Lown JW (1994) Biochem Pharmacol 47:373–385
28. (a) Tamaoki T, Nomoto H, Takahashi I, Kato Y, Morimoto M, Tomita F (1986) Biochem Biophys Res Commun 135:397–402; Reviews on Indolocarbazoles: (b) Nakano H, Omura S (2009) J Antibiot 62:17–26; (c) Prudhomme M (1997) Curr Pharm Des 3:265–290
29. Kishi T, Saito H, Sano H, Takahashi I, Tamaoki T (1990) Eur Patent 0,390,181, 30 March 1990
30. Broka CA (1991) Tetrahedron Lett 32:859–862
31. Hauser FM, Sengupta D, Corlett SA (1994) J Org Chem 59:1967–1969
32. (a) Coleman RS, Grant EB (1994) J Am Chem Soc 116:8795–8796; (b) Coleman RS, Grant EB (1995) J Am Chem Soc 117:10889–10904
33. (a) Merlic CA, Aldrich CC, Albaneze-Walker J, Saghatelian A (2000) J Am Chem Soc 122:3224–3225; (b) Merlic CA, Aldrich CC, Albaneze-Walker J, Saghatelian A, Mammen J (2001) J Org Chem 66:1297–1309
34. Nasini G, Merlini L, Andreetti GD, Bocelli G, Sgarabotto P (1982) Tetrahedron 38:2787–2796
35. Arnone A, Merlini L, Mondelli R, Nasini G, Ragg E, Scaglioni L, Weiss U (1993) J Chem Soc Perkin Trans 2:1447–1454
36. Kishi T, Tahara S, Tsuda M, Tanaka C, Takahashi S (1991) Planta Med 57:376–379 (shiraiachrome A is called hypocrellin B here)
37. O'Brien EM, Morgan BJ, Mulrooney CA, Carroll PJ, Kozlowski MC (2010) J Org Chem 75:57–68
38. Li X, Hewgley JB, Mulrooney C, Yang J, Kozlowski MC (2003) J Org Chem 68:5500–5511
39. (a) Mulrooney CA, Li X, DiVirgilio ES, Kozlowski MC (2003) J Am Chem Soc 125:6856–6857; (b) Mulrooney CA, Morgan BJ, Li X, Kozlowski MC (2010) J Org Chem 75:16–29
40. Kita Y, Tohma H, Hatanaka K, Takada T, Fujita S, Mitoh S, Sajurai H, Oka S (1994) J Am Chem Soc 116:3684–3691
41. (a) Olah GA, Laali K, Mehrotra AK (1983) J Org Chem 48:3360–3362; (b) Cohen T, Schambach RA (1970) J Am Chem Soc 92:3189–3190; (c) Barton DHR, Lacher B, Zard SZ (1985) Tetrahedron Lett 26:5939–5942
42. Dickstein JS, Mulrooney CA, O'Brien EM, Morgan BJ, Kozlowski MC (2007) Org Lett 9:2441–2444
43. O'Brien EM, Morgan BJ, Kozlowski MC (2008) Angew Chem Int Ed 47:6877–6880
44. Morgan BJ, Mulrooney CA, O'Brien EM, Kozlowski MC (2010) J Org Chem 75:30–43
45. Morgan BJ, Dey S, Johnson SW, Kozlowski MC (2009) J Am Chem Soc 131:9413–9425
46. Reviews: (a) Muci AR, Buchwald SL (2002) Top Curr Chem 219:131–209; (b) Hartwig JF (2008) Nature 455:314–322; (c) Carril M, SanMartin R, Dominguez E (2008) Chem Soc Rev 37:639–647
47. (a) Torraca KE, Huang X, Parrish CA, Buchwald SL (2001) J Am Chem Soc 123:10770–10771; (b) Vorogushi AV, Huang X, Buchwald SL (2005) J Am Chem Soc 127:8146–8149; (c) Anderson KW, Ikawa T, Tundel RE, Buchwald SL (2006) J Am Chem Soc 128:10694–10695
48. (a) Garner P, Anderson JT, Dey S (1998) J Org Chem 63:5732–5733; (b) Chimiak A, Przychodzen W, Rachon J (2002) Heteroatom Chem 13:169–194

49. (a) Ohno K, Tsuji J (1968) J Am Chem Soc 90:99–107; (b) Abu-Hasanayn F, Coldman ME, Goldman AS (1992) J Am Chem Soc 114:2520–2524; (c) Kreis M, Palmelund A, Bunch L, Madsena R (2006) Adv Synth Catal 348:2148–2154
50. Morgan BJ, Mulrooney CA, Kozlowski MC (2010) J Org Chem 75:44–56
51. Kobayashi E, Ando K, Nakano H, Iida T, Ohno H, Morimoto M, Tamaoki T (1989) J Antibiot 42:1470–1474
52. (a) Myers AG, Tanaka D, Mannion MR (2002) J Am Chem Soc 124:11250–11251; (b) Tanaka D, Romeril SP, Myers AG (2005) J Am Chem Soc 127:10323–10333

Chapter 8
Phomactin A

Yu Tang, Kevin P. Cole, and Richard P. Hsung

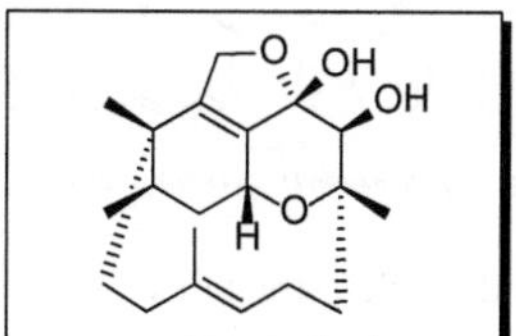

8.1 Introduction

8.1.1 Isolation

Marine organisms have provided a highly diversified natural product structures with a wide range of biological properties [1], leading to intense isolation, synthesis, and medicinal chemistry efforts. In particular, marine fungi have received much recent interest due to the radically different environment from which they have adapted, relative to their terrestrial counterparts. To identify new platelet-activating factor (PAF) antagonists from marine sources, Sugano et al. reported the structure of (+)-phomactin A in 1991 (Fig. 8.1) [2]. (+)-Phomactin A was extracted and purified from the culture filtrate of a marine fungus, *Phoma* sp. (SANK 11486), which is a parasite collected off crab shells, *Chionoecetes opilio*, found off the coast of the Fukui Prefecture in Japan. It was determined to have a moderate PAF aggregation inhibitory ability (IC_{50} = 10 mM).

Y. Tang (✉) • K.P. Cole • R.P. Hsung
School of Pharmacy and Department of Chemistry, University of Wisconsin – Madison, Madison, WI 53706, USA

J.J. Li and E.J. Corey (eds.), *Total Synthesis of Natural Products*,
DOI 10.1007/978-3-642-34065-9_8, © Springer-Verlag Berlin Heidelberg 2012

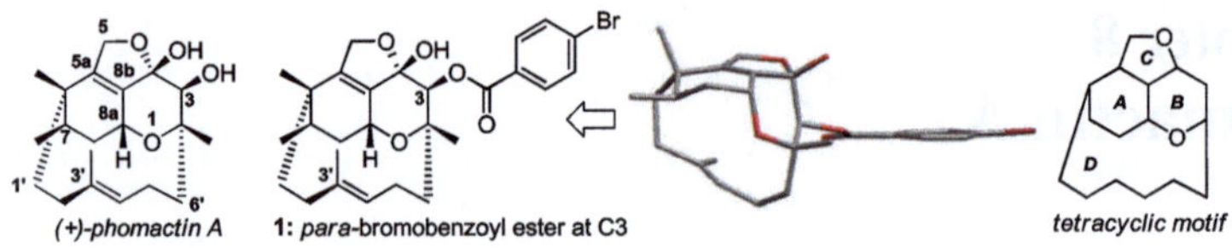

Fig. 8.1 (+)-Phomactin A and its C3-*para*-bromobenzoyl ester derivative

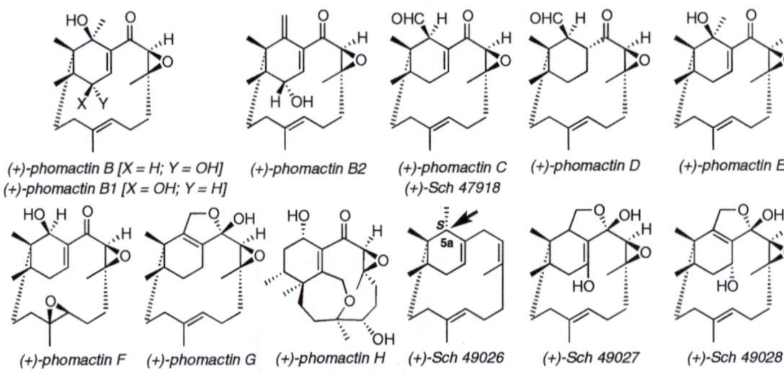

Fig. 8.2 The phomactin family and the Sch structures

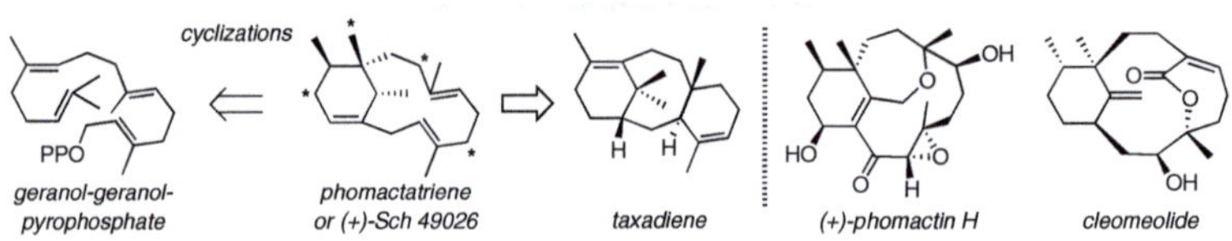

Fig. 8.3 A biosynthetic sketch

The structure of (+)-phomactin A was determined using both NMR and crystallographic methods (Fig. 8.1). Although the crystal structure is of low quality, it clearly revealed the unusual ABCD-tetracyclic topology as well as the absolute stereochemistry. Subsequently, nine additional phomactins were isolated from various fungal sources with many of them displaying anti-PAF activity [3–5]: B [3], B1 [3], B2 [3], C [3] (or Sch 47918 [6]), D [3], E [4], F [4], G [4], and finally, H [5] (Fig. 8.2).

Chu et al. at Schering-Plough independently reported the isolation of Sch 47918 in 1992, which is identical to (+)-phomactin C, although in the opposite antipode [6]. In 1993, Chu et al. also reported structures of (+)-Sch 49026, 49027, and 49028 [7]. However, based on Pattenden's work [8], Sch 49028 was misassigned, and it should actually be phomactin A! In addition, when Oikawa [9] isolated phomactatriene (see Fig. 8.3), a biosynthetic precursor to all phomactins, the C5a stereochemistry of (+)-Sch 49026 was corrected as (*S*) and not (*R*). The entire family possesses an unusual and highly oxygenated diterpenoid molecular architecture, encompassing a novel bicyclo[9.3.1]pentadecane core as punctuated in (+)-Sch 49026.

8.1.2 Biosynthesis

By using an elegant ^{13}C-labeling study that involved incubation of Phoma sp. with 1-^{13}C and 1,2-^{13}C acetate, Oikawa et al. [9]. were able to isolate the proposed biosynthetic intermediate phomactatriene (or Sch 49026), with ^{13}C incorporation from singly labeled acetate units as indicated by "*" in Fig. 8.3. Phomactatriene is strikingly reminiscent of taxadiene, a biosynthetic intermediate for Taxol^{TM}. The net biosynthesis for both involves geranylgeranyl diphosphate (GGDP) cyclization [9]. It is noteworthy that prior to isolation of phomactins, the only known related structure is cleomeolide, a diterpene from the herb *Cleome viscosa* [10] that remarkably resembles phomactin H.

8.1.3 Medicinal Chemistry

PAF is a proinflammatory phospholipid mediator found in many different cell types in the human body, referring to a class of compounds with the general structure of 1-*O*-alkyl/acyl-2-acetylglycero-3-phosphocholine. PAF mediates an incredibly diverse array of biological processes including wound healing, inflammation, bronchoconstriction, vascular permeability, apoptosis, angiogenesis, reproduction, and long-term potentiation [11]. PAF is also implicated in many inflammatory, respiratory-related problems, including asthma and pulmonary edema, and cardiovascular diseases [12–14]. The phomactins are able to inhibit the binding of PAF to its receptors and hence could potentially treat some of the disorders associated with excessive PAF accumulation [2–5]. Although the exact mechanism is unknown, it is known that the mode of action is unique relative to other known inhibitors, thereby representing a new class of PAF inhibitors [2]. Only limited structure activity relationships have been reported by Sugano et al. focusing on derivatization of phomactin D, the most potent member of the family in inhibiting PAF aggregation ($IC_{50} = 0.80$ mM) [12], and by Rawal et al. on some simple phomactin analogs [15].

8.1.4 Synthetic Challenges

Phomactin A is the most challenging family member architecturally. The fragments that are most challenging are highlighted in Fig. 8.4. In Box-A, the highly sensitive hydrated furan is prone to dehydration under acidic or basic conditions, and any total synthesis almost certainly must save introduction of this fragment until the end game. Box-B relates to the strained and somewhat twisted electron-rich double bond. This trisubstituted olefin is extremely reactive toward electrophilic oxidants.

Additionally, we believe that the olefin is not free to rotate and rotamer issues may arise in an attempted synthesis. In Box-C, the cis-methyl groups at C6 and C7

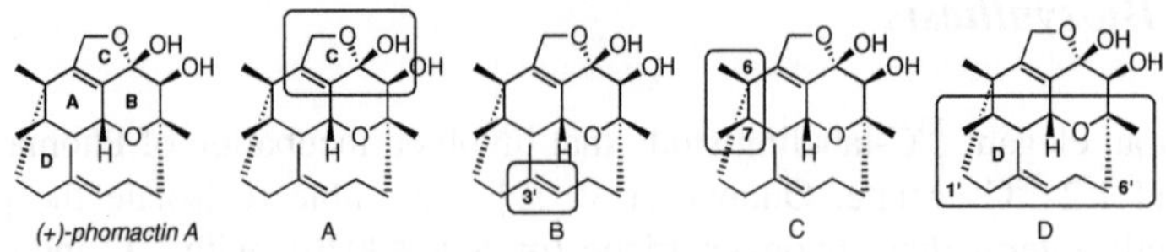

Fig. 8.4 Challenging features in (+)-phomactin A

are highlighted with C6 being the extremely hindered quaternary center. This center must be installed at an early time and will hinder attempts to functionalize centers nearby. It is noteworthy that many conformations of phomactin A will always have one methyl group assuming an axial position, severely hindering any reagent approach to the A-ring from the top face, while the belt (we will refer to the C1′–C6′ portion as the belt) effectively blocks the bottom face. Lastly, Box-D showcases the unusual 12-membered (D-ring) bridged/fused ring system, which only makes the system more challenging.

In nearly two decades, challenging structured of phomactins coupled with the interesting biological activity has elicited an impressive amount of synthetic efforts [16–18]. (+)-Phomactin D was first synthesized by Yamada in 1996 [19], and Wulff [20] reported the synthesis of (±)-phomactin B2 in 2007. However, (+)-phomactin A has been the most popular target because of its unique topology. To date, two monumental total syntheses have been accomplished: Pattenden's [21] racemic synthesis and Halcomb's [22] asymmetric synthesis in 2002 and 2003, respectively. Both syntheses are beautifully done but also mimicked Yamada's synthesis of D, thereby underscoring the remarkable influence of Yamada's earlier work on the phomactin chemistry. Upon completing their synthesis of (±)-phomactin A, Pattenden [23] completed (±)-phomactin G via a modified route used for A.

8.2 The Architecturally Distinctive ABD-Tricycle

8.2.1 Retrosynthetic Analysis

We commenced our own approach toward (±)-phomactin A with an intent to feature an *oxa*-[3 + 3] annulation strategy [24–28] that was developed in our lab [29–32] and, in particular, an intramolecular annulation. While *oxa*-[3 + 3] annulations or related reaction manifolds [33, 34] are known and can be traced back more than six decades [35], an intramolecular variants of this reaction were not known [24–28]. There were no applications of intramolecular oxa-[3 + 3] annulations in natural product synthesis [27, 28, 36] until our approach toward phomactin A was disclosed. Subsequently, an account toward (±)-likonide B was reported by Trauner [36e].

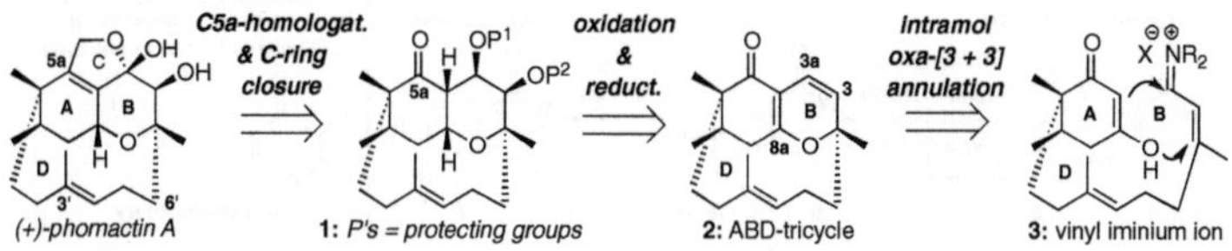

Fig. 8.5 Our plan to phomactin A

More specifically as shown in Fig. 8.5, we envisioned that phomactin A could be elaborated through C5a-homologation and C-ring closure of **1**, which could be attained via a sequence of oxidative and reductive transformations of **2**. From the onset, we recognized the unique structural topology of **2**. A Spartan^{TM} model reveals a highly strained caged motif with the plane of the AB-ring being in close proximity with the D-ring olefin at C3′. Our retrosynthetic analysis appeared to be reasonable except the C5a-homologation, which would involve a quite hindered C5a-carbonyl group. Whatever concerns we had regarding the end game did not deter us from pursuing phomactin A because ABD-tricycle **2** may be readily constructed via an intramolecular *oxa*-[3 + 3] annulation of vinyl iminium ion **3**. However, we did not anticipate the difficulties we would encounter with the regioselectivity of the annulation.

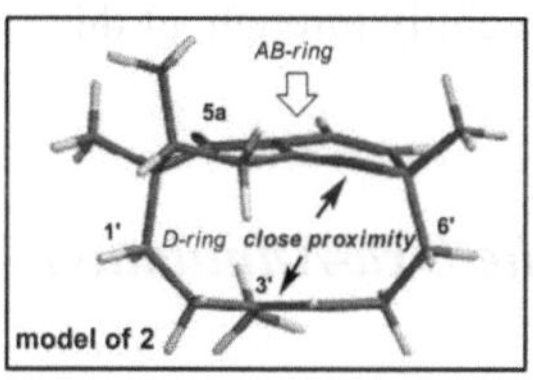

Mechanistically, the intermolecular *oxa*-[3 + 3] annulation of diketone **4** [37] with α,β-unsaturated iminium ion **5** involves a tandem cascade [38] of Knoevenagel condensation–6π-electron electrocyclic ring-closure of 1-oxatrienes **6** [39–41] to give pyranyl heterocycles **7** (Fig. 8.6) [42]. This tandem sequence can be considered as a stepwise formal [3 + 3] cycloaddition [43–45] or annulation in which two σ bonds are formed along with a new stereogenic center adjacent to the heteroatom. It reveals the synthetic power of oxa-6π pericyclic ring-closures that could find its precedent six decades ago [35] while representing a biomimetic strategy in natural product synthesis [36, 40]. The challenge in this and other related annulations is the regiochemical control [24–28], head-to-head versus head-to-tail (see the box), which can be unpredictable and lead to complex mixtures [46]. However, the use of α,β-unsaturated iminium salts in our annulations has led to head-to-head regioselectivity almost exclusively [24–28]. The intramolecular version of this annulation could entail a very different sequence consisting of *O*-1,4-addition followed by *C*-1,2-addition/β-elimination to give the desired ABD-tricycle **2**.

In our early designs, we indeed ignored the perfectly plausible regiochemical orientation shown as **3B**, which could lead to a disaster. Furthermore, our tunnel vision was enhanced by the fact that all the synthetic approaches reported at the

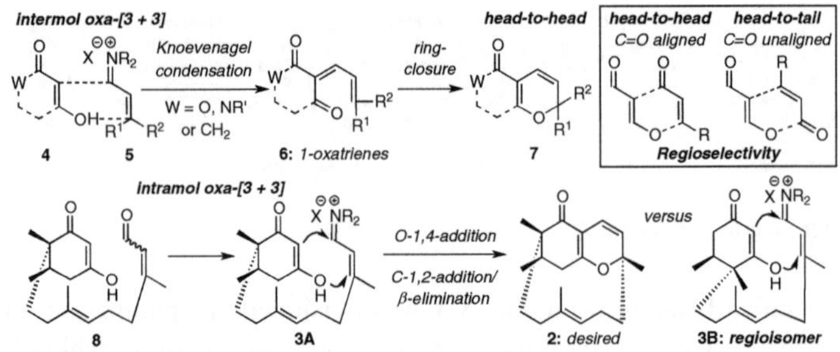

Fig. 8.6 Regiochemical issues in our synthetic approach

time [16, 17], the macrocyclic D-ring, was constructed: (1) through ring-closing metathesis (RCM), sulfone alkylation, acetylenic addition, Julia olefination, and Nozaki–Hiyama–Kishi (NHK), Stille, or Suzuki cross coupling, which is different from our strategy; and (2) at late stages of the synthesis. We thus had a unique approach to the D-ring that would be assembled at an early stage of the synthesis, thereby further accentuating the synthetic prowess of the *oxa*-[3 + 3] annulation strategy. Consequently, it was without recognizing such a possible disaster and relevant calculations of **3A** versus **3B**, we commenced on our long journey and found ourselves struggling mightily throughout this endeavor [47–50].

8.2.2 Approaches to the Oxa-Annulation Precursor

The first approach to *oxa*-[3 + 3] annulation precursor **8** would rely on late-stage oxidation of the cyclohexanone motif in **10** to the diketone, and **10** can be accessed through a sequence of 1,4-conjugate addition of a Me-nucleophile followed by trapping with a homoallyl electrophile **9** to set the A-ring stereochemistry (Scheme 8.1). Unfortunately, while the addition of Me_2CuLi proceeded smoothly, no alkylation using iodide **9** was observed; instead, only a mixture of 2,3-dimethylcyclohexanones and recovered iodide were isolated.

Fortunately, we were able to rapidly revamp the synthesis and envisioned a clever solution by using vinylogous ester **12** (Scheme 8.2). Vinylogous ester **12** would solve the regioselective deprotonation problem, as there is only one α-proton. We also hoped that the existing stereocenter would strongly influence the diastereoselectivity of the reaction. At the time, we were unaware of Pattenden's success with similar systems. Vinylogous ester **12** could be readily attained from commercially available 5-methyl-1,3-cyclohexanedione. However, we quickly realized that the reaction suffered from major problems. First, the ratio of diastereomers **13** was *always* 2.7:1 though in favor of the desired isomer, and no dramatic change in this ratio was observed under any optimizations. Secondly, these isomers were separable only partially after repeated MPLC, and the mixture in the ensuing steps did not result separable samples. Most disconcertingly, the

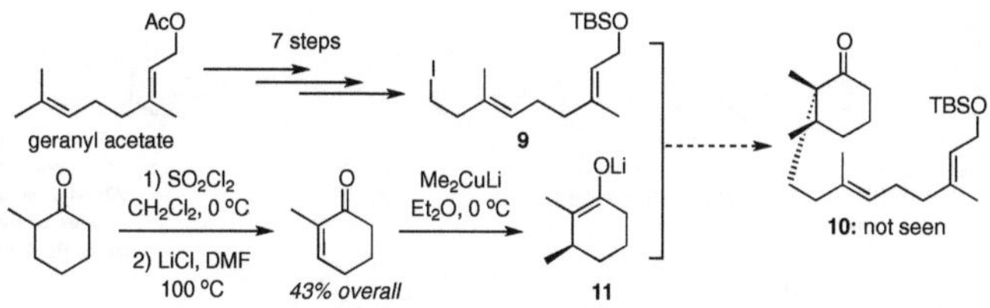

Scheme 8.1 Failed 1,4-addition and trapping with iodide **9**

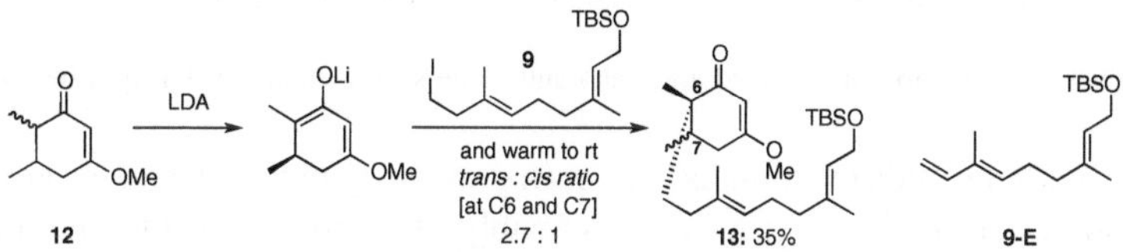

Scheme 8.2 An alternative: α-alkylations with iodide **9**

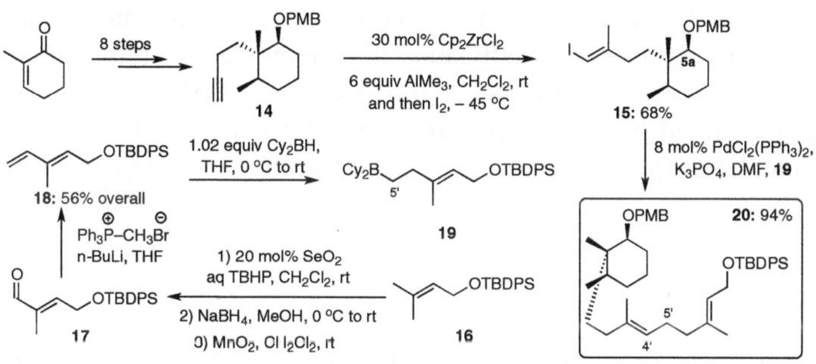

Scheme 8.3 Suzuki–Miyaura cross coupling

yield was seldom greater than 30 %. Frequently, no alkylation took place, and in many cases, alkene **9-E** was seen resulting from competing E2-elimination of **9**.

The third and the most operative route at the time involved a key $PdCl_2(PPh_3)_2$ catalyzed *B*-alkyl Suzuki–Miyaura coupling of alkyl borane **19** and iodide **15**, leading to the C4′–C5′ cross-coupled alkene **20** in high yield as summarized in Scheme 8.3. With alkene **20** in hand, we were ready to pursue the formation of the 1,3-diketone motif in the A-ring, which turned out to be significantly more challenging than it appeared on paper [47b].

As shown in Scheme 8.4, when using LDA/PhSeBr followed by exposing to H_2O_2, enone **21** was attained and could be epoxidized to epoxy ketone **22** as a single diastereomer. The stereoselectivity can be explained by the exclusive axial attack of the hydroperoxide anion. However, treatment of this epoxide intermediate with catalytic $Pd(PPh_3)_4$ gave only traces of the desired diketone **23**. The reaction provided mostly recoved starting **22** along with β-hydroxy ketone resulting from ring-opening of the epoxide, which was inseparable from diketone **23**. To the best

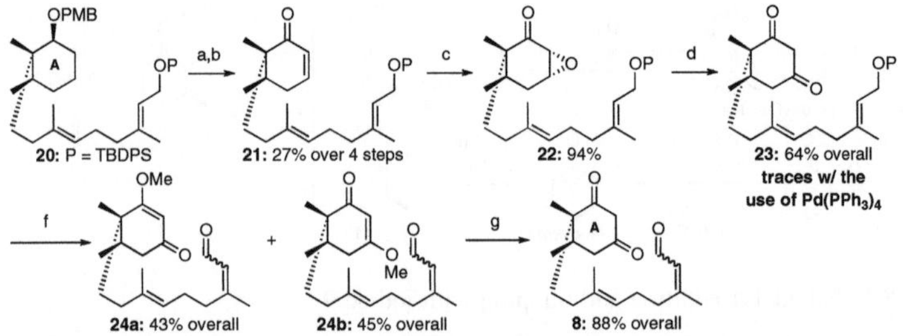

a) 1. DDQ, CH_2Cl_2, H_2O, 2. TPAP, NMO; b) 1. LDA, HMPA, – 78°C, PhSeBr, 2. ; H_2O_2, THF, 0 °C; c) K_2CO_3, H_2O_2; d) 1. SmI_2, THF/MeOH, – 90°C, 2. PCC; e) 1. K_2CO_3, Me_2SO_4, reflux, 2. TBAF, rt, 3. DMP [O]; h) 3 *N* HCl; acetone, rt.

Scheme 8.4 Completion of the third approach and some struggles in the A-ring transformation

of our knowledge, this reaction has not been previously reported, and we are unsure of the source of the hydrogen, but presumably, it is from phosphine reduction of water that was not excluded. PCC oxidation of a small amount of this β-hydroxy ketone worked reasonably well to afford the desired diketone **23**. Thus, SmI_2 was used to directly access β-hydroxy ketone with the ensuing PCC oxidation to give diketone **23**. Lastly, desilylation turned out to be problematic due to the diketone acidity. To resolve this issue, methylation of diketone **23** was required, and exposing these regioisomeric methyl vinylogous esters to TBAF successfully afforded the corresponding alcohols, which were oxidized to enals **24a/b**. Upon hydrolysis, the *oxa*-annulation precursor **8** was at last attained in 22 steps from 2-methylcyclohexanone.

8.2.3 An Improved Synthesis of Oxa-Annulation Precursor

We recognized that the A-ring synthesis should incorporate as much oxygenation as possible to avoid the difficult process of installing it at a later time. In addition, we felt that it would be best to form the diketone moiety in an already protected fashion and unmask it at the end. Consequently, our fourth-generation approach to the *oxa*-[3 + 3] annulation precursor **8** began with the use of a Diels–Alder cycloaddition to form the A-ring with the added possibility of achieving an asymmetric manifold.

With Rawal's diene, the Diels–Alder cycloaddition with tiglic aldehyde occurred with remarkable ease at 50 °C in THF, and the reaction was observed to slowly occur even at temperature as low as –10 °C (Scheme 8.5). This high level of reactivity toward a rather unreactive and/or hindered dienophile is presumably due to the dimethylamino group being more electron donating (significantly raising its HOMO level) relative to Danishefsky's diene [51]. In this fashion, high yields of the desired amino cyclohexene **25** were obtained. This key reaction sets the stereochemistry and provides us with extensive A-ring oxygenation. It was also observed that **25** was reasonably stable to moisture and could be manipulated to a certain extent, but it was unstable to chromatography and underwent clean

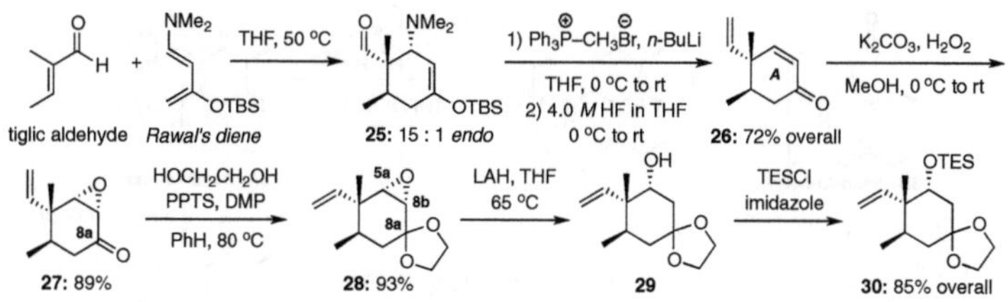

Scheme 8.5 Regiochemical issues in our synthetic approach

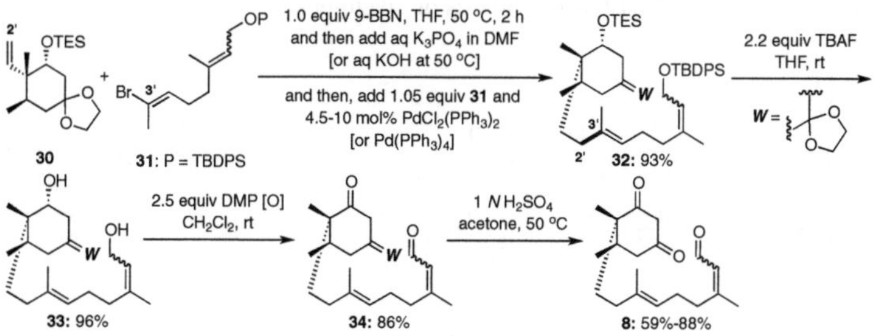

Scheme 8.6 Completion of a 12-step synthesis of **8**

elimination to give the enone under acidic conditions. We were pleased to find that Wittig olefination of the crude cycloadduct **25** proceeded with ease to give an alkene that was not isolated but exposed to HF to cleanly afford enone **26**. Standard reactions would lead to alkene **30** in anticipation of the Suzuki–Miyaura coupling.

Again, we turned our attention to the *B*-alkyl Suzuki–Miyaura coupling. As shown in Scheme 8.6, coupling of vinyl bromide **31** was successful with alkene **30**. Ultimately, TBAF removal of both silicon ethers in **32** and an ensuing double Dess–Martin periodinane oxidation gave enal **34**. Finally, treatment of **34** with 1 N H_2SO_4 in acetone at 50 °C removed the ketal moiety and afforded the desired diketo-enal **8** as a mixture of enal isomers. The hydrolysis reaction was slow at room temperature, and when heated, the reaction was usually stopped with some starting material still present, due to the formation of an unknown by-product. Overall, at 12 linear steps, this route to diketo-enal **8** was much shorter than the third-generation synthesis (22 steps) and was easily capable of supplying **8** in $\geq$10 g quantities.

8.2.4 *Key Oxa-Annulation and the D-Ring Atropisomerism*

We attempted the key intramolecular *oxa*-[3 + 3] annulation reaction of diketo-enal **8** under high-dilution conditions at room temperature using piperidinium acetate salt in THF. We were delighted upon the discovery of the formation of the desired

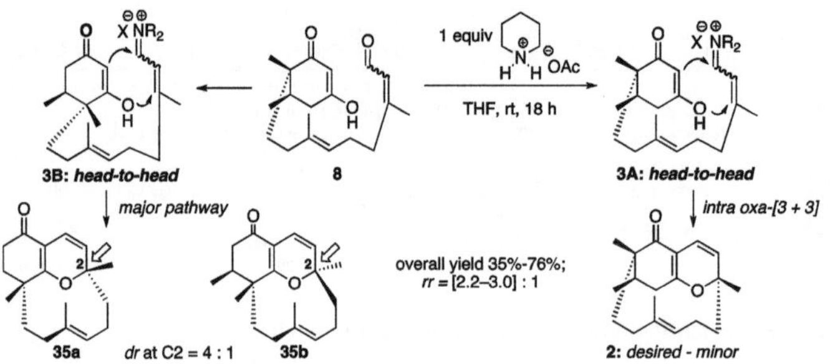

Scheme 8.7 Completion of a 12-step synthesis of diketo-enal **8**

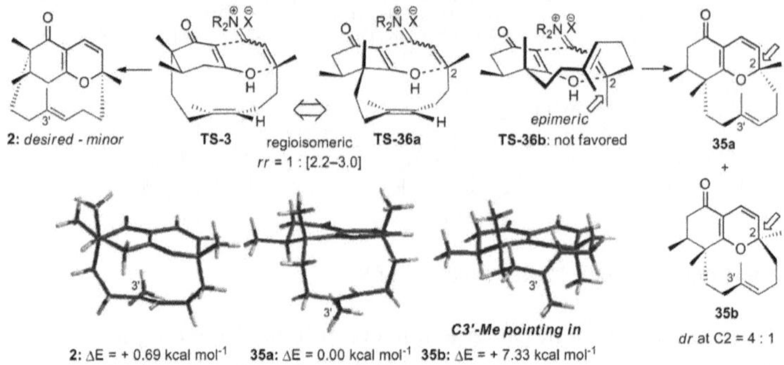

Fig. 8.7 Three TS for the key oxa-[3 + 3] annulation

annulation product **2**, along with regioisomeric products **35a** and **35b** arriving from annulation of **3B** as an inseparable 4:1 mixture (Scheme 8.7). The overall ratio for **2**: (**35a** + **35b**) ranged between 1:2.2 and 1:3.0, thereby implying that the desired ABD-tricycle in this annulation is actually the minor product.

Although the overall yield of the reaction was as good as 76 %, we were disappointed by the poor ratio of desired cycloadduct to the isomeric products and that the overall yield of the annulation varied with low end being 35 % [48]. The three annulation products, **2**, **35a**, and **35b** would presumably be derived from the three respective transition states shown in Fig. 8.7. However, when we attained relative energies of **2**, **35a**, and **35b** employing HF6-31G*/B3LYP6-31G* calculations using Spartan^{TM}, we found something disconcerting. That is, while the relative energetics appeared to be in the right order, the proposed minor diastereomer **35b** with the allylic (C3′) Me group pointing inward possesses a very high energy. This promoted us to examine very carefully the assignment of isomers **35a** and **35b**, as we recognize the significant implication of being correct in these assignments to our ultimate total synthesis efforts.

While the discovery of this novel D-ring atropisomerism reveals the unique and intricate architecture of these tricycles, it alerted to us the actual alignment of the C3′ allyl methyl group in the desired ABD-tricycle **2** and, more importantly, the

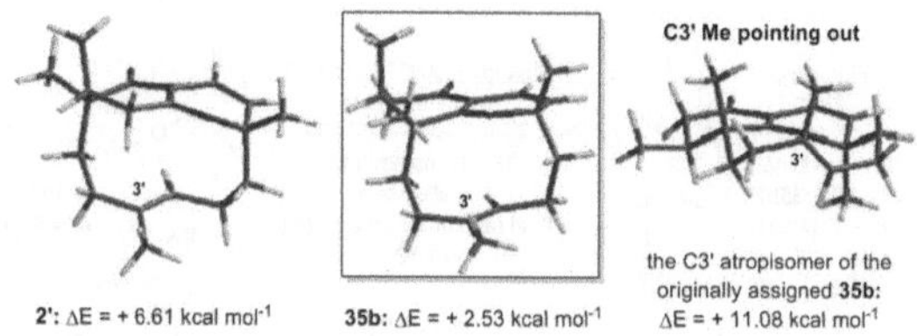

Fig. 8.8 Assignment of 2 and energetics of possible atropisomers

vigor of its assignment. Thus, we returned our attention to the desired ABD-tricycle **2** and the initial assignment of **2**, which was not a trivial matter. The desired ABD-tricycle **2** existed as a single belt olefin rotamer and was observed to have a large R_f difference (*less* polar) from the mixture of undesired **35a** and **35b**. This can be explained by the presence of the quaternary center adjacent to the most Lewis-basic site (the carbonyl) in the case of **2**. While no crystal structure was attained, interesting correlations using nOe [47b] on the single isomer **2** supported our structural assignment and its respective transition state **TS-3** for the annulation (Fig. 8.8). In addition, we also observed a long range through ^1H–^1H COSY interaction between the C3a vinyl proton and the C8 allylic protons, an interaction that is impossible in **35a** and **35b**, due to the position of the quaternary center.

Further HF6-31G*/B3LYP6-31G* calculations would suggest that the C3′ Me group of **2** is also pointing in as in the assignment of **35a** and **35b**. Atropisomer of **2** (see model **2′** in Fig. 8.8) is 6.61 kcal mol^{-1} less stable. Additionally, the C3′ atropisomer of the originally assigned **31b** possesses an even higher energy (11.08 kcal mol^{-1}!!); the actual **35b**, atropisomeric with **35a**, is only 2.53 kcal mol^{-1}. Consequently, we were pleased at the knowledge that we had obtained the three most stable isomers in the annulation according to our earlier calculations. Whether these other atropisomers were ever formed or were simply too costly or unstable to form at room temperature, we did not screen for their existence.

The amount of anhydride used did turn out to be very important to the regioselectivity. When lesser amounts of anhydride were used, the ratio of products **2**:**35a**:**35b** was noted to be 1.7:1:1 (the originally observed ratio). When the amount of anhydride was increased, this ratio could be as high as 5:4:1. Most critically, we uncovered that the undesired isomers could be equilibrated back to the originally observed mixture as shown in Scheme 8.8. Mechanistically, the equilibration likely proceeds through pericyclic ring-opening [41] of **35a** and **35b** to give 1-oxatriene **36B**, and the overall isomerization from **36B** to 1-oxatriene **36A** would require a sequence of 1,4-addition of piperidine and elimination. We now believe that this number represents a thermodynamic ratio, while the 5:4:1 ratio is kinetic. This discovery was immense in the efficient accumulation of annulation product **2**, which was needed for the major challenges awaiting.

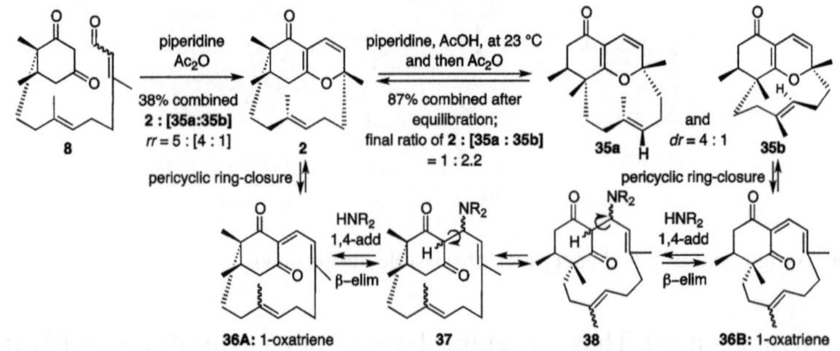

Scheme 8.8 An attempt to improve the regioisomeric ratio and a successful equilibration

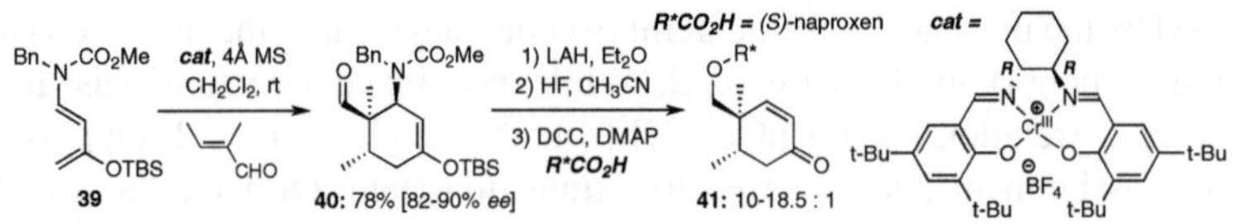

Scheme 8.9 An asymmetric Diels–Alder cycloaddition

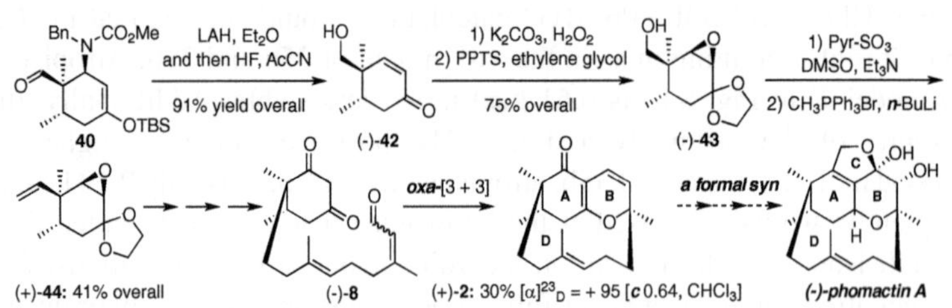

Scheme 8.10 A formal asymmetric synthesis of (–)-phomactin A

8.2.5 A Formal Synthesis of (–)-Phomactin A

An added bonus of our new route to diketo-enal **8** was that the A-ring could be constructed in an enantioselectively manner via Rawal's asymmetric Diels–Alder route [51b, 52, 53]. As shown in Scheme 8.9, after much experimentation [49], using diene **39** and the (*R*,*R*)-Cr(III)-salen catalyst [54] with BF_4 as counter anion, we obtained **36** in 78 % yield with *ee* up to 90 %, as determined by ^{1}H NMR analysis of the (*S*)-naproxen ester **41**. Rawal [52] had reported a related cycloaddition of tiglic aldehyde that gave 62 % yield with 93 % *ee* but with SbF_6 as counter anion. As summarized in Scheme 8.10, the optically enriched diketo-enal (–) **8** could be generated and subjected to the optimized annulation conditions to give ABD-tricycle (+)-**2** in an overall 13 steps from diene **39**, thereby constituting a formal synthesis of (–)-phomactin A.

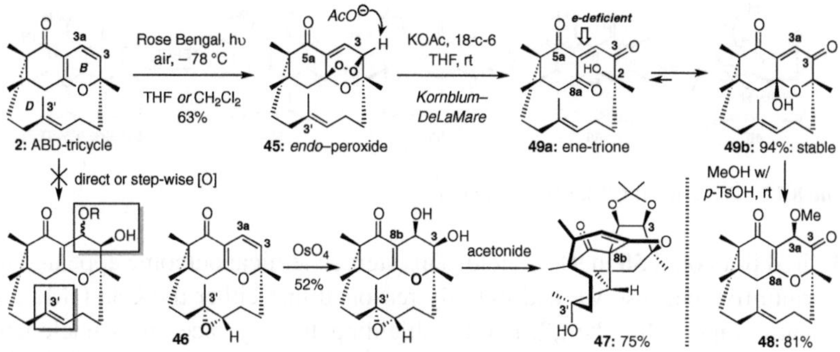

Scheme 8.11 Endoperoxide ring opening and C3a oxygenation

8.3 Lessons Learned from the Challenging Structural Topology

8.3.1 Oxidations of C3 and C3a in B-Ring

The hardest part was over—so we thought. All we need are four maybe at most five steps for oxidation of B-ring, reduction of AB-ring junction, and homologation and C-ring formation!

We began our journey by examining oxidation of the C3-3a olefin in B-ring, which would require selectivity over the C3′-olefin in D-ring. Unfortunately, after months of struggle, no direct or stepwise oxidative process would be selective for the C3-3a olefin over the D-ring olefin. Using buffered peracetic acid or *m*-CPBA would give epoxide **46a** instead (Scheme 8.11). An ensuing OsO_4 dihydroxylation followed by standard acetonide protection led to a complete skeleton rearrangement to give **47**! Fortuitously and remarkably, upon irradiation with a 300-W lamp of a THF solution of **2** at −78 °C containing Rose Bengal as sensitizer and an air bubbler, the formation of endoperoxide **45** was observed with no sever competing ene or [2 + 2] pathway.

In another interesting turn of event, while we were attempting to add acetate anion in a 1,4-fashion into the enone system of **45**, we isolated ene-dione **49b** when using KOAc in the presence of 18-crown-6 (Scheme 8.11). While this was not what we had anticipated, it turned out to be a known protocol developed by Kornblum and DeLaMare to ring-open endoperoxides through a deprotonation pathway [55]. It is noteworthy that ene-dione **49b** did not equilibrate to the ene-trione **49b** which could suffer from the aforementioned HDA pathway as we had also suffered from [47c].

Although a major drawback to this reaction was the destruction of the C3 stereochemistry, it represented the first successful attempt to open the peroxide in a productive manner. To regain the C3 stereocenter, we treated vinylogous ester **48** (from acid-promoted solvolysis of **49b**) $NaBH_4$ but only to find alcohols were isolated in a 4:1 ratio (Scheme 8.12). Fortunately, L-Selectride® gave exclusively

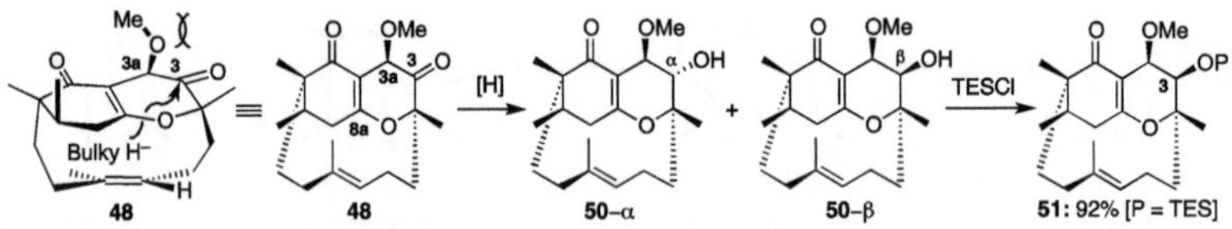

Scheme 8.12 The critical C3-ketone reduction

the desired β-alcohol **50** in 91 % yield. This stereochemical outcome appeared to be counterintuitive because we had initially reasoned that either the C3a-OMe group could out-compete the "belt," thereby shielding the top face from nucleophiles more effectively and leading to the hydride delivery, especially a small hydride, from the bottom face. However, in hindsight, by examining the Spartan™ model, it would appear that the C3a-OMe group is *pseudo*-axially oriented. Small hydrides could get by with relative ease from the top face, but with a more bulky hydride, the C3a-OMe group was able to completely prevent the reagent approaching from the top face. This success completes the task of oxidizing C3 and C3a, and more importantly, the C3 stereochemistry was reset despite losing it in the Kornblum–DeLaMare process.

8.3.2 Reduction of C8a and C8b at the AB-Ring Junction

To demethylate, we exposed **51** to BBr_3 and obtained the desired alcohol **52** (Scheme 8.13). This success is highly remarkable because (1) the TES ether could just as easily had been cleaved and (2) the C3a-OMe group could also be cleaved to form an oxocarbenium ion through arrow pushing from the B-ring oxygen atom. With the free alcohol **52** in hand, we were set to investigate its oxidation to diketone **53**. We were initially apprehensive because of the possibility of forming the oxocarbenium species upon initial activation of the allylic alcohol by an oxidizing species. Swern oxidation failed presumably due to generation of "the unwanted" oxocarbenium species through a sulfonium intermediate such as **54**. While TPAP and PCC were slow and messy, Dess–Martin periodinane rapidly and quantitatively provided access to diketone **53**.

Literature search showed the dihydro-chromandione motif in **53**, or other related system, is relatively unknown and/or under-investigated, and the proposed conjugate reduction to this system was unprecedented. Fortuitously, our attempt employed $NaBH_4$ in MeOH at 0 °C, and while a new spot was rapidly produced, the conversion was slow. When the solution was warmed to ambient temperature and additional aliquots of $NaBH_4$ were added, a mixture of diketone **55** and hydroxy ketone **56** was isolated. In subsequent trials, **55** could be reduced cleanly to **56** especially at elevated temperatures.

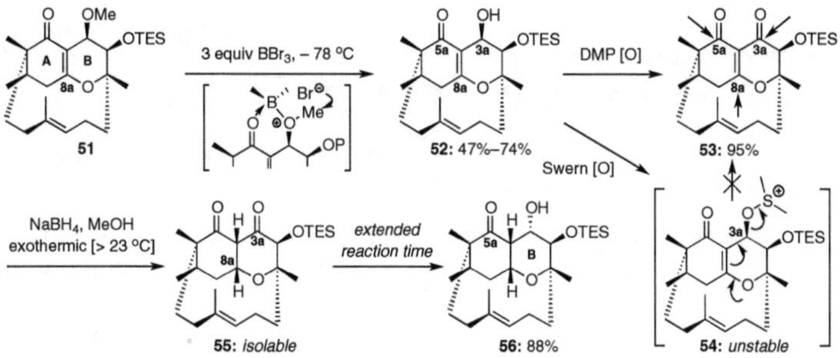

Scheme 8.13 Successful demethylation and oxidation

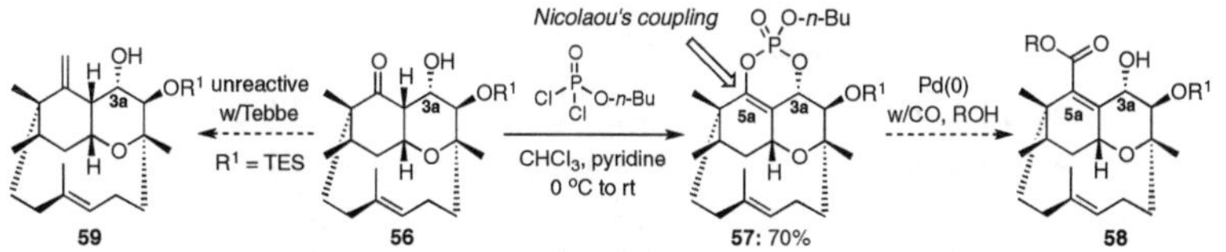

Scheme 8.14 Unsuccessful direct C5a-homologation

8.3.3 Homologation at C5a in the A-Ring

8.3.3.1 Initial Failures

The third task is to install the 20th and final carbon via homologation C5a. Given the resistance of the C5a-carbonyl to attack by hydride, smallest conceivable nucleophile, during the formation of **56**, we recognized there might be severe difficulties in the need of adding a larger carbon nucleophile. The entire synthetic plan would be futile if our efforts were in vain due to the steric hindrance surrounding C5a.

Toward C5a-homologation, after failing in Tebbe's olefination, we explored idea of cyclic enol-phosphate **57** inspired by Nicolaou's [56] cross-coupling chemistry (Scheme 8.14). The desired enol-phosphate **57** was obtained from hydroxyl ketone **56** as a single diastereomer, although the phosphorus stereochemistry was not determined. However, exposure of the enol-phosphate **57** to Pd(0) under a CO atmosphere did not result in carbonylation. As suggested in a conversation with Professor Larry Overman, it is possible that the phosphate is simply too hindered to undergo oxidative insertion by palladium, and thus, we abandoned this approach and elected to investigate the use of small one-carbon nucleophiles.

Unfortunately, upon treating hydroxy ketone **56** with MeLi for extended reaction times, some sort of fragmentation reaction appeared to occur (Scheme 8.15). While the desired addition product **63** was not seen, the fragmentation that took place may be attributed to a retro-aldol reaction through intermediates such as **60** and **61**. The tentatively assigned structure of final product **62** is shown, although it was not fully

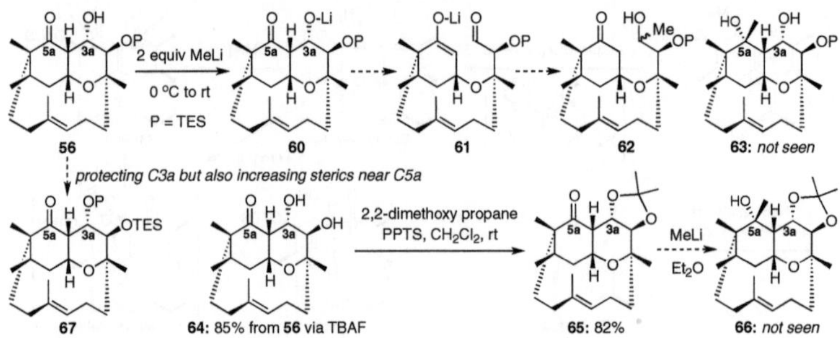

Scheme 8.15 Unsuccessful attempts at MeLi addition

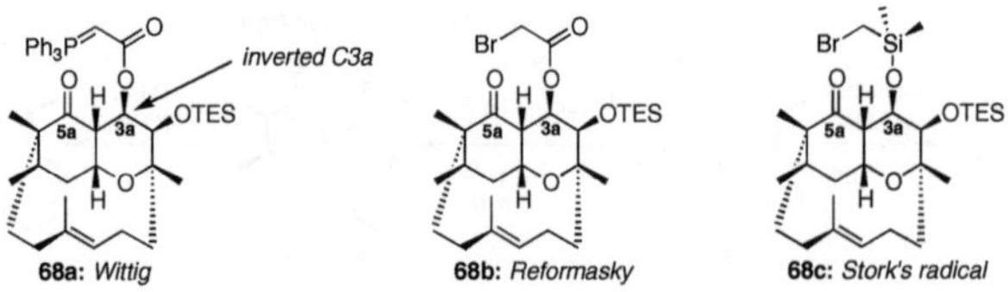

Scheme 8.16 Thoughts on possible intramolecular delivery

characterized. We then considered silylation of C3a-OH but that would lead to additional steric congestion near C5. Thus, we elected to pursue acetonide **65** via removal of the TES group. Unfortunately again, exposure of acetonide **65** to MeLi did not afford the desired addition product. In these studies, MeLi has become the "gold standard" to judge whether or not a compound was reactive toward a nucleophile.

While these failures signify that our approach to phomactin A might be doomed, they provoked us to think deeply how we could retool our plans. Among many thoughts, interesting designs such as an intramolecular delivery using systems such as **68a–c** (Scheme 8.16) were recognized. None was enticing enough to warrant any actual experimental efforts, but the thought exercise led us to recognize potential conformation issues with respect to this unique structural topology. This turn of event began with a critical NMR observation because many exhibit unusual broadenings in NMR likely due to conformational exchange. Thus, there could exist a more reactive conformation, allowing access to the C5a-carbonyl.

8.3.3.2 A Meticulous and Critical NMR Observation

When diketone **55** was reduced to hydroxyl ketone **56**, an interesting change occurred in the ^{1}H NMR that gave us significant insight into for our proposed conformations of these systems. A selected region of the ^{1}H NMR for compounds **2** and **56** is shown in Fig. 8.9.

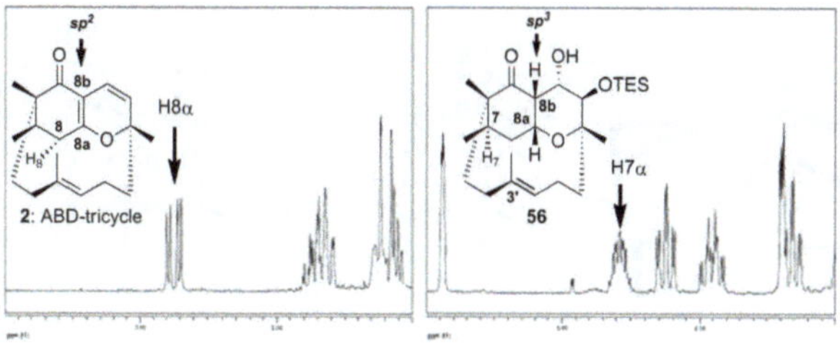

Fig. 8.9 ^{1}H NMR of ABD-tricycle **2** and hydroxy ketone **56**

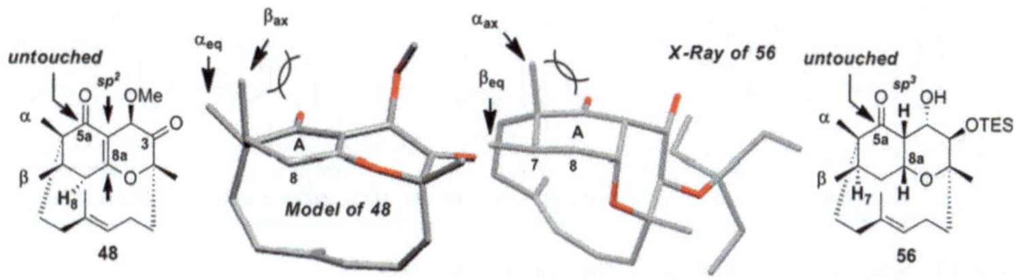

Fig. 8.10 Conformation analysis: neither could be homologated

The resonance at δ 2.90 ppm in the NMR spectrum of ABD-tricycle **2** was assigned as the H8α proton (α denoting proton being down), noting the rather highly deshielded nature of its chemical shift. At the time, we believed that this proton was *pseudo*-axial because in addition to deshielding from the α,β-unsaturated carbonyl, it could experience anisotropic deshielding from the belt olefin at C3′. We observed that when the C8a carbon is no longer sp^2-hybridized but sp^3-hybridized as in **56**, H8α no longer experiences this dramatic deshielding effect. Instead, the H7α methyne proton is now likely *pseudo*-axial and experiences anisotropic effect from the C3′-olefin, thereby shifting its NMR signal significantly downfield. This is very distinct in hydroxy ketone **56** with H7a at 2.8 ppm and in the natural product as well as all compounds where C8a is sp^3-hybridized including endoperoxide **45**. These unique and meticulous observations led us to examine the A-ring conformation more carefully because we felt that the A-ring conformation likely plays a crucial role in the reactivity of the C5a-carbonyl toward nucleophiles.

8.3.3.3 A Strategic Conformational Analysis

We first examined ketone **48** and hydroxy ketone **56** because (1) no reduction of C5a-carbonyl in **48** occurred when using L-Selectride™ to reduce the C3-ketone and (2) $NaBH_4$ did not touch the C5a-carbonyl group in hydroxy ketone **56**. By using minimized Spartan™ model of vinylogous ester **48** and the X-ray structure of hydroxy ketone **56**, we found unique conformational elements (Fig. 8.10) in addition to fully validating the aforementioned NMR analysis of possible positions for H8α when C8a is sp^2-hybridzied and H7 when C8a is sp^3-hybridized. In **48**,

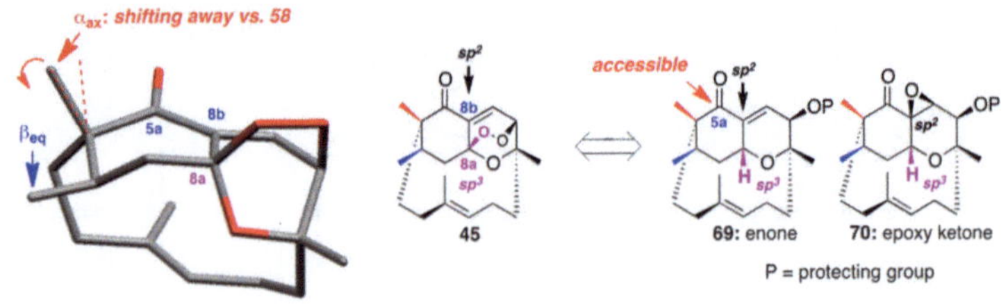

Fig. 8.11 Rationale for choosing **65** and **66**

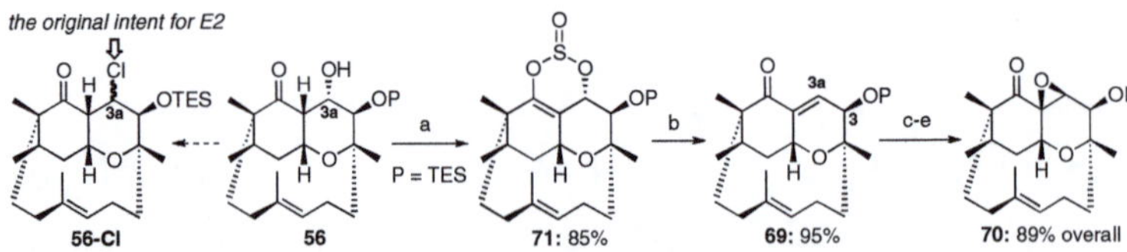

a) $SOCl_2$, pyridine, 0 °C; b) Tol, reflux; c) *p*-TsOH, MeOH, 50 °C; d) K_2CO_3, H_2O_2, MeOH, rt; e) TESCl, imid, CH_2Cl_2, rt.

Scheme 8.17 Serendipitous success in the enone synthesis

α-Me in the A-ring is *pseudo*-equatorial with β-Me being *pseudo*-axial, thereby blocking any incoming nucleophiles toward the C5a-carbonyl group.

We were mesmerized by the X-ray structure of hydroxy ketone **56** with its AB-ring junction being both sp^3-hybridized, revealing a very different conformation from that of **48**. The structure showed near-perfect chairs for the *cis*-fused 1-*oxa*-decalinic. With this conformational preference, the β-Me group is *pseudo*-equatorial with the α-Me group now being *pseudo*-axial position. We believe this new α-Me orientation is responsible for the fact that the C5a-carbonyl is not reduced by $NaBH_4$: it is sterically inaccessible, even by a nucleophile as small as a hydride.

In contrast, the AB-ring junction of endoperoxide **45** has respective sp^3- and sp^2-hybridizations at C8a and C8b (Fig. 8.11), which forces the A-ring into a twisted boat and bend both Me groups back toward *pseudo*-equatorial positions. This A-ring conformation is very similar to that of phomactin A itself (see Fig. 8.1). *Most critically here, while the β-Me group remains pseudo-equatorial as in **45**, the α-Me group shifts away versus its respective position in **45***. We hoped that this minor but not so insignificant shift would also be present in enone **69** and epoxy ketone **70**, in which C8a is sp^3-hybridized and C8b is sp^2-hybridized (*pseudo*-sp^2 for **70**) and that there would be enough exposure of the π*–C=O orbital at C5a for the nucleophile to enter.

8.3.3.4 The Experimental Realization

With the above conformational analysis, syntheses of enone **65** and epoxy ketone **70** became our focal points, but this would once again represent another challenging experimental endeavor (Scheme 8.17). Serendipitously, when we decided to

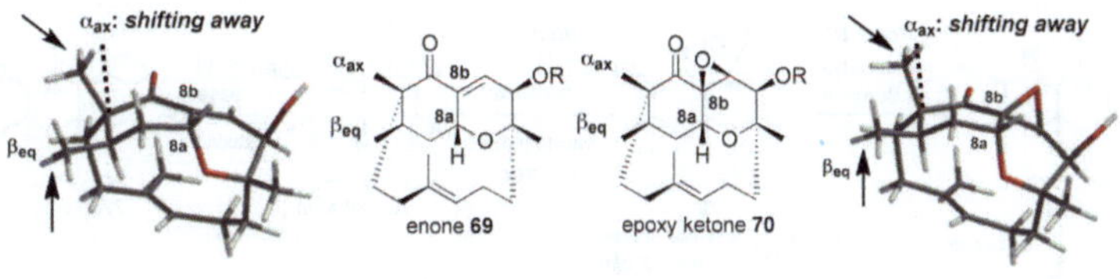

Fig. 8.12 Reaffirming the conformations in **69** and **70**

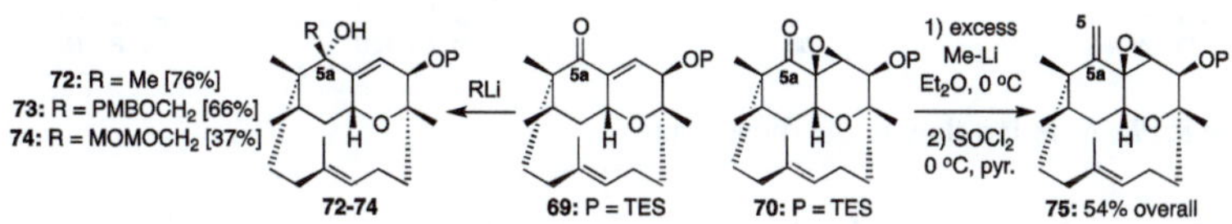

Scheme 8.18 Successful C5a-homologations

eliminate the secondary alcohol to construct the enone through chlorination with $SOCl_2$ and pyridine (see **56-Cl**), we isolated instead the crystalline cyclic enol-sulfite **71** as a single diastereomer and the sulfur stereochemistry was not assigned.

Interestingly, we were intrigued by the ESI mass spectrum of the compound, as the observed base peak consisted of $[M–SO_2+Na]^+$. This led us to explore a thermal retro-Diels–Alder reaction that could afford the desired enone **69**. It is noteworthy that the chemistry of cyclic enol-sulfites would appear to be an under-explored area with a few references reporting their isolation being found [57]. At last, we were also able to prepare epoxy ketone **70** from **69** in three steps, albeit epoxidation did not take place unless the TES group was removed. Spartan^{TM} models reaffirmed our initial conformational assessment of enone **69** and epoxy ketone **70**, which contain sp^3-hybridized C8a and sp^2-hybridized C8b (pseudo-sp^2-hybridized C8b for **70**) at the AB-ring junction (Fig. 8.12) and displayed the desired twisted-boat conformation in A-ring.

Although an initial attempt using methyl Wittig reagent generated from Ph_3PCH_3Br and *n*-BuLi failed, enone **69** proved to be reactive to MeLi!; the desired tertiary alcohol **72** was isolated in good yield as a single diastereomer with the new methyl group having come from the sterically less hindered α-face of **69** (Scheme 8.18). Encouraged by this outcome, we employed an oxygenated methyl lithium source using the PMB-stannane [58] and stannane of MOM ether [59]. At last, we could also homologate epoxy ketone **70** via addition of MeLi followed by elimination using $SOCl_2$ to give vinyl epoxide **75**. This punctuates our success in homologating a difficult C5a-carbonyl through a thoughtful conformational analysis that was made possible from meticulous observations in the proton NMR spectrum.

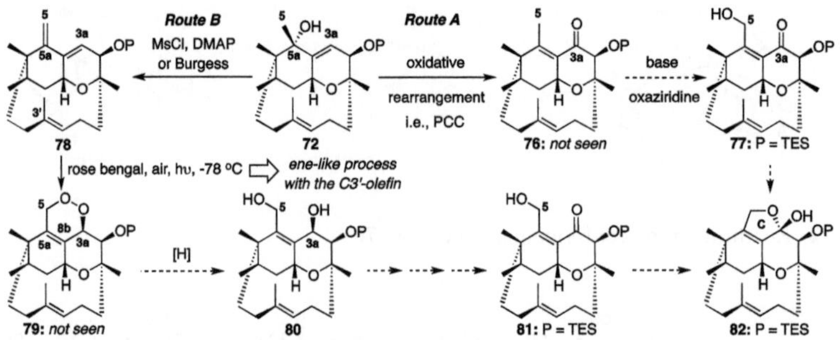

Scheme 8.19 A design flaw in the diene route

8.4 Completion of the Total Synthesis

At this point, we had exerted an immense amount of effort to overcome a number of difficult transformations and their unprecedented challenges: (1) oxidation at C3 and C3a, (2) reduction at C8a, and (3) homologation at C5a. We learned to truly respect the unique structural topology that ABD-tricycle **2** possesses. What remained for the total synthesis was the C-ring construction, which posed serious challenge of its own.

8.4.1 The Diene Route

We had two possible routes in which alcohol **72** could be used (Scheme 8.19). Route A would involve rearrangement of tertiary alcohol **72** to enone **76**. Deprotonation at C5 and generation of the enolate followed by exposure to an oxaziridine or other oxygen electrophile equivalents might directly afford the hydrated furan C-ring of phomactin A (see **82**) via hydroxy enone **81**. We had also hoped to make use of a chromium-mediated oxidative rearrangement of tertiary allylic alcohols. Unfortunately, treatment of **72** to PCC produced only unidentified baseline materials, thereby quickly eliminating this route.

Route B would involve another $^1O_2$ Diels–Alder reaction using diene **78**. We envisioned that reduction of cycloadduct **79** could afford 1,4-diol **80**, which could lead to the C-ring formation again through hydroxyl enone **77**. Toward this goal, we investigated the elimination of tertiary alcohol **72** to diene **78**. $POCl_3$ in pyridine was unreactive, while MsCl in DMAP/pyridine slowly provided the diene as an insepa-rable mixture with some unidentified side-products. The Burgess reagent is well known for its ability to eliminate a wide range of alcohols in a mild manner [60]. Consequently, treatment of alcohol **72** with a small excess of Burgess reagent in toluene at 70 °C rapidly induced elimination to diene **78**, although not in high purity. Nevertheless, we exposed this impure diene briefly to singlet oxygen and were dismayed to note that there was no desired endoperoxide **79**, only

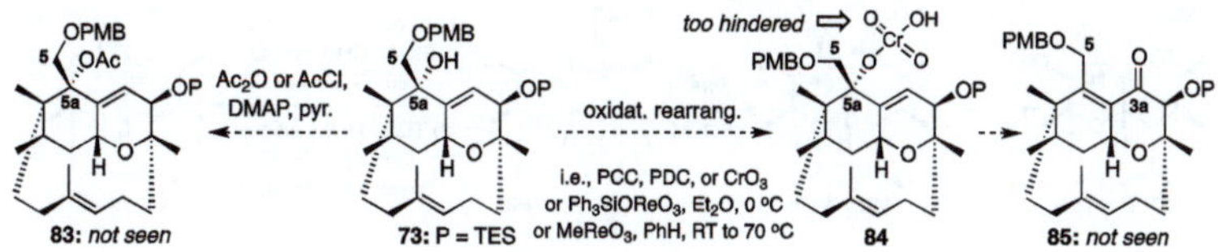

83: *not seen* 73: P = TES 84 85: *not seen*

Scheme 8.20 Initial thoughts on the allyl alcohol route

decomposition along with possible ene-reaction products. We quickly recognized a possible design flaw here. The diene motif in **78** is not coplanar, with terminal carbons of the diene (C5 and C3a) at a dihedral angle or 59° (see model of **78** in the left inset), thereby prohibiting it from participating in any form of cycloaddition as a 4π-component. Consequently, we abandoned the diene route altogether.

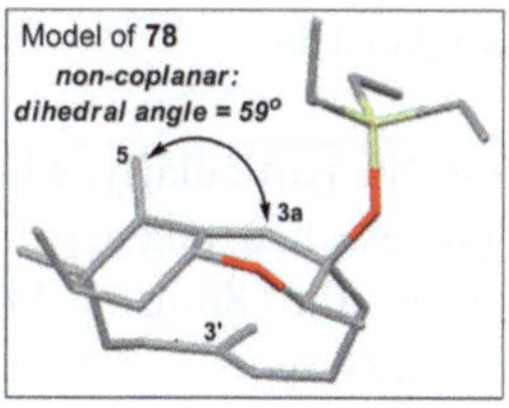

8.4.2 *The Allyl Alcohol Route*

On paper, the allylic alcohol route employing PMB-ether **73** or MOM-ether **74** appeared to be even more enticing than the diene route. As shown in Scheme 8.20, we again attempted oxidative rearrangement using PMB-ether **73** but more aggressively this time in an attempt to construct PMB-protected hydroxy enone **83**, which should be steps away from phomactin A. However, with Cr [61, 62], or later with Re [63], based reagents (at the suggestion of Professors Daesung Lee and Matt McIntosh) only resulted in either slow decomposition of the starting material or removal of the PMB group and subsequent decompositions.

8.4.3 *The Vinyl Epoxide Route*

The key to the success of the vinyl epoxide route would reside in a nucleophilic ring-opening of vinyl epoxide **75**. If this ring-opening proceeds regioselectively at C5 via a S_N2' pathway, it would have been the most welcome at this stage of our efforts (Scheme 8.21). However, an array of conditions, including Pd(0)-mediated and Lewis acidic conditions, were screened over a period of almost 2 years, but we never observed the desired 1,4-diol **80**. Instead, we could at various times see small

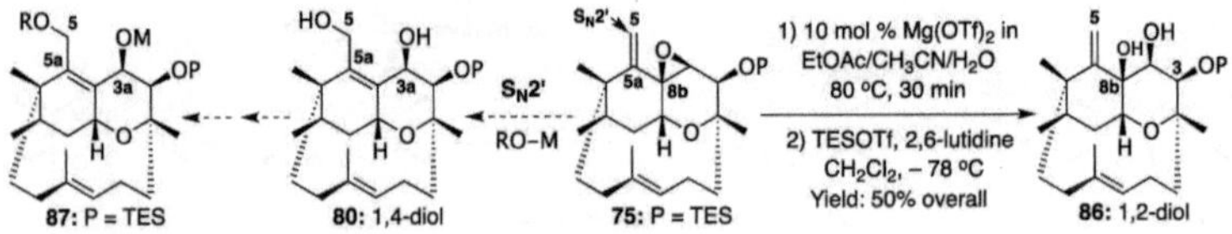

Scheme 8.21 An enticing S_N2' addition approach

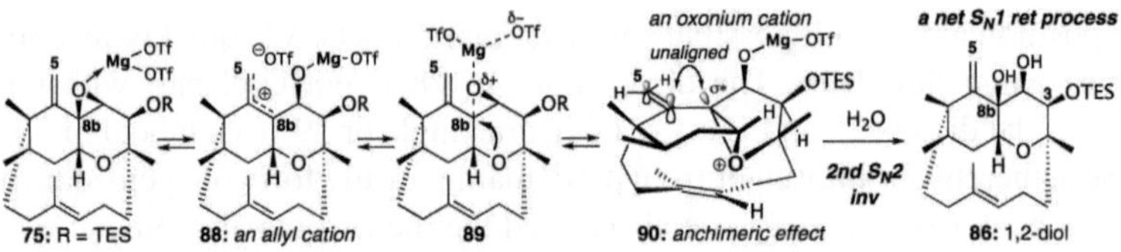

Scheme 8.22 Possible pathways to 1,2-diol **86**

amounts of product resembling **86** particularly when using $Mg(OTf)_2$, thereby suggesting that an S_N1-like process had taken place. Consequently, we seized the opportunity and synthesized 1,2-diol **86** using $Mg(OTf)_2$ in wet CH_3CN [64]. Re-silylating the C3-OH was necessary as it did not survive the conditions. The formation of 1,2-diol **86** via a stereochemically retentive S_N1-like process intrigued us. The formation of an allyl cation species, such as **88** (Scheme 8.22) via activation of **75** that allows for a straightforward retentive S_N1 addition to give **86**, represented one possibility, although true cationic species are not often considered as part of intermediates even in Lewis acid-promoted nucleophilic ring-openings. Even if the ring-opening occurs, it would be highly reversible in this context.

Perhaps a more plausible pathway would involve an intermediate in which anchimeric assistance takes place through the pyranyl oxygen atom (**89** → **90**). This would render an overall process of S_N2 addition in an inverted fashion. In addition, from the model, it appeared that if such an anchimeric assistance indeed existed, S_N2 addition could proceed faster, whereas the S_N2' addition pathway could be slowed or even suppressed because the $\sigma^*_{C\text{–}O}$ (of the oxonium bond) is not aligned with the system of the C5a–C5 *exo*-cyclic olefin in the conformation assumed by **90** (see model). The following experiment would actually support this assertion.

As shown in Scheme 8.23, the S_N2' addition pathway was indeed possible when using Li-I, and in the presence of $Sc(OTf)_3$, the desired allyl iodide **91** could be attained. Likewise, when using LiBr follow by DMP oxidation, bromo-enone **92** could be isolated in 50 % overall yield. Both **87** and **88** represented possible solutions to our problem. However, attempts to displace the bromide using Ag salts and an oxygen nucleophile only led to α-hydroxy ketone **93** in low yields, while allyl iodide **91** led to 1,2-diol **86**. Despite our high level of confidence for a straightforward S_N2/S_N1 addition, the S_N2' addition pathway was unexpectedly dominant here. Models again revealed the delicate balance between various competing pathways dictated by sterics and/or stereoelectronics. In **86**, the $\sigma^*_{C\text{–}O}$ of the

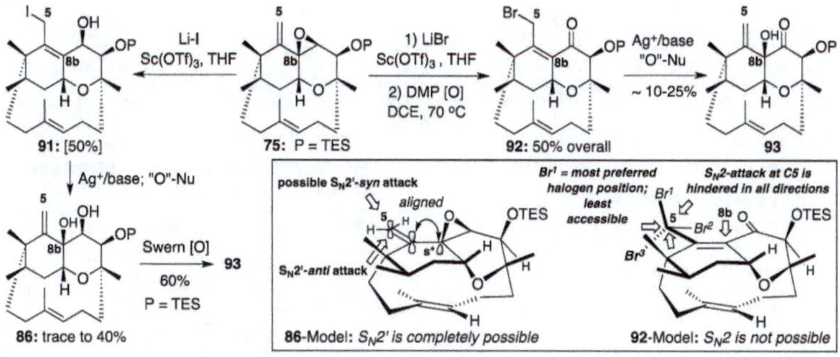

Scheme 8.23 A successful S_N2' addition with halides

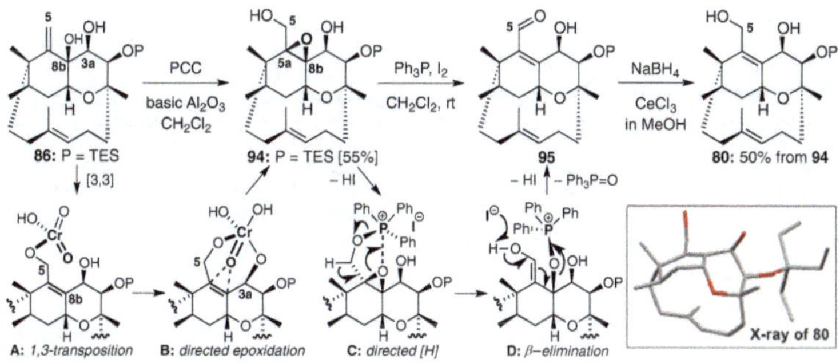

Scheme 8.24 A mystery in the epoxidation

epoxide is now perfectly aligned with the p-system of the C5a-5 *exo*-cyclic olefin, thereby allowing the S_N2' addition to proceed either in an *anti* and/or *syn* manner. On the other hand, in bromo-enone **92**, the S_N2 addition pathway (attack at C5) is impeded by surrounding sterics regardless of the position of the bromide leaving group (though it would likely prefer position $\boldsymbol{Br^1}$, which provides the least allylic strain). Consequently, the less hindered S_N2' addition pathway through C8b would come to dominate.

The above result prompted us to examine 1,2-diol **86**, and we were very pleased to find that 1,3-allylic alcohol transposition was successful using PCC on alumina, although epoxidation had again taken place on the initially transposed olefin at C8b and C5a (Scheme 8.24). Such an epoxidation had been reported by Dauben [61], Herz [62], and more recently by McIntosh [65]. Subsequent treatment of **94** with $Ph_3P\text{-}I_2$ [66] afforded hydroxy enal **95**. Although it might appear to be counterintuitive, intermediates **A–D** would provide a very reasonable description of events that took place in both of these transformations. An ensuing Luche reduction of hydroxy enal **95** gave 1,4-diol **80**, which was confirmed through its X-ray structure. We had finally transposed 1,2-diol **86** to 1,4-diol **80**, which represents the equivalent of the original intent of adding an oxygen nucleophile via S_N2', when examined vinyl epoxide **75**.

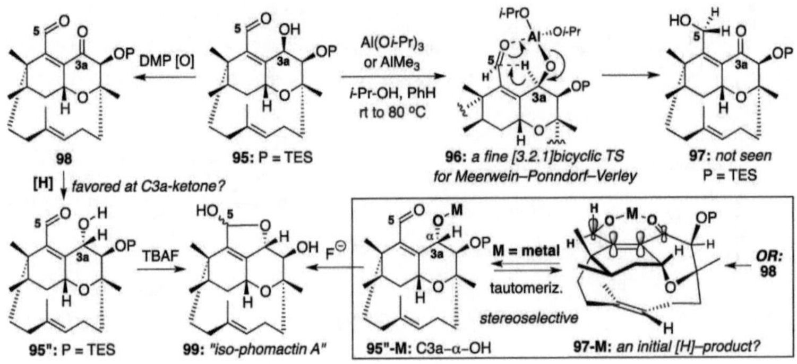

Scheme 8.25 Attempts to correct the oxidation states

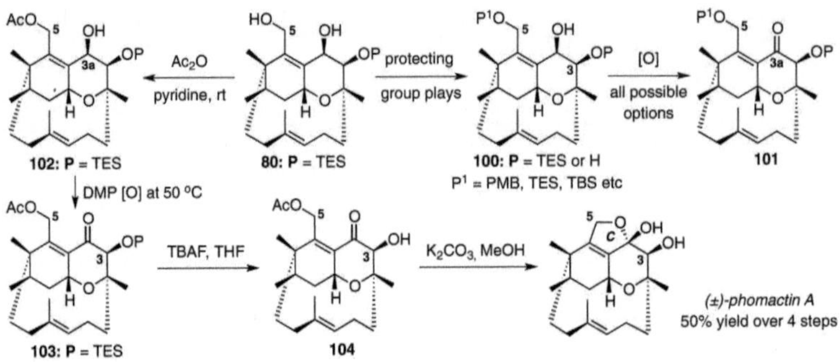

Scheme 8.26 The final obstacle: protecting groups

In hindsight, we probably should have focused on 1,4-diol **80** and simply struggled through the less glamorous chemistry of protection and deprotection, but we were fascinated with hydroxy enal **95**. As shown in Scheme 8.25, we envisioned a Meerwein–Ponndorf–Verley process of self-redox to access hydroxy enone **97** because the proposed transition state looked almost too perfect to pass on. When that failed with Al(O*i*-Pr)$_3$ or AlMe$_3$, we turned to keto aldehyde **98** in the hope of being able to reduce the presumably more reactive aldehyde in chemoselective manner. A number of hydride sources were tried, and occasionally, we were able to see a key NMR resonance that appeared to match phomactin A. One time, when using Dibal-H, we were able to gain a very small sample of semi-pure product after removal of the TES group. It turned out that this material was not phomactin A but appeared to be "*iso*-phomactin A" **99**.

We then returned to 1,4-diol **80**, and after failing in a number of protection strategies, we managed to selectively acylate C5-OH (Scheme 8.26). Subsequently, we found that oxidation of the C3a-OH group employing Dess–Martin periodinane reagent at a temperature above room temperature gave enone **103** with good fidelity. With the concern of the equilibrium issue described above, we desialylated prior to deacetylation in the hope that lactonization would simply occur to form the lactol C-ring. This indeed occurred, and at last, we completed our total synthesis of (±)-phomactin A in 24 steps from ABD-tricycle **2** and 35 steps from Rawal's amino diene.

8.5 Conclusion

We began our interest in phomactin A with an attempt to showcase our intramolecular *oxa*-[3 + 3] annulation strategy to construct its unique ABD-tricyclic manifold. This resulted a distinctly new approach with the 12-membered D-ring of phomactin A being assembled simultaneously with the 1-oxadecalin at an early stage. We believed that completing this total synthesis would serve to greatly elevate the visibility of our *oxa*-[3 + 3] annulation strategy. However, we learned much more than that because the ABD-tricycle represents a unique structural topology that would pose a number of unprecedented challenges.

Each operation represented a major lesson in synthesis, and with each passing lesson, we learned to respect intricacies of this de novo tricyclic system. While certainly not the most elegant way to complete a synthesis, we have no regrets in pursuing this exercise. In this case, while turning a 15-step synthetic design into a 34-step, colossal task should have no business in drug discovery; it proved to be a fertile ground for learning to solve any given problems, to think and deduce logically, to observe meticulously, to be persistent when needed, and to be daring when there is no other way. That was a perfect exercise perfectly suited for a perfect academic setting.

References

1. For reviews, see: (a) Molinski TF, Dalisay DS, Lievens SL, Saludes JP (2009) Nat Rev Drug Dis 8:69; (b) Blunt JW, Copp BR, Munro MGH, Northcotec PT, Prinsep MR (2003) Nat Prod Rep 20:1; (c) Faulkner DJ (2000) Nat Prod Rep 17:1; (d) Simmons TL, Andrianasolo E, McPhail K, Flatt P, Gerwick WH (2005) Mol Cancer Ther 4:333
2. Sugano M, Sato A, Iijima Y, Oshima T, Furuya K, Kuwano H, Hata T, Hanzawa H (1991) J Am Chem Soc 113:5463
3. Sugano M, Sato A, Iijima Y, Oshima T, Furuya K, Haruyama H, Yoda K, Hata T (1994) J Org Chem 59:564
4. Sugano M, Sato A, Iijima Y, Oshima T, Furuya K, Kuwano H, Hata T (1995) J Antibiot 48:1188
5. Koyama K, Ishino M, Takatori K, Sugita T, Kinoshita K, Takahashi K (2004) Tetrahedron Lett 45:6947
6. Chu M, Truumees I, Gunnarsson I, Bishop WR, Kreutner W, Horan AC, Patel MG, Gullo VP, Puar MS (1993) J Antibiot 46:554
7. Chu M, Patel MG, Gullo VP, Truumees I, Puar MS, McPhail AT (1992) J Org Chem 57:5817
8. Cheing JWC, Goldring WPD, Pattenden G (2003) Chem Commun 2788
9. Tokiwano T, Fukushi E, Endo T, Oikawa H (2004) Chem Commun 1324

10. Mahato SB, Pal BC, Kawasaki T, Miyahara K, Tanaka O, Yamasaki K (1979) J Am Chem Soc 101:4720
11. Stafforinif DM, McIntyre TM, Zimmerman GA, Prescott SM (2003) Crit Rev Clin Lab Sci 40:643
12. Sugano M, Sato A, Saito K, Takaishi S, Matsushita Y, Iijima Y (1996) J Med Chem 39:5281
13. Braquet P, Touqui L, Shen TY, Varaftig BB (1987) Pharm Rev 39:97–145
14. Nagase T, Ishii S, Kume K, Uozumi N, Izumi T, Ouchi Y, Shimizu T (1999) J Clin Invest 104:1071
15. Zhu X, Lambertino AT, Houghton TJ, McGilvra JD, Xu C, Rawal VH, Leff AR (2003) Life Sci 73:3005
16. For an excellent review on synthetic efforts toward phomactins, see: Goldring WPD, Pattenden G (2006) Acc Chem Res 39:354
17. Also see: Cole KP, Hsung RP (2003) ChemTracts 16:811
18. For approaches, see: (a) Foote KM, Hayes CJ, Pattenden G (1996) Tetrahedron Lett 37:275; (b) Chen D, Wang J, Totah NI (1999) J Org Chem 64:1776; (c) Seth PP, Totah NI (1999) J Org Chem 64:8750; (d) Seth PP, Totah NI (2000) Org Lett 2:2507; (e) Kallan NC, Halcomb RL (2000) Org Lett 2:2687; (f) Chemler SR, Danishefsky SJ (2000) Org Lett 2:2695; (g) Seth PP, Chen D, Wang J, Gao X, Totah NI (2000) Org Lett 56:10185; (h) Foote K, John M, Pattenden G (2001) Synlett 365; (i) Mi B, Maleczka R (2001) Org Lett 3:1491; (j) Chemler SR, Iserloh U, Danishefsky SJ (2001) Org Lett 3:2949; (k) Houghton T, Choi S, Rawal VH (2001) Org Lett 3:3615; (l) Mohr PJ, Halcomb RL (2002) Org Lett 4:2413; (m) DiBlasi CM, Hamblett CL, Leighton JL (2002) Abstracts of Papers, 224th ACS-National Meeting, Boston, MA, 2002; ORGN-364; (n) DiBlasi CM (2003) Ph.D. Dissertation. Thesis Advisor: Professor James L. Leighton, Columbia University; (o) Balnaves AS, McGowan G, Shapland PDP, Thomas EJ (2003) Tetrahedron Lett 44:2713; (p) Cheing JWC, Goldring WPD, Pattenden G (2003) Chem Commun 2788; (q) Ryu K, Cho Y-S, Jung S-I, Cho C-G (2006) Org Lett 8:3343; (r) Huang J, Wang H, Wu C, Wulff WD (2007) Org Lett 9:2799; (s) Seth PP, Chen D, Wang J, Gao X, Totah NI (2007) Tetrahedron Lett 48:4605; (t) Peng W, Lee C-S (2008) Synlett 142; Sugano M, Sato A, Iijima Y, Furuya K, Hata T, Kuwano H (1995) J Antibiot 48:1188; (u) Ciesielski J, Canterbury DP, Frontier AJ (2009) Org Lett 11:4374; (v) Shapland PDP, Thomas EJ (2009) Tetrahedron 65:4201; (w) Blackburn TJ, Helliwell M, Kilner M, Lee ATL, Thomas EJ (2009) Tetrahedron Lett 50:3550; (x) Schwartz KD, White JD (2011) Org Lett 13:248; (y) Huang S, Du G, Lee C-S (2011) J Org Chem 76:6534
19. For total synthesis of (+)-phomactin D, see: Miyaoka H, Saka Y, Miura S, Yamada Y (1996) Tetrahedron Lett 37:7107
20. For total synthesis of (±)-phomactin B2, see: Huang J, Wu C, Wulff WD (2007) J Am Chem Soc 129:13366
21. For total synthesis of (±)-phomactin A, see: (a) Goldring WPD, Pattenden G (2002) Chem Commun 1736; For detailed full papers, see: (b) Diaper CM, Goldring WPD, Pattenden G (2003) Org Biomol Chem 1:3949; (c) Foote KM, Hayes CJ, John MP, Pattenden G (2003) Org Biomol Chem 1:3917
22. For total synthesis of (+)-phomactin A, see: Mohr PJ, Halcomb RL (2003) J Am Chem Soc 125:1712
23. For total synthesis of (±)-phomactin G, see: Goldring WPD, Pattenden G (2004) Org Biomol Chem 2:466
24. Harrity JPA, Provoost O (2005) Org Biomol Chem 3:1349
25. Hsung RP, Kurdyumov AV, Sydorenko N (2005) Eur J Org Chem 1:23
26. Hsung RP, Wei L-L, Sklenicka HM, Shen HC, McLaughlin MJ, Zehnder LR (2001) Trends Heterocycl Chem 7:1–24
27. Hsung RP, Cole KP (2004) In: Harmata M (ed) Strategies and tactics in organic synthesis, vol 4. Elsevier Science, Pergamon, Oxford, UK, p 41
28. Buchanan GS, Feltenberger JB, Hsung RP (2010) Curr Org Syn 7:363
29. Hsung RP, Shen HC, Douglas CJ, Morgan CD, Degen SJ, Yao LJ (1999) J Org Chem 64:690

30. Hsung RP, Wei L-L, Sklenicka HM, Douglas CJ, McLaughlin MJ, Mulder JA, Yao L (1999) Org Lett 1:509
31. Shen HC, Wang J, Cole KP, McLaughlin MJ, Morgan CD, Douglas CJ, Hsung RP, Coverdale HA, Gerasyuto AI, Hahn JM, Liu J, Wei L-L, Sklenicka HM, Zehnder LR, Zificsak CA (2003) J Org Chem 68:1729
32. Kurdyumov AV, Lin N, Hsung RP, Gullickson GC, Cole KP, Sydorenko N, Swidorski JJ (2006) Org Lett 8:191
33. For recent studies, see: (a) Sagar R, Park J, Koh M, Park SB (2009) J Org Chem 74:2171; (b) Yamamoto Y, Itonaga K (2009) Org Lett 11:717; (c) Hubert C, Moreau J, Batany J, Duboc A, Hurvois J-P, Renaud J-L (2008) Adv Synth Catal 350:40; (d) Brioche JCR, Goodenough KM, Whatrup DJ, Harrity JPA (2008) J Org Chem 73:1946; (e) Brioche JCR, Goodenough KM, Whatrup DJ, Harrity JPA (2007) Org Lett 9:3491; (f) Epstein OL, Rovis T (2006) J Am Chem Soc 128:16480
34. Also see: (a) Appendino G, Cravotto G, Tagliapietra S, Nano GM, Palmisano G (1990) Helv Chim Acta 73:1865; (b) Schuda PF, Price WA (1987) J Org Chem 52:1972; (c) de March P, Moreno-Mañas M, Casado J, Pleixats R, Roca JL (1984) J Heterocycl Chem 21:85; (d) Tietze LF, v. Kiedrowski G, Berger B (1982) Synthesis 683; (e) de Groot A, Jansen BJM (1975) Tetrahedron Lett 16:3407
35. For the earliest work on *oxa*-[3+ 3] annulations, see: Ikawa M, Stahmann MA, Link KP (1944) J Am Chem Soc 66:902
36. For applications in natural product syntheses, see: (a) Yamashita S, Iso K, Kitajima K, Himura M, Hirama M (2011) J Org Chem 76:2408; (b) Brioche JCR, Goodenough KM, Whatrup DJ, Harrity JPA (2007) Org Lett 9:3941; (c) Kurdyumov AV, Hsung RP (2006) J Am Chem Soc 128:6272; (d) Liu H, Siegel DR, Danishefsky SJ (2006) Org Lett 8:423; (e) Olson BS, Trauner D (2005) Synlett 700; (f) Malerich JP, Maimone TJ, Elliott GI, Trauner D (2005) J Am Chem Soc 127:6276; (g) Sunazuka T, Handa M, Nagai K, Shirahata T, Harigaya Y, Otoguro K, Kuwajima I, Õmura S (2004) Tetrahedron 60:7845; (h) Kurdyumov AV, Hsung RP, Ihlen K, Wang J (2003) Org Lett 5:3935; (i) Hsung RP, Cole KP, Zehnder LR, Wang J, Wei L-L, Yang X-F, Coverdale HA (2003) Tetrahedron 59:311; (j) Malerich JP, Trauner D (2003) J Am Chem Soc 125:9554; (k) Zehnder LR, Hsung RP, Wang J, Golding GM (2000) Angew Chem Int Ed 39:3876
37. For a key review on chemistry of 1,3-dicarbonyls, see: Simon C, Constantieux T, Rodriguez J (2004) Eur J Org Chem 4957
38. For beautiful reviews on tandem reactions, see: (a) Tietze LF (1996) Chem Rev 96:115; (b) Tietze LF, Beifuss U (1993) Angew Chem Int Ed 32:131; (c) Tietze LF (1990) J Heterocycl Chem 27:47
39. For reviews for pericyclic ring-closures, see: (a) Marvell EN (1980) Thermal electrocyclic reactions. Academic, New York; (b) Okamura WH, de Lera AR (1991) In: Trost BM, Fleming I, Paquette LA (eds) Comprehensive organic synthesis, vol. 5. Pergamo, New York, pp 699–750
40. For informative reviews on ring-closure in natural product synthesis, see: (a) Beaudry CM, Malerich JP, Trauner D (2005) Chem Rev 105:4757; (b) Pindur U, Schneider GH (1994) Chem Soc Rev 409
41. For leading references on electrocyclic ring-closures involving 1-oxatrienes, see: (a) Shishido K, Shitara E, Fukumoto K (1985) J Am Chem Soc 107:5810; (b) Shishido K, Hiroya K, Fukumoto K, Kametani T (1986) Tetrahedron Lett 27:971
42. For a leading review, see: Tang Y, Oppenheimer J, Song Z, You L, Zhang X, Hsung RP (2006) Tetrahedron 62:10785
43. The term "formal [3 + 3]" was used to describe [3 + 3] carbo-cycloadditions. See: Seebach D, Missbach M, Calderari G, Eberle M (1990) J Am Chem Soc 112:7625
44. For earlier studies on *carbo*-[3 + 3] formal cycloadditions and annulations, see: Landesman HK, Stork G (1956) J Am Chem Soc 78:5129

45. For leading reviews on step-wise *carbo*-[3 + 3] formal cycloadditions, see: (a) Filippini M-H, Rodriguez J (1999) Chem Rev 99:27; (b) Butkus E (2001) Synlett 1827; (c) Feist H, Langer P (2007) Synthesis 327
46. For a seminal study, see: de March P, Moreno-Mañas M, Casado J, Pleixats R, Roca JL, Trius A (1984) J Heterocyclic Chem 21:1369
47. (a) Cole KP, Hsung RP (2003) Org Lett 5:4843; (b) Buchanan GS, Cole KP, Li G, Tang Y, You L, Hsung RP (2011) Tetrahedron 67:10105; (c) Buchanan GS, Cole KP, Tang Y, Hsung RP (2011) J Org Chem 76:7027
48. For an initial communication on recognition of the unique structural topology of ABD-tricycle **2**, see: Cole KP, Hsung RP (2005) Chem Commun 5784
49. You L, Hsung RP, Bedermann AA, Kurdyumov AK, Tang Y, Buchanan GS, Cole KP (2008) Adv Synth Catal 350:2885
50. For an initial communication on completion of our total synthesis of (±)-phomactin A, see: Tang Y, Cole KP, Buchanan GS, Li G, Hsung RP (2009) Org Lett 11:1591
51. (a) Kozmin SA, Iwama T, Huang Y, Rawal VH (2002) J Am Chem Soc 124:4628; (b) Kozmin SA, Green MT, Rawal VH (1999) J Org Chem 64:8045
52. (a) Huang Y, Iwama T, Rawal VH (2000) J Am Chem Soc 122:7843; (b) Huang Y, Iwama T, Rawal VH (2002) Org Lett 4:1163; Also see: (c) Kozmin SA, Iwama T, Huang Y, Rawal VH (2002) J Am Chem Soc 124:4628
53. Huang Y, Iwama T, Rawal VH (2002) J Am Chem Soc 124:5950
54. (a) Jacobsen EN (2000) Acc Chem Res 33:421; (b) For application of Cr(III)-salen catalysts in enantioselective hetero-Diels-Alder reactions: Schaus S, Brånalt J, Jacobsen EN (1998) J Org Chem 63:403
55. Kornblum N, DeLaMare HE (1951) J Am Chem Soc 73:880
56. Nicolaou KC, Shi GQ, Gunzer JL, Gartner GP, Yang Z (1997) J Am Chem Soc 119:5467
57. Sheppard WA (1968) J Org Chem 33:3297
58. Ferezou J-P, Julia M, Li Y, Liu L-W, Pancrazi A, Porteu F (1994) Bull Soc Chim Fr 131:865
59. Shiozaki M (1991) J Org Chem 56:528
60. Burgess EM, Penton HR, Taylor EA (1970) J Am Chem Soc 92:5244
61. Dauben WG, Michno DM (1977) J Org Chem 42:682
62. Sundararaman P, Werner Herz W (1977) J Org Chem 42:813
63. For some examples, see: (a) Gordon MS, Jacob J, Espenson JH, Jensen JH (1998) Organometallics 17:1835; (b) Wang G, Jimtaisong A, Luck RL (2004) Organometallics 23:4522; (c) Osborn JA, Bellemin-Laponnaz S, Gisie H, Le Ny J-P (1997) Angew Chem Int Ed 36:976; (d) Grubbs RH, Morrill C (2005) J Am Chem Soc 127:2842; (e) Osborn JA, Le Ny J-P, Bellemin-Laponnaz S (2000) Tetrahedron Lett 41:1549
64. For related protocols, see: (a) Boyer F-D, Hanna I (2005) J Org Chem 70:1077; (b) Iranpoor N, Shekarriz M, Shiriny F (1998) Synth Commun 28:347
65. For a recent example, see: (a) Chai Y, McIntosh MC (2004) Tetrahedron Lett 45:3269; Also see: (b) Sundararaman P, Herz W (1977) J Org Chem 42:813; (c) Chu A, Mander LN (1988) Tetrahedron Lett 29:2727
66. Wydra H, Paryzek Z (1984) Tetrahedron Lett 25:2601

Chapter 9
(+)-11,11′-Dideoxyverticillin A

Justin Kim and Mohammad Movassaghi

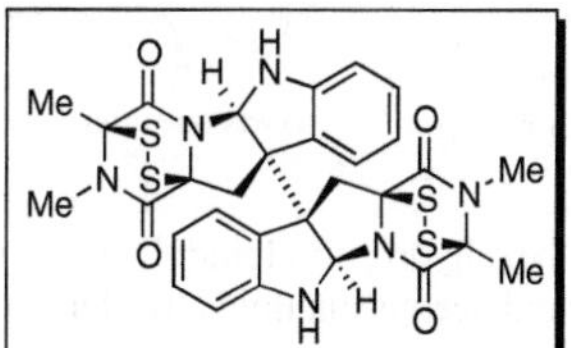

(+)-11,11′-dideoxyverticillin A

9.1 Introduction and Classification

The dimeric epipolythiodiketopiperazine alkaloids are a family of fungal metabolites renowned for their structural and functional complexity [1]. Isolated across eight genera of fungi, over 40 members of this class of natural products have been discovered since the isolation of their inaugural members, (+)-chaetocin A (2) [2] and (+)-verticillin A [3], in 1970 (Fig. 9.1). As evidenced by the large number of constituents comprising this family, there are many avenues by which each member derives its identity. For instance, while all members possess a C3–C3′ dimeric linkage, a thiolated diketopiperazine moiety, and a pentacyclic subunit arising from a tryptophan-containing cyclic dipeptide, there is considerable diversity in the second amino acid that is incorporated into these natural products—alanine, serine, valine, glycine, and threonine are all prevalent components. In addition, these variable amino acids, along with the integral tryptophan residue, are often observed

J. Kim • M. Movassaghi (✉)
Department of Chemistry, Massachusetts Institute of Technology, 77 Massachusetts Avenue, Cambridge, MA 02139, USA
e-mail: movassag@mit.edu

J.J. Li and E.J. Corey (eds.), *Total Synthesis of Natural Products*,
DOI 10.1007/978-3-642-34065-9_9, © Springer-Verlag Berlin Heidelberg 2012

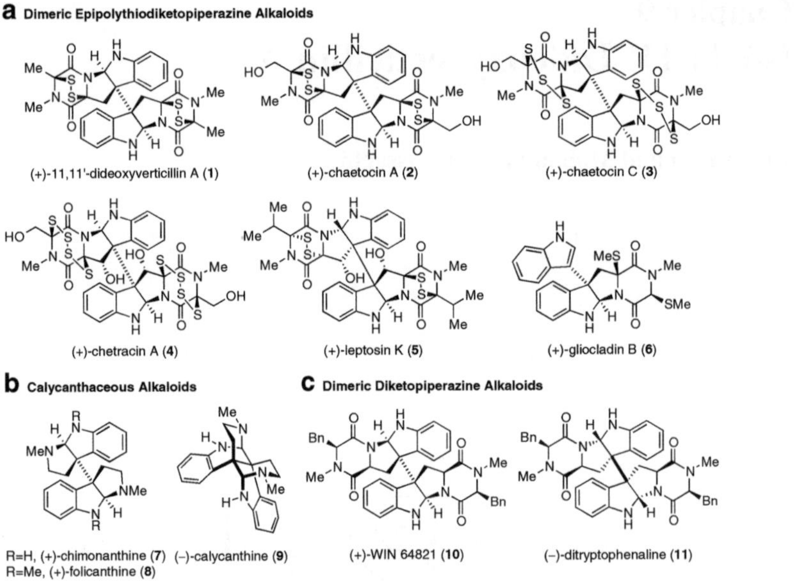

Fig. 9.1 Representative dimeric hexahydropyrroloindole alkaloids. (**a**) Dimeric epipolythiodiketopiperazine alkaloids. (**b**) Calycanthaceous alkaloids. (**c**) Dimeric diketopiperazine alkaloids

with oxidative modifications such as hydroxylation of their β-positions. Beyond the constitutional variations, there are also variations with respect to their stereochemical construction with some members displaying a pseudo-C_2-symmetric axis about their C3–C3′ dimeric linkage, while yet others have a pseudo-*meso*-core; both *endo* and *exo* epipolysulfides have also been reported. Most importantly, as the name epi*poly*thiodiketopiperazine alkaloid implies, a great source of diversity arises from the variations in the degree of sulfuration. Di-, tri-, and tetrasulfides exist in a variety of combinations across the two conjoined monomeric subunits. Congeners in which the polysulfide has been reductively methylated or eliminated altogether also exist and are incorporated within this family.

The dimeric epipolythiodiketopiperazine alkaloids are characterized by a densely functionalized core structure possessing up to 12 stereogenic centers, 6 of which are fully substituted and nearly all of which are embedded in a contiguous fashion. (+)-11,11′-Dideoxyverticillin A (**1**) [4], specifically, is in possession of eight stereogenic centers, all of which are obligatory among this class of natural products. While these imposing molecular architectures, replete with vicinal quaternary centers and the steric congestion they afford, would pose a unique challenge to their synthesis in their own right, the intricacy of this class of molecules is further augmented by the unique array of functional groups adorning their structure. The eponymous epipolythiodiketopiperazine motif, oft-recognized to be highly acid-, base-, and redox-sensitive [5], in combination with a plethora of labile centers of the (hemi)aminal oxidation state, renders many of these natural products fragile to a wide range of chemical conditions.

These epipolythiodiketopiperazine alkaloids, together with the calycanthaceous alkaloids (Fig. 9.1b), form a superfamily of natural products termed the dimeric hexahydropyrroloindole alkaloids [6–8]. The main dichotomy within this superfamily arises from the biogenetic elaboration of tryptamine versus tryptophan building blocks. The tryptamine-based calycanthaceous alkaloids, boasting members such as chimonanthine (**7**), calycanthine (**9**), and folicanthine (**8**), are largely plant derived and have a long and rich history in the context of natural product synthesis [7, 9].

9.2 Pharmacology

Beyond their chemical complexity, the dimeric epipolythiodiketopiperazines are well appreciated for the wide range of potent biological activities that they display. While there seems to be a general consensus of antibacterial and cytotoxic activity among a large number of the dimeric epipolythiodiketopiperazine constituency, in accordance with their structural diversity, their collective biological profile spans a range of properties including antiviral, antileukemial, antinematodal, antitumor, antibiotic, and proapoptotic activities. (+)-11,11′-Dideoxyverticillin A (**1**), the focus of this chapter, for example, has demonstrated antiangiogenic activity, displayed efficacy against several cancer cell lines, and exhibited potent inhibitory activity toward the epidermal growth factor receptor (IC_{50} = 0.14 nM) [10]. Meanwhile, its congener (+)-chaetocin A (**2**), in addition to being cytotoxic toward HeLa cells (IC_{50} = 0.025 μg/mL) and *Staphylococcus aureus* (IC_{50} = 0.1 μg/mL) [11], is the first lysine-specific histone methyltransferase inhibitor of the SU(VAR) 3–9 type and has shown great utility in the study of heterochromatin-mediated gene repression [12].

Structure–activity relationship studies have demonstrated the essential nature of the epipolythiodiketopiperazine functional group for the biological action of these natural products [1c, 11, 13]. Indeed, reports on monomeric epidithiodiketopiperazine alkaloids such as gliotoxin and sporidesmin have identified two different mechanisms of action, both of which assign a central role to the obligatory epidisulfide. The first mode of toxicity involves redox cycling [14] whereby native cellular reductants such as glutathione reduce the epidithiodiketopiperazine moiety to its dithiol form [15]. The thiols are highly reactive toward molecular oxygen, auto-oxidizing back to the epidisulfide and generating reactive oxygen species such as superoxide and hydrogen peroxide in the process. The cytotoxic effects of sporidesmin appear to be modulated by reactive oxygen species either directly or indirectly as secondary messengers in signaling pathways. The prevalence and generality of this mechanism is far from being established, however, as subsequent studies on gliotoxin-treated cells have challenged the role of the reactive oxygen species-mediated mechanism in the induction of apoptosis [16].

The second mode of toxicity is postulated to involve the direct interaction of the epidithiodiketopiperazine motif with target proteins, forming mixed disulfides with cysteine residues in various proteins. Gliotoxin, for example, has been demonstrated to form a 1:1 covalent complex with alcohol dehydrogenase [13b, 17]. Epidithiodiketopiperazines can also catalyze the formation of disulfide bonds between proximally located cysteine residues in proteins such as in creatine kinase [18]. Recently, epidithiodiketopiperazines have also been implicated in a zinc ejection mechanism, whereby the epidisulfide can shuffle disulfide bonds in the CH1 domain of proteins, coordinate to the zinc atoms that are essential to the tertiary structure of that domain, and remove the metal cation [12d, 19].

The toxicity effects of epidithiodiketopiperazine alkaloids are often significantly enhanced in cells with highly reducing environments as a result of the redox-uptake mechanism [20]. While cells often have active transporters involved in the rapid efflux of epidisulfide-containing secondary metabolites, reduction of this moiety by glutathione or other native cellular reductants prevents the small molecule from crossing the plasma membrane. Through this mechanism, epidithiodiketopiperazines can accumulate in cells to a concentration several orders of magnitude greater than the applied concentration. The cytotoxic effects are thus highly amplified within those cells.

Finally, it has been reported that there is a positive correlation between the degree of sulfur incorporation on dimeric epipolythiodiketopiperazine alkaloids and their cytotoxic and antibiotic activities. Epitrithiodiketopiperazine alkaloid (+)-chaetocin C (**3**) ($IC_{50} = 0.025$ μg/mL) is four times as potent as the epidithiodiketopiperazine-containing (+)-chaetocin A (**2**) toward *Staphylococcus aureus* and twice as cytotoxic against HeLa cells ($IC_{50} = 0.02$ μg/mL). The trend is inclusive of the epitetrasulfide-containing chetracin A as well as the mixed epidi-/epitrisulfide-containing (+)-chaetocin B. It is postulated that the bioactivity of these natural products parallels the instability of the episulfide bridge toward reduction [3]; this proposal is consistent with mechanisms invoking dithiol or thiolate species as active pharmacophores.

9.3 Biosynthesis

A comprehensive compilation of data derived from feeding and other biosynthetic studies performed by Kirby [21], Sammes [22], and Taylor [23], genomic studies conducted by Howlett [1c, 24], as well as our own synthetic studies probing the innate reactivity of various intermediates (vide infra) culminated in our formulation of the biosynthetic hypothesis delineated below (Scheme 9.1). The biosynthesis features two distinct phases: an oxidative dimerization phase and a thiolation phase in which diketopiperazine structures are elaborated to afford the defining epipolysulfide motifs (Scheme 9.1) [25, 26].

The biosynthetic hypothesis commences with the action of a nonribosomal peptide synthetase on natural amino acid precursors. While the incorporation of

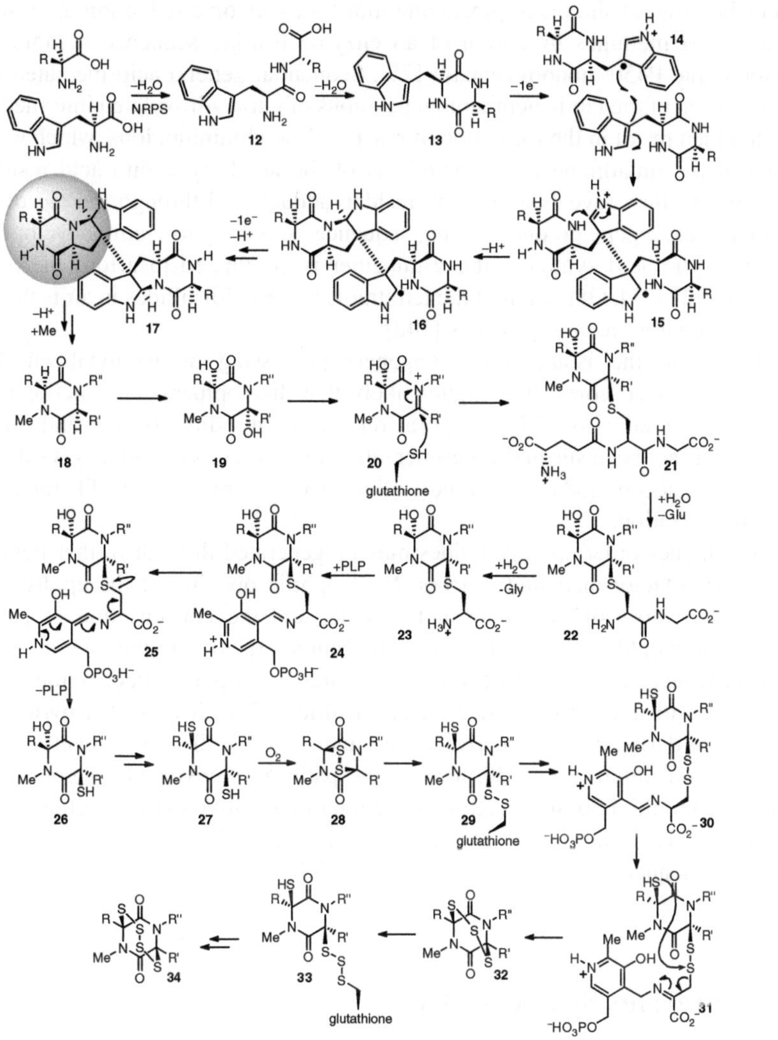

Scheme 9.1 Postulated biosynthetic pathway for dimeric epipolythiodiketopiperazine alkaloids

L-tryptophan is compulsory, the biosynthetic machinery displays wide latitude in its ability to condense a second auxiliary amino acid—L-alanine in the case of (+)-11,11′-dideoxyverticillin A (1)—to afford a tryptophan-derived diketopiperazine intermediate **13**. Mirroring Woodward and Robinson's biogenetic hypothesis for the calycanthaceous alkaloids, single-electron oxidation of the electron-rich tryptophan residue would likely initiate an oxidative dimerization of the diketopiperazine precursor with concomitant cyclization to yield the octacyclic intermediate **17**. Subsequent *N*-methylation of the amides would then yield an unembellished skeletal core of the dimeric epipolythiodiketopiperazine alkaloids. The first step en route to elaboration of the diketopiperazine motif may involve oxidation at each of the four C_α–H methines by action of an enzyme bearing sequence homology to a cytochrome P450 monooxygenase [25]. Sequential general acid-mediated ionization at each of the labile hemiaminal positions of tetrahydroxylated intermediate **19** would then result in the formation of reactive *N*-acyliminium ions, which would be trapped by glutathione [26]. Hydrolysis of the ancillary amino acid residues of glutathione then reveals a cysteine residue adduct, and through the aid of a PLP cofactor and a gene product bearing sequence homology to 1-aminocyclopropane-1-carboxylic acid synthase, the α-thioether is eliminated to afford the requisite tetrathiol **27** [27]. Oxidation of the tetrathiols to the epidisulfides in air is thought to afford the target natural products [14d].

Based on the isolation of congeners possessing unfunctionalized dimeric diketopiperazine cores, it is highly likely that diketopiperazine functionalization succeeds dimerization. The temporal relation of the dimerization and thiolation phases has not been verified experimentally; however, this hypothesis is sufficiently flexible to accommodate the dimerization phase at any stage of diketopiperazine functionalization.

The higher-order epipolysulfides may be generated through further iteration of the PLP cofactor-mediated process. Nucleophilic opening of the epidisulfide by glutathione, loss of glutamate and glycine, and the condensation of pyridoxal phosphate would provide Schiff base **30**. Tautomerization to the internal imine **31** may provide an electron sink for intramolecular nucleophilic attack of the thiol onto the mixed disulfide to generate an epitrisulfide **32** with loss of a pyridoxamine phosphate derivative. Alternatively, elimination of the disulfane–diketopiperazine moiety from intermediate **30**, akin to the pathway envisioned for conversion of sulfide **24** to thiol **26**, followed by oxidation of a putative thiol–disulfane diketopiperazine may result in the corresponding epitrisulfide **32**. Further iteration of these plausible sequences would afford epitetrasulfide **34**.

9.4 Previous Synthetic Work

Several of the principal challenges faced during the synthesis of the dimeric epipolythiodiketopiperazine alkaloids were identified at the outset of the work presented in this chapter: establishing the vicinal quaternary stereocenters of the C3–C3′ dimeric linkage, installation of an epidithiodiketopiperazine motif, and controlled extension of the episulfide bridge to higher order homologs. Some of these issues had been addressed previously in the literature on simpler systems with varying degrees of success. The results of these synthetic efforts are delineated below.

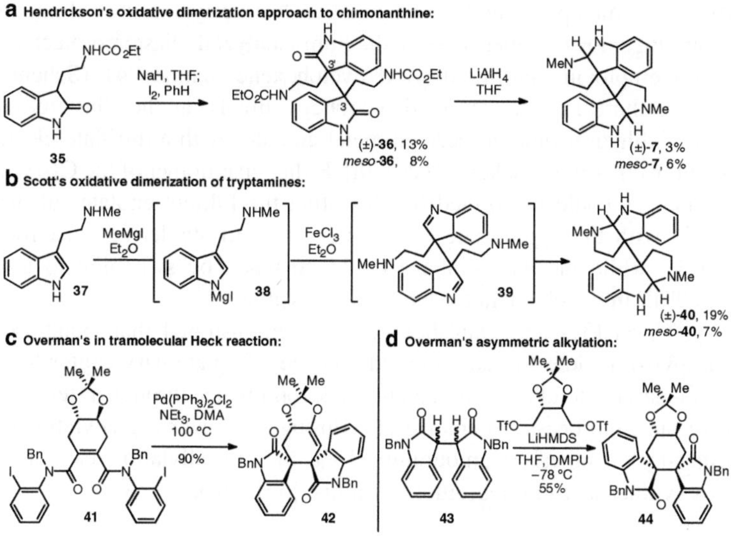

Scheme 9.2 Prior approaches to the C3–C3′ linked dimers. (**a**) Hendrickson's oxidative dimerization approach to chimonanthine. (**b**) Scott's oxidative dimerization of tryptamines. (**c**) Overman's intramolecular Heck reaction. (**d**) Overman's asymetric alkylation

9.4.1 Previous Approaches to the C3–C3′ Dimeric Linkages

The isolation of calycanthine (**9**) in 1888 by Eccles [28] and the subsequent proposition for its origins in the oxidative dimerization of tryptamine by Woodward [29] and Robinson [30] had prompted several key synthetic studies based on a biomimetic approach. Hendrickson was the first to experimentally verify the plausibility of forming the C3–C3′ linked dimers through an oxidative radical dimerization strategy (Scheme 9.2a). He demonstrated that the sodium enolate of a tryptamine-derived oxindole could be oxidized with iodine to afford a mixture of three possible stereoisomers. The racemic product was isolated in 13 % yield, while the meso product was isolated in 8 % yield. Global reduction of the oxindole and carbamates afforded the first synthetic samples of chimonanthine (**7**) [9a].

Scott was able to leverage the same type of methodology in an impressive display in which *N*-methyltryptamine was dimerized directly to afford chimonanthine (**7**) (Scheme 9.2b) [9c]. Deprotonation of the indole 1*H* proton with methyl Grignard followed by treatment with $FeCl_3$ accomplished the single-electron oxidation and dimerization of the indole moiety. The racemic and meso stereoisomeric products were obtained as a mixture in 19 % and 7 % yields, respectively. Takayama later found hypervalent iodine to be a superior oxidant, affording yields of 17 % and 30 %, respectively [9j]. In both cases, however, as in the case of Hendrickson's example, stereocontrol could not be achieved.

In 1999, Overman provided the first enantioselective synthesis of chimonanthine through an ingenious sequence of palladium-catalyzed diastereoselective Heck cyclizations employing tartrate-derived cyclohexene diamide **41** (Scheme 9.2c) [9h]. Shortly thereafter, he reported a strategy involving the diastereoselective alkylation of dihydroisoindigo-derived metal enolates with a ditriflate electrophile synthesized from tartrate (Scheme 9.2d) [9i]. Either enantiomer of the C_2-symmetric C3–C3′ linkages could be formed by alkylating the lithium enolates of dihydroisoindigo **43** with the appropriate enantiomer of a tartrate-derived electrophile in THF/DMPU, while the *meso*-linkage could be made by switching to a sodium enolate in THF in which chelate control was possible.

Finally, Crich [31] and Danishefsky [32] demonstrated that enantioselective hexahydropyrroloindole formation could precede C3-quaternary center formation. The carbon–carbon bond formation could be accomplished through a stereoretentive cationic or radical process as demonstrated through the reverse-prenylation reaction in Danishefsky's synthesis of amouramine [32a, b] or the allylation reaction en route to Crich's synthesis of (+)-debromoflustramine [33], respectively.

9.4.2 Previous Approaches to the Epidithiodiketopiperazine Motif

In the four-decade history preceding the synthesis of (+)-11,11′-dideoxyverticillin A (**1**), three unique approaches to the epidithiodiketopiperazine group were developed and applied to varying effect (Scheme 9.3). The first method was developed by Trown in 1968 and involved the bromination of sarcosine anhydride (**45**) with molecular bromine in refluxing 1,2-dichlorobenzene [34]. Subsequent S_N2 displacement of the secondary bromides with potassium thioacetate and acidic hydrolysis of the resultant thioesters revealed a dithiol intermediate, which could be oxidized to the epidisulfide with Ellman's reagent. Due to the tendency for alpha-alkylated amino acids to undergo elimination and further bromination, this method lacked a generality extending beyond the glycine or sarcosine anhydride-derived substrates. Thiolation strategies based on nucleophilic displacement using thioacetates or trithiocarbonates [35] also could not be applied to hindered alpha-substituted amino acid derivatives. Yet, the simplicity of Trown's synthesis proved highly enabling, serving as the opening sequence to several synthetic approaches centered on the elaboration of this epidithiodiketopiperazine core (vide infra) [36, 37].

A complementary method was reported 3 years later by Hino, in which a readily enolizable diketopiperazine **48** was directly converted to the epidisulfide by deprotonation with sodium hydride and exposure to sulfur monochloride [38]. As with the Trown method, this method was limited to a specific class of substrates, namely, ones possessing a 1,3-dicarbonyl motif at each of the reactive centers, yet it has also seen subsequent applications in total synthesis [39, 40]. In 1972, Schmidt was able to significantly broaden the scope of the enolate thiolation method by introducing elemental sulfur as the electrophilic agent [41]. In contrast to Hino's method in which formation of a highly reactive, unstable adduct requires readily

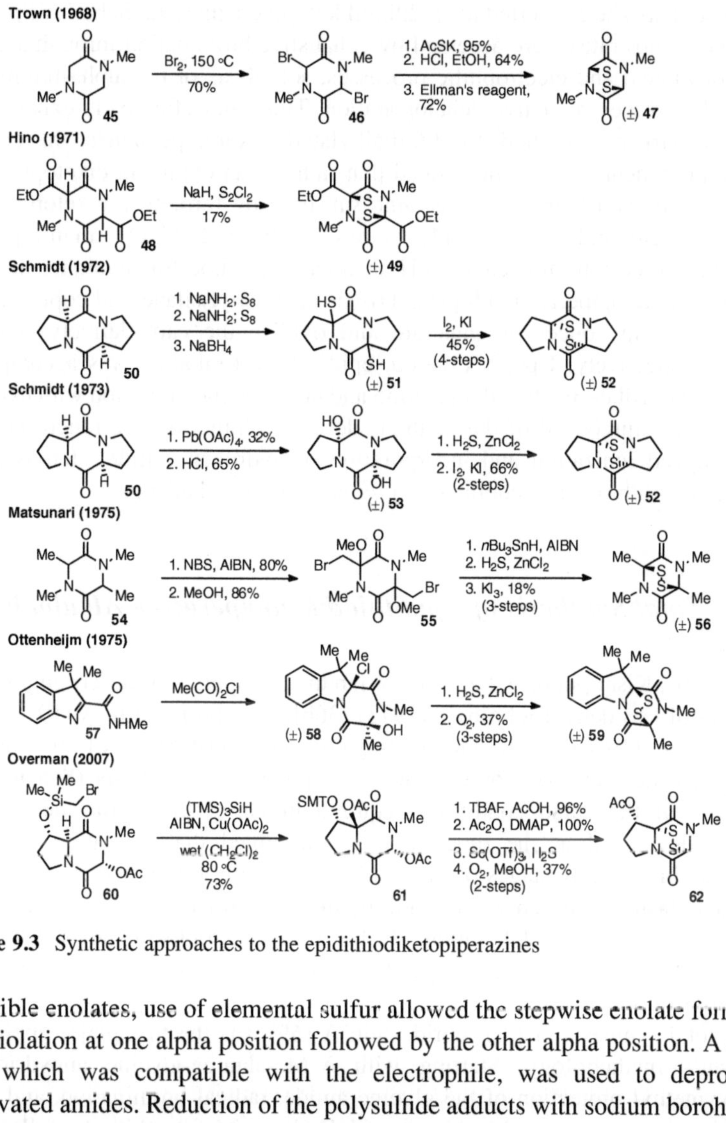

Scheme 9.3 Synthetic approaches to the epidithiodiketopiperazines

accessible enolates, use of elemental sulfur allowed the stepwise enolate formation and thiolation at one alpha position followed by the other alpha position. A strong base, which was compatible with the electrophile, was used to deprotonate unactivated amides. Reduction of the polysulfide adducts with sodium borohydride afforded a dithiol precursor **51**, rounding out the strategy.

The final method for constructing epidithiodiketopiperazine motifs relied on the nucleophilic thiolation of *N*-acyliminium ions. Access to alpha-oxidized diketopiperazine structures was central to this approach, and key developments were made in this regard. Schmidt first demonstrated the feasibility of this ionization approach in 1973 by conversion of proline anhydride to its diacetate using $Pb(OAc)_4$ [42]. Hydrolysis of the acetates, ionization of the hemiaminals with zinc chloride in the presence of hydrogen sulfide, and oxidation with iodine provided the epidisulfide of interest. In 1975, Matsunari reported access to alpha-methoxy diketopiperazines, which could also be converted to epidithiodiketopiperazines via Schmidt's protocol [43]. These substrates were accessed by exhaustive bromination through a combination of radical and electrophilic processes, solvolysis of the alpha-bromides in methanol, and then reductive dehalogenation. This work effectively expanded the scope of Trown's method to 3,6-dialkylated diketopiperazines. In contrast, Ottenheijm concurrently demonstrated that rather than oxidizing diketopiperazine structures, one could arrive at the same substrate by forming a diketopiperazine from alpha-keto acid chlorides [44]. More recently, in 2007, Overman reported a radical-promoted intramolecular C–H abstraction method for oxidizing the alpha position of diketopiperazine **60** [45]. Tris(trimethylsilyl)silane and copper-acetate were utilized in concert to generate and oxidize the captodatively stabilized radicals, respectively. Upon oxidation, epidisulfide formation was accomplished via scandium triflate-mediated ionization and nucleophilic thiolation with hydrogen sulfide. Interestingly, all of the methods described hitherto have relied on Lewis acid-catalyzed ionization and incorporation of hydrogen sulfide; stereoselective access to *cis*-dithiols thus continued to remain a major challenge.

9.4.3 Total Synthesis of Epidithiodiketopiperazine Alkaloids

In the 1970s, Kishi published a series of landmark papers [36] describing the total syntheses of (±)-dehydrogliotoxin (1973) [36b], (±)-sporidesmin A (1973) [36c], (±)-gliotoxin (1976) [36d], and (±)-hyalodendrin (1976) [36e] in which he employed a new method for epidithiodiketopiperazine synthesis (Scheme 9.4). Cognizant of the harsh conditions required in all of the sulfur incorporation methods developed at the time, it was determined that thiolation would be performed in the early stages of the syntheses. A dithiol intermediate obtained in a similar fashion to Trown's epidithiodiketopiperazine was protected as a dithioacetal, and after elaboration of this core diketopiperazine structure, the dithioacetal was unraveled under mild conditions in the final steps to afford the target epidisulfides.

The total synthesis of (±)-sporidesmin A (**69**) was the first major application of Kishi's methodology. Starting with *N*-Me-alanine-glycine anhydride **63**, methoxymethyl protection of the glycine amide, radical bromination, and potassium thioacetate workup led to thioester **64**. Hydrolysis of the thioester followed by boron trifluoride-mediated dithioacetal formation with the trithiane of anisaldehyde afforded the thiolated diketopiperazine core in 80 % yield as a 2:1 mixture of diastereomers. Deprotonation of the glycinyl methine and acylation of the carbanion with acid chloride **66** yielded the complete carbon skeleton of (±)-sporidesmin A (**69**) in 61 % yield. A straightforward sequence involving ketone reduction and protecting group manipulations was capped with an oxidative cyclization using iodosobenzene diacetate to afford (±)-sporidesmin A diacetate (**68**). After hydrolysis of the acetates, the dithioacetal was converted directly to the epidisulfide by its exposure to *m*CPBA and boron trifluoride.

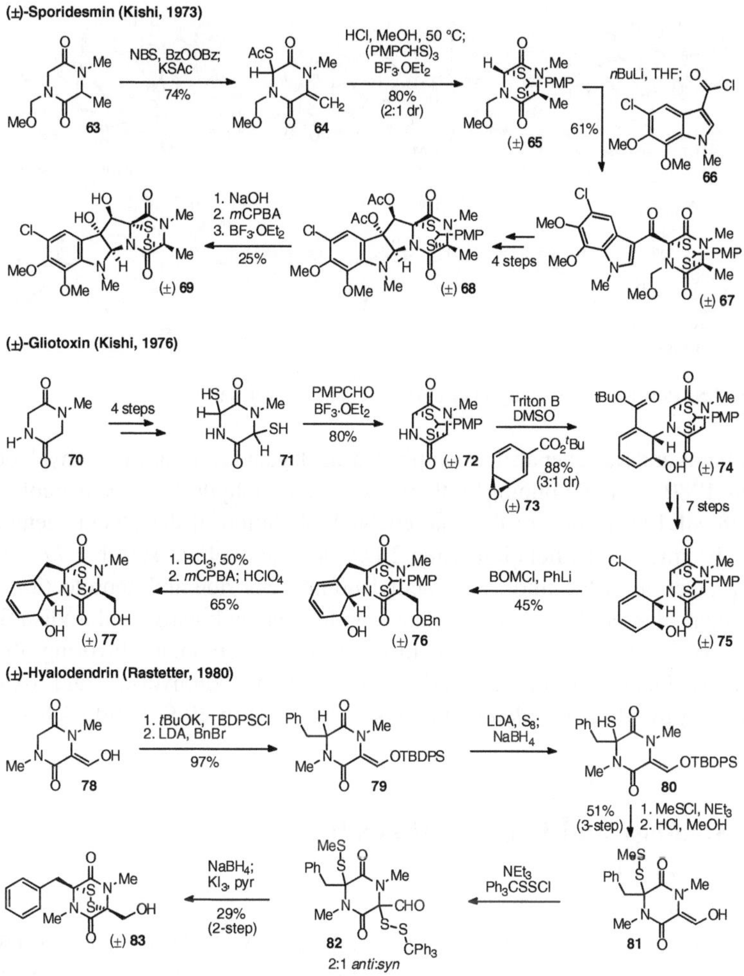

Scheme 9.4 Total synthetic of epipolythiodiketopiperazine alkaloids

In a similar fashion, (±)-gliotoxin (**77**) was synthesized from the dithiol adduct of sarcosine-glycine anhydride **71**. Lewis acid-catalyzed dithioacetal formation with anisaldehyde was followed by its conjugation to 4-carbo-*tert*-butoxybenzene oxide (**73**) using Triton B in DMSO. The Michael adduct was attained in 88 % yield with a 3:1 dr favoring the desired diastereomer. After a straightforward 7-step conversion of the *tert*-butyl ester to the allylic chloride, controlled addition of phenyllithium into a THF solution of **75** and excess chloromethyl benzyl ether afforded the carbon skeleton of gliotoxin in 45 % yield. Benzyl deprotection set up the system for a mild unveiling of the epidisulfide using *m*CPBA and perchloric acid in 65 % yield.

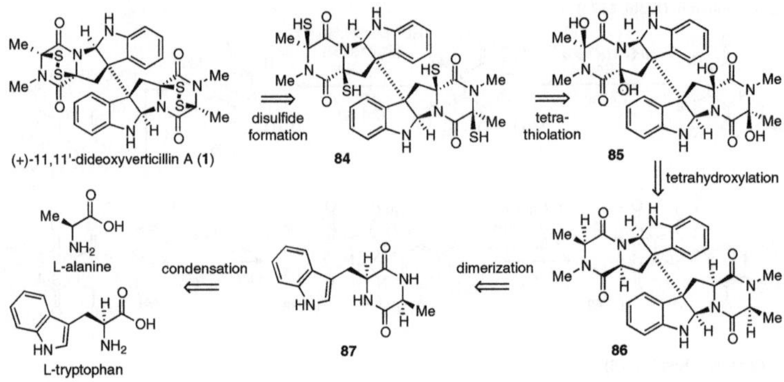

Scheme 9.5 Retrosynthetic analysis of (+)-11,11′-dideoxyverticillin A

Williams and Rastetter also accomplished an elegant synthesis of (±)-hyalodendrin (**83**) in 1980 [39]. Beginning with the sarcosine anhydride-derived enolic aldehyde **78**, silyl protection of the enal enabled alkylation of the glycine center with benzyl bromide and thiolation using LDA and monoclinic sulfur *a la* Schmidt. After protection of the thiol with methylsulfenyl chloride and deprotection of the silyl ether, the enol was sulfenylated with triphenylmethyl chlorodisulfide to afford bis(disulfide) **82** as a 2:1 mixture of diastereomers favoring the anti isomer. Reduction of the disulfides with sodium borohydride and oxidation with KI_3 in pyridine afforded (±)-hyalodendrin (**83**) in 29 % yield (Scheme 9.4).

9.5 Strategy and Retrosynthesis for (+)-11,11′-Dideoxyverticillin A

The retrosynthesis of (+)-11,11′-dideoxyverticillin A (**1**) [25] (Scheme 9.5) was largely inspired by plausible biosynthetic processes responsible for its production in nature [25]. The vast array of isolable congeners available in their various stages of complexity together with the encoded biosynthetic machinery teased out from the annotations of biosynthetic gene clusters presented a clear window into the assembly of the dimeric epidithiodiketopiperazine alkaloids. Always cognizant of genus-dependent versus genus-independent modifications, the level of variability that was tolerated in core modifications represented among members of the epipolythiodiketopiperazine family was taken as indication of the degree to which a transformation was substrate or enzyme directed. Careful analysis of gliotoxin-feeding studies by Kirby, Taylor, and Sammes; Howlett's gene cluster analyses of the same product; as well as our own hypotheses about the innate reactivities of intermediates led to the retrosynthesis described below [25, 26].

In the most ambitious disconnection that could be entertained, the natural product was dissected in half to afford a fully functionalized monomeric radical. While it was abundantly clear that the sensitivity of the epidisulfide moiety to radical conditions would preclude such a process, this disconnection was the idealized expression of our desire for the most convergent construction of this complex alkaloid and the willingness to face the challenges inherent to late-stage functionalization reactions. More conservatively, however, (+)-11,11′-dideoxyverticillin A (**1**) was postulated to arise from the auto-oxidation of a tetrathiol intermediate, which could itself be derived from the nucleophilic thiolation of an *N*-acyliminium ion species. These highly reactive electrophiles could potentially arise from the tetrahydroxylation of a dimeric tetracycle. Invocation of an unfunctionalized dimeric tetracycle was rooted in the observation of such species in isolates such as (+)-WIN 64821 (**10**) and (−)-ditryptophenaline (**11**) [46]. While these dimers would be accessed in the same spirit by which they are synthesized in nature—a single-electron oxidative dimerization of tryptophan-containing cyclic dipeptides—we hypothesized that decoupling the cyclization and dimerization steps would afford a degree of stereocontrol in C3–C3′ bond formation that is uniquely desired in the laboratory synthesis of these molecules [7]. As such, the dimers were to be constructed through a single-electron reductive dimerization of tetracyclic bromides, obtained from the two-electron oxidation of a tryptophan-containing cyclic dipeptide. This diketopiperazine intermediate would in turn be derived from the condensation of natural L-tryptophan and L-alanine amino acids.

The challenging tetrafunctionalization reactions notwithstanding, the proposed strategy would be highly contingent upon the execution of each reaction with a high level of stereocontrol. In a strategy reminiscent of Seebach's self-reproduction of chirality [47], the final stereochemical arrangement arises during thiolation with stereochemical induction from the C3 to C3′ vicinal quaternary stereocenters. These carbon stereocenters are derived from the amino acid stereochemistries during the bromocyclization step, bringing the process full circle. The constant relay and amplification of stereochemical information allows the entire three-dimensional structure of these epidithiodiketopiperazine alkaloids to be encoded in two natural amino acids.

9.5.1 Synthesis of (+)-11,11′-Dideoxyverticillin A

The synthesis of (+)-11,11′-dideoxyverticillin A (**1**) opened with the formation of an L-Trp-L-Ala dipeptide derivative using standard EDC peptide coupling conditions (Scheme 9.6). Boc-deprotection with trifluoroacetic acid, concentration, and subjection of the amine salt to basic polar protic conditions afforded diketopiperazine (−)-**91** in 84 % yield over the two steps. Brominative cyclization using molecular bromine in acetonitrile at 0 °C yielded a diastereomeric mixture of tetracyclic products in a 4:1 ratio with the desired *endo*-product isolated in 76 % yield. This sequence, inspired by the pioneering work on C3a-functionalized hexahydropyrroloindoles by Hino, Crich, and Danishefsky and originally extended to the diketopiperazine-containing alkaloids by our group in the context of our

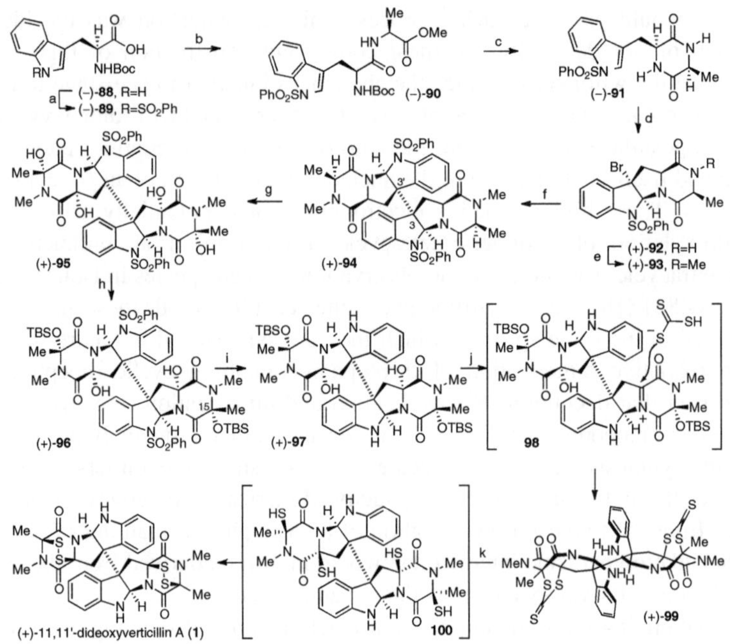

(a) LHMDS, $PhSO_2Cl$, THF, 71%; (b) L-Ala-OMe, EDC·HCl, HOBt, Et_3N, CH_2Cl_2, 94%; (c) TFA, CH_2Cl_2; morpholine, *t*BuOH, 84%; (d) Br_2, MeCN, 76%; (e) MeI, K_2CO_3, acetone, 77%; (f) $CoCl(PPh_3)_3$, acetone, 46%; (g) Py_2AgMnO_4, CH_2Cl_2, 63%; (h) TBSCl, PPY (5 mol%), Et_3N, DMF, 55%; (i) 5% Na(Hg), NaH_2PO_4, MeOH, 87%; (j) K_2CS_3, TFA, CH_2Cl_2, 56%; (k) ethanolamine, acetone; KI_3, pyr, 62%.

Scheme 9.6 Total synthesis of (+)-11,11′-dideoxyverticillin A (1)

(+)-WIN 64821 (**10**) and (−)-ditryptophenaline (**11**) syntheses [7], not only effectively differentiated the two amide moieties but also most importantly marked the first in a series of stereochemical transfer steps in which the stereochemistry of the constituent L-amino acids was relayed to ultimately define each of the relative and absolute stereochemical configurations at all eight stereogenic centers found in the target compound. Finally, *N*-methylation of the base-sensitive amide in 77 % yield using methyl iodide and potassium carbonate in acetone completed the 5-step synthesis of our key tetracyclic bromide monomer starting from commercially available amino acid derivatives.

The octacyclic dimer (+)-**94** could be obtained in short order from the tetracyclic bromide (+)-**93** via a Co(I)-mediated reductive dimerization protocol first implemented in our prior syntheses of (+)-chimonanthine (**7**), (+)-folicanthine (**8**), and (−)-calycanthine (**9**) [7]. Simple exposure of intermediate (+)-**93** to tris (triphenylphosphine)cobalt(I) chloride [48] in acetone under anaerobic conditions rapidly afforded dimer (+)-**94** in 46 % yield. While higher yields (52 % yield) could be obtained in tetrahydrofuran on small scale, performing the reaction in acetone reproducibly afforded higher yields on gram scales. Notably, the product was obtained in similar efficiency on multi-gram scale (43 % yield on 8-g scale)

Scheme 9.7 Epidithiodiketopiperazine formation on monomeric model systems

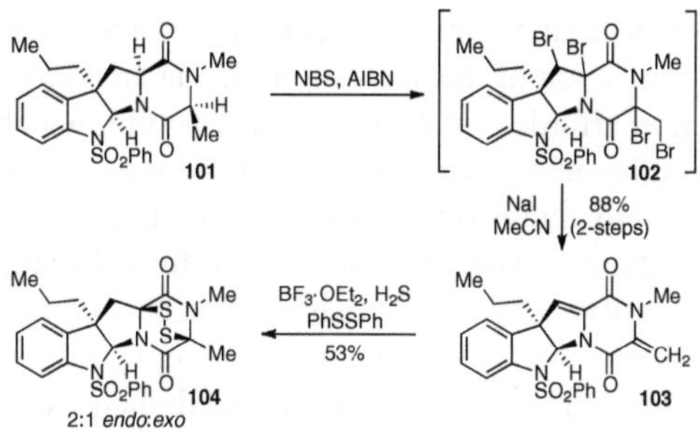

even despite forgoing the challenging chromatographic separation of the dimerization precursor from its 11*R* epimer (5:1 ratio) generated during the preceding *N*-methylation step. The dimerization event is believed to operate through a single-electron reduction mechanism in which the tertiary benzylic bromides are rapidly reduced to their corresponding radicals; radical recombination produces the desired dimer. While the radical centers were potentially stereochemically labile, preference for *cis*-fusion on [3.3.0]-bicyclic ring systems led to the preservation of the original bromide stereochemistry and established the C3–C3′ dimeric linkage in a stereoretentive manner.

With the unadorned dimeric skeleton of (+)-11,11′-dideoxyverticillin A (**1**) in hand, we initially set out to oxidize the C_α positions of each of the amino acid residues via enol/enolate oxidation chemistry; however, even highly optimized conditions utilizing KHMDS and Davis oxaziridine in a DMPU/THF solvent mixture only afforded partially oxidized products, various diastereomeric products, as well as decomposition byproducts. Tetraol products could be obtained but in single digit yields and as diastereomeric mixtures. Neither the use of bis-lithiated amide bases nor direct thiolation of enolates fared any better. In retrospect, beyond the reactivity issues, the base sensitivity of the resultant hemiaminal products (vide infra) proved to be the bane of this approach. Furthermore, nonbasic conditions such as those used in amide activation strategies were ineffective for the alpha-functionalization of the amino acid residues as the diketopiperazine substructure seemed particularly inert to these conditions. At this critical juncture, recognizing the potential for generating mixtures of up to ten different stereoisomers in the oxidative transformations and the challenges associated with their analysis, we moved to a monomeric model system.

Coincident to our shift to a model system, the unique push–pull arrangement of the electron-rich amide nitrogen and the electron-withdrawing amide carbonyl about the C_α-position of the amino acid residues piqued our interest in the potential for radical-mediated oxidation of those positions. Indeed, capitalizing on the potential for captodative stabilization of radicals on the amino acid residues, exposure of tetracyclic substrate **101** to *N*-bromosuccinimide and AIBN in CCl_4 led to its tetrabromination (Scheme 9.7). In the propagation phase, C_α–H abstraction by a bromine radical was followed by reaction with transiently generated molecular bromine to regenerate a bromine radical and afford an alkyl bromide. Ionization of the alkyl bromide via an *N*-acyliminium ion and subsequent elimination afforded an enamide substructure, which underwent further electrophilic bromination to generate a *vic*-dibromide. Iteration of this sequence on the second amino acid residue resulted in a tetrabromide [25]. Reductive debromination with sodium iodide in acetonitrile afforded diene **103** in 88 % yield over two steps. Gratifyingly, the diene proved to be a competent precursor to the epidithiodiketopiperazine group. Indeed, bubbling hydrogen sulfide through a solution containing diene **103**, boron trifluoride, and diphenyl disulfide as an internal oxidant afforded a 2:1 mixture of *endo*/*exo* epidithiodiketopiperazines in 53 % overall yield. Interestingly, the reactivity of the Br-centered radical was uniquely tuned for this reaction as a variety of C-centered radical species were incapable of intermolecular hydrogen atom abstraction, while O-centered radical species resulted in a range of decomposition products.

Efforts to apply the newfound conditions to the dimeric octacyclic bisdiketopiperazine (+)-**94** were unfortunately unfruitful owing to putative modes of C3–C3′ bond fragmentation and potentially unfavorable steric interactions arising from the conjoined monomeric subunit. Close inspection of the bond dissociation energies associated with the reactive species in the bromine radical-mediated C_α-abstraction method applied to the monomer gave rise to a powerful insight [49]. Provided that a bromine radical is capable of C_α–H abstraction on the diketopiperazines and that the average bond dissociation energy of a formyl C–H bond is similar to that of an H–Br bond, mild oxidants which oxidize formyl groups to their corresponding carboxylic acids via a hydrogen atom abstraction mechanism should be suitable in effecting the desired oxidative transformation. Cognizant of the fact that organic-soluble permanganate reagents are believed to oxidize aldehydes to carboxylic acids through a C–H abstraction process, such conditions were sought out and applied to the tetracyclic monomer. Exposure of diketopiperazine **101** to tetra-*n*-butylammonium permanganate in pyridine [50] for 2 h afforded a tetracyclic diol as a single diastereomer in 78 % yield.

Oxidation of the dimer, however, proved far more challenging. Using the tetra-*n*-butylammonium permanganate conditions explored on the monomer, a tetraol was obtained in 9 % yield. The yield was able to be improved to 40 % by optimizing the isolation procedure for this highly polar and elimination-prone compound, but the tetrahydroxylation reaction was still fraught with competitive epimerization of the hemiaminals and the formation of complex reaction mixtures containing incompletely oxidized intermediates. Suspecting the involvement of pyridine in the epimerization of the hemiaminal-containing structures, the solvent was replaced by dichloromethane. After further experimentation, bis(pyridine)silver(I) permanganate (Py_2AgMnO_4) [51] was found to be the optimal oxidant. Exposure of dimer (+)-**94** to Py_2AgMnO_4 (4.8 equiv) in dichloromethane at 23 °C for 2 h produced the dimeric octacyclic tetraol (+)-**95** in 63 % yield as a single diastereomer. Single crystal X-ray diffraction analysis of the product revealed that the oxidation occurred with complete retention of stereochemistry at each of the C_α–H methines.

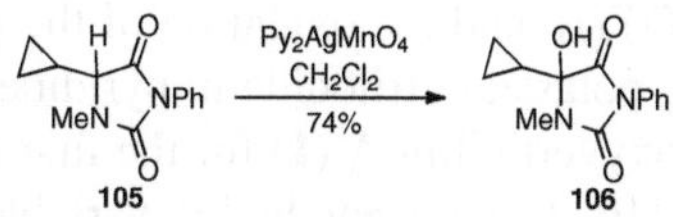

Scheme 9.8 Radical clock-based mechanistic probe

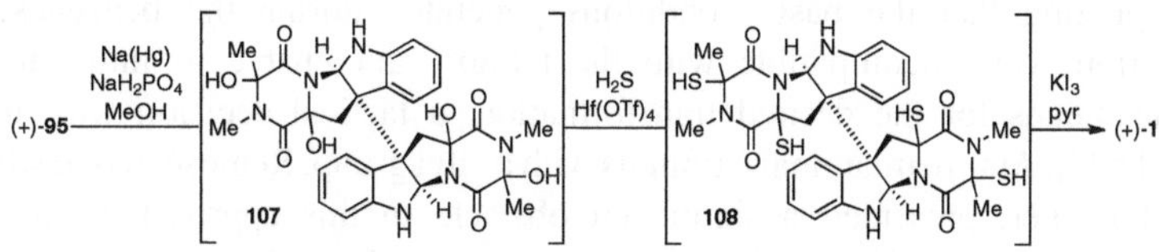

Scheme 9.9 First generation route to (+)-11,11′-dideoxyverticillin A (1)

In a *single step* and on *multi-gram scale*, four C–H bonds were replaced by four C–O bonds in a highly stereoretentive manner with an efficiency approaching ca. 90 % yield per hydroxylation event.

The oxidation is believed to proceed through a fast abstraction-rebound mechanism consistent with mechanistic and computational studies of permanganate-mediated oxidations by Mayer [52] and Houk [53], respectively. Radical-clock hydantoin **85**, upon subjection to the optimal oxidation conditions, afforded the hydroxylated cyclopropyl hydantoin **105** in 74 % yield (Scheme 9.8), demonstrating that a radical-mediated rebound process would be faster than the rate of cyclopropyl ring opening. Further studies on a thermodynamically more stable *cyclo*-D-Trp-L-Ala derivative revealed that only the C_α(L-Ala) position is oxidized under the reaction conditions, leaving the C_α(D-Trp) position unchanged. Whether due to a nonoptimal C–H conformation for abstraction or a more sterically impacted transition state in the oxidation of the tryptophan residue, this observation stresses the important consequences of stereochemistry when choosing the amino acid precursors.

The tetraols were found to be highly sensitive toward acidic and basic conditions. Under Bronsted acidic conditions, the hemiaminals readily eliminated to generate a tetraene, while under basic conditions, the tetraol either decomposed or epimerized to generate a mixture of diastereomers. It is speculated that the base-mediated epimerization proceeds through ring–chain tautomerization involving a putative alpha-keto amide derivative. It is also of note that simple dissolution of tetraol (+)-**95** in methanol also leads to its degradation to a complex mixture of products.

Nonetheless, with facile access to the tetraol, a first generation synthesis of (+)-11,11′-dideoxyverticillin A (**1**) was developed (Scheme 9.9). In said strategy, the key tetraol intermediate (+)-**95** was subject to a rapid three-step sequence involving (1) benzenesulfonyl deprotection using sodium amalgam in dibasic sodium phosphate buffered methanol, (2) thiolation of the crude diaminotetraol product **100** using condensed hydrogen sulfide in a sealed tube with a highly oxophilic Lewis acid $Hf(OTf)_4$, and (3) oxidation of the putative tetrathiol **108** to bisepidisulfide (+)-**1** using potassium triiodide in pyridine. This sequence afforded the target (+)-11,11′-dideoxyverticillin A (**1**) for the first time in yields of 2–15 % over three steps (Scheme 9.9). The low and highly variable yields together with the use of pressurized toxic hydrogen sulfide gas, however, necessitated the reevaluation of the final stages of the synthesis.

Recognizing that the basic conditions generated during the benzenesulfonyl deprotection were incompatible with the tetraol, and unable to find alternative nonbasic means for the desired transformation, a tactical decision was made to protect the C_α(Ala)-hemiaminal carbinols with a silyl group. Regioselective silylation issues, however, presented a significant obstacle to this approach as the steric constraints of a *cis*-configured diol on each monomer tolerated only a single silylation event per diketopiperazine. Incidentally, protection of the alanine-derived hydroxyl groups was necessary in imparting greater base stability to the tetrahydroxylated compound. Despite efforts to capitalize on potential substrate-controlled selectivities, initial variations on the silyl groups failed to provide the desired bis-silylated compound in yields in excess of 36 %. Cognizant of the role the nucleophilic catalyst plays in the transition state, both quinuclidine and pyridine-type catalysts were also evaluated. After significant optimization, Fu's (*R*)-(+)-4-pyrrolidinopyridinyl (pentamethylcyclopentadienyl)iron catalyst (5 mol %) [54], *tert*-butyl(chloro) dimethylsilane, and triethylamine in DMF effected the regioselective protection of the C_α(Ala)-hydroxyls to give diol (+)-**96** in 55 % yield (Scheme 9.6). Notably, the other enantiomer of the catalyst provides the desired product in only 36 % yield.

Benzenesulfonyl deprotection of diol (+)-**96** proceeded smoothly using sodium amalgam in buffered methanol, and the octacyclic diaminodiol (+)-**97** was obtained in 87 % yield as a prelude to the thiolation event. Our thiolation strategy was originally inspired by the Woodward–Prévost method for the *cis*-dihydroxylation of alkenes [55]; a successful thiol variant would enable stereocontrolled formation of a *cis*-dithiol. In accordance with this strategy, it was hypothesized that dithioacetic acid could trap an *N*-acyliminium ion generated from the first acid-induced hemiaminal ionization event. A second ionization event on the same diketopiperazine subunit would result in the intramolecular trapping of the *N*-acyliminium ion to generate a thiocarbenium ion, which upon aqueous quenching would reveal a *cis*-dithiol intermediate. Execution of this sequence, however, proved difficult, often resulting in intractable mixtures. In an effort to temporally partition the discrete thiol incorporation and hydrolysis steps, diprotic nucleophiles were explored.

After extensive experimentation, it was discovered that exposure of silyl ether (+)-**97** to potassium trithiocarbonate and trifluoroacetic acid in dichloromethane affords a 25:7:1 mixture of *endo*/*endo*:*endo*/*exo*:*exo*/*exo* bisdithiepanethione products, reflecting a ca. 5:1 preference for nucleophilic approach from the *endo*-face of each diketopiperazine moiety. Resubjection of the isolated bisdithiepanethione diastereomers to the original reaction conditions did not result in their equilibration, indicating that the products were a result of kinetic trapping.

Scheme 9.10 Schmidt's structural elucidation of a dithiol intermediate

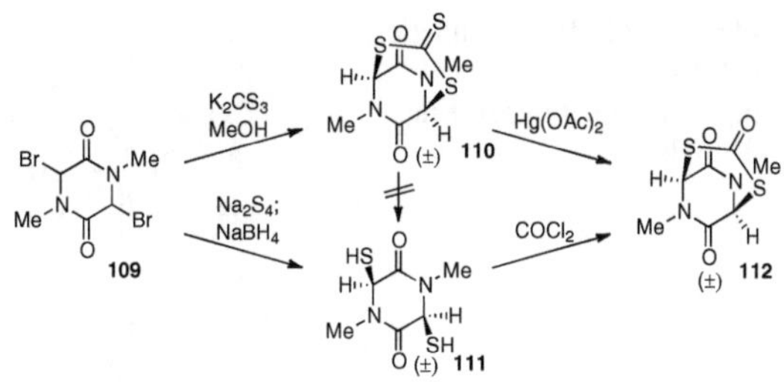

Significantly, the desired *endo*/*endo* diastereomer was isolated as the major product in 56 % yield.

In the thiolation event, it is believed that ionization of the less sterically hindered C_α(Trp)-hydroxyl groups precedes that of the C_α(Ala)-silyl ethers. Accordingly, *N*-acyliminium ion formation and nucleophilic trapping at this position transpires with stereoinduction from the proximally located C3 or C3′ stereocenter during this critical diastereodetermining step. Subsequent ionization of the C_α(Ala)-silyl ether then results in intramolecular closure to the bisdithiepanethione structure. The high level of diastereoselection observed for the dimeric tetraol relative to that which is observed for its monomeric counterpart (*ca.* 2:1 *endo*/*exo*, vide supra) is consistent with the proposed ionization sequence. The additional steric shielding afforded by the quaternary carbon center of the pendant monomer at the C3 position is effective in directing the nucleophile to the concave face of the [3.3.0]-bicycle. It is also of particular note that even despite the occurrence of a stereoablative hemiaminal ionization process, memory of the original amino acid stereochemistry at the alpha carbons, embedded in the stereochemical configuration at C3, is retrieved and re-encoded into the C11 and C15 stereocenters.

This synthetic approach represents a nuanced yet not insignificant development with respect to stereocontrolled formation of *cis*-dithiols on diketopiperazine structures. Only after extensive experimentation was it discovered that trifluoroacetic acid is a uniquely effective medium in which the trithiocarbonate salt was simultaneously soluble and stable in spite of the acidic conditions. Due to the rapid decomposition of trithiocarbonate salts in acidic solutions of the reagent, these reagents were exclusively used under alkaline conditions and consequently limited to S_N2-displacement reactions prior to our work. In addition, while Schmidt had reported the installation of a trithiocarbonate bridge onto sarcosine anhydride in 1971 during the structural elucidation of the dithiols obtained in Trown's epidisulfide synthesis [34] and through his own methodology employing Na_2S_4, he stopped short of accessing an epidisulfide (Scheme 9.10) [35]. Remarking on the difficulties of unraveling the bridged structure, the trithiocarbonate was converted to a dithiocarbonate, and independent synthesis of this structure by reaction of phosgene with Trown's dithiol enabled their direct comparison.

Consistent with these observations, the use of highly basic reagents for the deprotection of the thiols proved futile. As with tetraol (+)-**95**, base sensitivity of the tetrathiol product necessitated the use of neutral reagents. After significant

Scheme 9.11 Retrosynthetic analysis for the epipolythiodiketopiperazine alkaloids

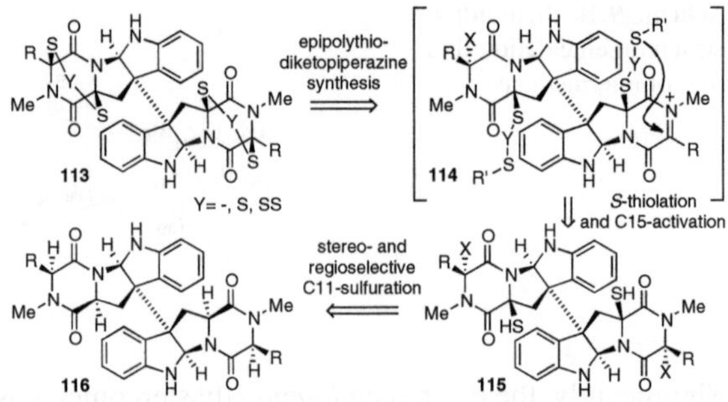

experimentation, it was realized that tethered dinucleophiles such as ethanolamine could unravel the bis(dithiepanethione) (+)-**99** efficiently to yield putative tetrathiol **100**, which could then be oxidized to bis(epidisulfide) (+)-**1** by exposure to air. After optimization, tetrathiol formation was followed immediately by an aqueous hydrochloric acid wash. Immediate treatment of the organic layer with potassium triiodide efficiently afforded (+)-11,11′-dideoxyverticillin A (**1**) in 62 % yield from bis(dithiepanethione) (+)-**99**.

9.5.2 Generalization to the Epipolythiodiketopiperazine Alkaloids

Upon completion of the synthesis of (+)-11,11′-dideoxyverticillin A (**1**), it was recognized that there were only a handful of reports of monomeric epitri- or epitetrathiodiketopiperazine syntheses, nearly all of which were accomplished with lack of sulfide chain length control. In the limited cases where selectivity was achieved, the results were highly substrate dependent or the method lacked substrate scope [56].

Our retrosynthesis for the epipolythiodiketopiperazine alkaloids by and large observes the basic strategic framework laid out in the synthesis of (+)-11,11′-dideoxyverticillin A (**1**); the main deviation is aptly in the final stages of thiol incorporation (Scheme 9.11).

The key precept for any strategy would involve complete stereochemical control and precision in the degree of sulfidation. Accordingly, we envisioned that the epipolysulfides could arise from the ionization of a C15 hemiaminal derivative and subsequent cyclization of a polysulfane onto the resultant *N*-acyliminium ion. The polysulfane would be derived from the corresponding thiol accessed by regioselective functionalization of the C_α(Trp) position.

This general approach to the dimeric epipolythiodiketopiperazine alkaloids, as portrayed through our syntheses of (+)-chaetocins A (**2**) and C (**3**) and (+)-12,12′-dideoxychetracin A [57], represents the first strategy for the chain length-controlled access to the epipolysulfide homologs. As exemplified by our synthesis of (+)-**1**, selectivity—regio, chemo, and stereo, as well as precision in the degree of sulfuration—proved to be the prime driving force of innovation throughout the work, and in spite of the exceedingly narrow reactivity profiles imposed by such strict constraints, deft navigation resulted in a successfully convergent and efficient synthesis.

9.6 Conclusion

The synthesis of (+)-11,11′-dideoxyverticillin A (**1**) represents the first total synthesis of a dimeric epidithiodiketopiperazine alkaloid since the discovery of this family of natural products nearly four decades prior. Our synthetic route was greatly inspired by biosynthetic considerations and presented highly efficient and chemoselective solutions to challenging transformations [1]. Included among these challenges were the implementation of a highly convergent radical dimerization strategy, the execution of a highly efficient and stereoselective tetrahydroxylation reaction, and the application of a *cis*-selective thiolation procedure in the late stages of the synthesis.

References

1. (a) Waring P, Eichner RD, Mullbacher A (1988) Med Res Rev 8:499; (b) Waring P, Beaver J (1996) Gen Pharmacol 27:1311; (c) Gardiner DM, Waring P, Howlett BJ (2005) Microbiology 151:1021; (d) Patron NJ, Waller RF, Cozijnsen AJ, Straney DC, Gardiner DM, Nierman WC, Howlett BJ (2007) BMC Evol Biol 7:174; (e) Huang R, Zhou X, Xu T, Yang X, Liu Y (2010) Chem Biodivers 7:2809; (f) Iwasa E, Hamashima Y, Sodeoka M (2011) Isr J Chem 51:420
2. Hauser D, Weber HP, Sigg HP (1970) Helv Chim Acta 53:1061
3. Katagiri K, Sato K, Hayakawa S, Matsushima T, Minato H (1970) J Antibiot (Tokyo) 23:420
4. Son BW, Jensen PR, Kauffman CA, Fenical W (1999) Nat Prod Res 13:213
5. Fukuyama T, Nakatsuka S-I, Kishi Y (1981) Tetrahedron 37:2045
6. (a) Hino T, Nakagawa M (1989) In: Brossi A (ed) The alkaloids: chemistry and pharmacology, vol 34, Academic, New York, NY, pp 1–75; (b) Anthoni U, Christopherson C, Nielsen PH (1999) In: Pelletier SW (ed) Alkaloids: chemical and biological perspectives, vol 13, Pergamon, New York, NY pp 163–236
7. Movassaghi M, Schmidt MA (2007) Angew Chem Int Ed 46:3725
8. Movassaghi M, Schmidt MA, Ashenhurst JA (2008) Angew Chem Int Ed 47:1485
9. (a) Hendrickson JB, Rees R, Goschke R (1962) Proc Chem Soc:383; (b) Hino T, Yamada S (1963) Tetrahedron Lett 4:1757; (c) Scott AI, McCapra F, Hall ES (1964) J Am Chem Soc 86:302; (d) Nakagawa M, Sugumi H, Kodato S, Hino T (1981) Tetrahedron Lett 22:5323; (e) Fang C-L, Horne S, Taylor N, Rodrigo R (1994) J Am Chem Soc 116:9480; (f) Link JT, Overman LE (1996) J Am Chem Soc 118:8166; (g) Somei M, Osikiri N, Hasegawa M, Yamada F (1999) Heterocycles 51:1237; (h) Overman LE, Paone DV, Stearns BA (1999) J Am Chem Soc 121:7702; (i) Overman LE, Larrow JF, Stearns BA, Vance JM (2000) Angew Chem Int Ed 39:213; (j) Ishikawa H, Takayama H, Aimi N (2002) Tetrahedron Lett 43:5637; (k) Matsuda Y, Kitajima M, Takayama H (2005) Heterocycles 65:1031; (l) Verotta L, Orsini F, Sbacchi M, Scheidler MA, Amador TA, Elisabetsky E (2002) Bioorg Med Chem 10:2133

[1] We acknowledge financial support by NIH-NIGMS (GM089732).

10. (a) Zhang Y-X, Chen Y, Guo X-N, Zhang X-W, Zhao W-M, Zhong L, Zhou J, Xi Y, Lin L-P, Ding J (2005) Anti-Cancer Drug 16:515; (b) Chen Y, Zhang Y-X, Li M-H, Zhao W-M, Shi Y-H, Miao Z-H, Zhang X-W, Lin L-P, Ding (2005) J Biochem Biophys Res Commun 329:1334; (c) Chen Y, Miao Z-H, Zhao W-M, Ding J (2005) FEBS Lett 579:3683
11. Saito T, Suzuki Y, Koyama K, Natori S, Iitaka Y, Knoshita T (1998) Chem Pharm Bull 36:1942
12. (a) Greiner D, Bonaldi T, Eskeland R, Roemer E, Imhof A (2005) Nat Chem Biol 1:143; (b) Isham CR, Tibodeau JD, Jin W, Timm MM, Bible KC (2007) Blood 109:2579; (c) Cherrier T, Suzanne S, Redel L, Calao M, Marban C, Samah B, Mukerjee R, Schwartz C, Gras G, Sawaya BE, Zeichner SL, Aunis D, Van Lint C, Rohr O (2009) Oncogene 28:3380; (d) Cook KM, Hilton ST, Mecinovic J, Motherwell WB, Figg WD, Schofield CJ (2009) J Bio Chem 284:26831; (e) Lakshmikuttyamma A, Scott SA, DeCoteau JF, Geyer CR (2010) Oncogene 29:576
13. Kwon-Chung KJ, Sugui JA (2009) Med Mycol 47:S97
14. (a) Munday R (1982) Chem Biol Interact 41:361; (b) Munday R (1987) J Appl Toxicol 7:17; (c) Chai CLL, Heath GA, Buleatt PB, O'Shea GA (1999) J Chem Soc Perk T 2:389; (d) Chai CLL, Waring P (2000) Redox Rep 5:257; (e) Scharf DH, Remme N, Heinekamp T, Hortschansky P, Brakhage AA, Hertweck C (2010) J Am Chem Soc 132:10136
15. Bernardo PH, Chai CLL, Deeble GJ, Liu X-M, Waring P (2001) Bioorg Med Chem Lett 11:483
16. Waring P, Mamchak A, Khan T, Sjaarda A, Sutton P (1995) Cell Death Differ 2:201
17. Waring P, Sjaarda A, Lin QH (1995) Biochem Pharmacol 49:1195
18. Hurne AM, Chai CLL, Waring P (2000) J Biol Chem 275:25202
19. Block KM, Wang H, Szabo LZ, Polaske NW, Henchey LK, Dubey R, Kushal S, Lazlo CF, Makhoul J, Song Z, Meuillet EJ, Olenyuk BZ (2009) J Am Chem Soc 131:18078.
20. (a) Jordan TW, Pedersen JS (1986) J Cell Sci 85:33; (b) Bernardo PH, Brasch N, Chai CLL, Waring P (2003) J Biol Chem 278:46549; (c) Orr JG, Leel V, Caermon GA, Marek CJ, Haughton EL, Erick LJ, Trim JE, Hawksworth GM, Halestrap AP, Wright MC (2004) Hepatology 40:232
21. Kirby GW, Robins DJ (1980) In: Steyn PS (ed) The biosynthesis of mycotoxins: a study in secondary metabolism, Academic, New York, NY, p 301
22. Sammes PG (1975) In: Herz W, Grisebach H, Kirby GW (ed) Progress in the chemistry of organic natural products, Springer, New York, NY, p 51
23. Taylor A (1971) In: Kadis S, Ciegler A (ed) Microbial toxins, Academic, New York, NY, p 337
24. (a) Fox EM, Howlett BJ (2008) Mycol Res 112:162; (b) Gardiner DM, Howlett BJ (2005) FEMS Microbiol Lett 248:241
25. Kim J, Ashenhurst JA, Movassaghi M (2009) Science 324:238
26. Kim J, Movassaghi M (2009) Chem Soc Rev 38:3035
27. Scharf DH, Remme N, Habel A, Chankhamjon P, Scherlach K, Heinekamp T, Hortschansky P, Brakhage AA, Hertweck C (2011) J Am Chem Soc 133:12322
28. Eccles GR (1888) Proc Am Pharm Assoc 84:382
29. Woodward RB, Yang NC, Katz TJ, Clark VM, Harley-Mason J, Ingleby RFJ, Sheppard N (1960) Proc Chem Soc:76
30. Robinson R, Teuber HJ (1954) Chem Ind:783
31. Crich D, Banerjee A (2007) Acc Chem Res 40:151
32. (a) Marsden SP, Depew KM, Danishefsky SJ (1994) J Am Chem Soc 116:11143; (b) Depew KM, Marsden SP, Zatorska D, Zatorski A, Bornmann WG, Danishefsky SJ (1999) J Am Chem Soc 121:11953; (c) Schkeryantz JM, Woo JCG, Siliphaivanh P, Depew KM, Danishefsky SJ (1999) J Am Chem Soc 121:11964
33. Bruncko M, Crich D, Samy R (1994) J Org Chem 59:5543
34. Trown PW (1968) Biochem Biophys Res Commun 33:402
35. Poisel H, Schmidt U (1971) Chem Ber 104:1714

36. (a) Kishi Y, Fukuyama T, Nakatsuka S (1973) J Am Chem Soc 95:6490; (b) Kishi Y, Fukuyama T, Nakatsuka S (1973) J Am Chem Soc 95:6492l; (c) Kishi Y, Nakatsuka S, Fukuyama T, Havel M (1973) J Am Chem Soc 95:6493; (d) Kishi Y, Fukuyama T (1976) J Am Chem Soc 98:6723; (e) Fukuyama T, Nakatsuka S, Kishi Y (1976) Tetrahedron Lett 38:3393
37. (a) Strunz GM, Kakushima M (1974) Experientia 30:719; (b) Srinivasan A, Kolar AJ, Olsen RK (1981) J Heterocycl Chem 18:1545; (c) Shimazaki N, Shima I, Hemmi K, Tsurumi Y, Hashimoto M (1987) Chem Pharm Bull 35:3527; (d) Wu Z, Williams LJ, Danishefsky SJ (2000) Angew Chem Int Ed 39:3866
38. Hino T, Sato T (1971) Tetrahedron Lett 12:3127
39. Williams RM, Rastetter WH (1980) J Org Chem 45:2625
40. Coffen DL, Katonak DA, Nelson NR, Sancilio FD (1977) J Org Chem 42:948
41. Ohler E, Poisel H, Tatruch F, Schmidt U (1972) Chem Ber 105:635
42. Ohler E, Tataruch F, Schmidt U (1973) Chem Ber 106:396
43. Yoshimura J, Nakamura H, Matsunari K, Sugiyama Y (1974) Chem Lett:559
44. Ottenheijm HCJ, Kerkhoff GPC, Bijen JWHA (1975) J Chem Soc Chem Comm:768
45. Overman LE, Sato T (2007) Org Lett 9:5267
46. (a) Springer JP, Buchi G, Kobbe B, Demain AL, Clardy J (1977) Tetrahedron Lett 18:2403; (b) Barrow CJ, Cai P, Snyder JK, Sedlock DM, Sun HH, Cooper R (1993) J Org Chem 58:6016
47. Seebach D, Boes M, Naef R, Schweizer WB (1983) J Am Chem Soc 105:5390
48. (a) Aresta M, Rossi M, Sacco A (1969) Inorg Chim Acta 3:227; (b) Yamada Y, Momose D-I (1981) Chem Lett:1277; (c) Bagal SK, Adlington RM, Baldwin JE, Marquez R (2004) J OrgChem 69:9100
49. Rauk A, Yu D, Taylor J, Shustov GV, Block DA, Armstrong DA (1999) Biochemistry 38:9089
50. Sala T, Sargent MV (1978) J Chem Soc Chem Commun:253
51. Firouzabadi H, Vessal B, Naderi M (1982) Tetrahedron Lett 23:1847
52. Gardner KA, Mayer JM (1995) Science 269:1849
53. Strassner T, Houk KN (2000) J Am Chem Soc 122:7821
54. Ruble JC, Fu GC (1998) J Am Chem Soc 120:11532
55. Woodward RB, Brutcher FV (1958) J Am Chem Soc 80:209
56. For prior epitrithiodiketopiperazine syntheses, see: (a) Brewer D, Rahman R, Safe S, Taylor A (1968) Chem Commun 24:1571; (b) Murdock KC, Angier RB (1970) Chem Comm:55; (c) Safe S, Taylor A (1970) J Chem Soc C 3:432; (d) Ref. [39]; (e) Ohler E, Tatruch F, Schmidt U (1972) Chem Ber 105:3658; (f) Ottenheijm HCJ, Herscheid JDM, Kerkhoff GPC, Spande TF (1976) J Org Chem 41:3433; (g) Curtis PJ, Greatbanks D, Hesp B, Cameron AF, Freer AA (1977) J Chem Soc Perk T I 2:180; (h) Coffen DL, Katonak DA, Nelson NR, Sancilio FD (1977) J Org Chem 42:948; (i) Kirby GW, Rao GV, Robins DJ, Stark WM (1986) Tetrahedron Lett 27:5539; For prior epitetrathiodiketopiperazine syntheses, see: (j) Refs. [39, 61b, c, 61g–i]; (k) Poisel H, Schmidt U (1971) Angew Chem Int Ed 10:130; (l) Yoshimura J, Sugiyama Y, Matsunari K, Nakamura H (1974) Bull Chem Soc Jpn 47:1215; (m) Strunz GM, Kakushima M, Stillwell MA (1975) Can J Chem 53:295; (n) Kirby GW, Robins DJ, Stark M (1983) J Chem Soc Chem Commun:812; (o) Waring P, Eichner RD, Tiwari-Palni U, Mullbacher A (1987) Aust J Chem 40:991
57. Kim J, Movassaghi M (2010) J Am Chem Soc 132:14376

Chapter 10
Retigeranic acid

David R. Adams and Tomáš Hudlický

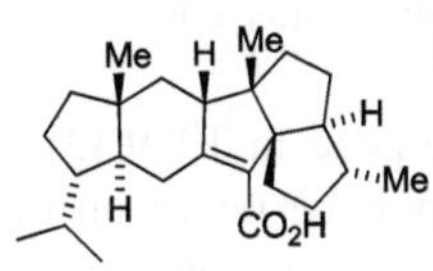

retigeranic acid A

10.1 Introduction

Retigeranic acid (**1**) belongs to the class of sesterterpenes, which comprise five isoprene units. The chemistry of this group of terpenes, representative examples of which are shown in Fig. 10.1, was the subject of a recent review [1]. The unique structure of retigeranic acid was undoubtedly of interest to all those who, in the late 1970s and early 1980s, participated in the development of methods for the synthesis of cyclopentanoid natural products. This particular discipline dominated an entire decade, 1978–1988, as many chemists were targeting various triquinane-containing natural products. The triquinane era probably began in 1977 when Zalkow reported the isolation of isocomene [2], a unique sesquiterpene containing an angular triquinane moiety. Retigeranic acid also contains this subunit but was not officially recognized as a triquinane at that time. The report of isocomene structure initiated vigorous activity in the synthetic community, with Oppolzer being the first to

D.R. Adams • T. Hudlický (✉)
Chemistry Department and Centre for Biotechnology, Brock University, 500 Glenridge Avenue, St. Catharines, Ontario L2S 3A1, Canada
e-mail: thudlicky@brocku.ca

DOI 10.1007/978-3-642-34065-9_10, © Springer-Verlag Berlin Heidelberg 2012

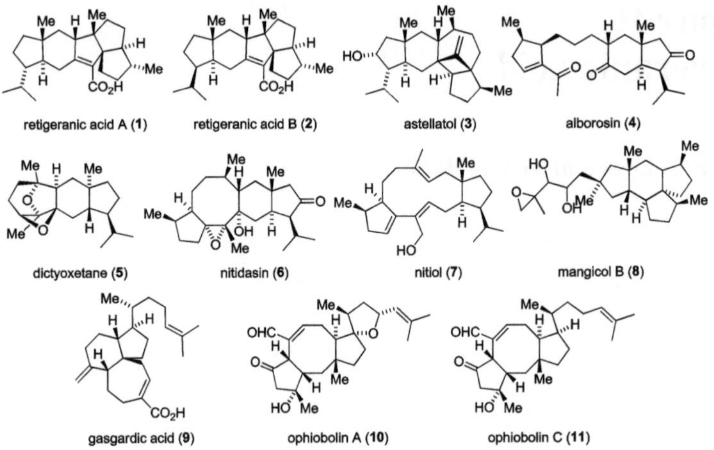

Fig. 10.1 Representative examples of sesterterpenes

synthesize this unique sesquiterpene in 1979 [3]. The publication of Oppolzer's synthesis appeared 1 month after the report on the total synthesis of isocomene published by Chatterjee [4], a report that was later proven to be completely fabricated [5].

The synthesis of other angularly fused triquinanes as well as linearly fused sesquiterpenes such as hirsutene and capnellene quickly followed. Many general methods for the synthesis of cyclopentanoid natural products emerged as a result of the target-oriented effort [6]. These accomplishments have been reviewed extensively on numerous occasions [7]. This chapter reviews the history of retigeranic acid from its isolation and structure determination to published approaches to its synthesis and the four total syntheses accomplished to date.

10.2 Isolation and Structure

In 1965 Seshardri et al. described the isolation of an unknown terpenoid acid "B" obtained from the lichens of *Lobaria retigera* in the western Himalayas [8]. (Note that this annotation bears no relation to the subsequent nomenclature later defined by Corey and Shibata.) Four collections had been made in the summer of 1962 from under the "Rutba" plants in the Valley of Flowers (12,500 feet) and on the way to Hemkund Lokpal (13,500 feet) and from underneath rocks and from pine trees in Ganghariya (10,000 feet). The samples were subjected to a series of increasingly polar extractions (petroleum ether, diethyl ether, acetone). The unknown terpenoid acid B was present in all petroleum ether extracts, except that of the sample obtained from the Valley of Flowers, in compositions ranging from 0.47 % to

2.75 %, as a colorless crystalline solid. Recrystallization of the terpenoid acid B from dioxane–acetone provided colorless, needle-like crystals with a melting point of 218–221 °C and $[\alpha]_D = -59$ ($c = 1.0$, $CHCl_3$). An orange-red color was reported for the Lieberman–Burchard test [9, 10], an assay typically used to identify cholesterol and derivatives by treatment with acetic anhydride and sulfuric acid to generate a chromophore. The IR spectrum (1,667 cm^{-1}) and UV absorption ($\lambda_{max} = 239$ nm) supported the Lieberman–Burchard evidence for the presence of an α,β-unsaturated carboxylic acid. A more detailed study of retigeranic acid (previously referred to as "compound B") by the same group was published in 1972 and included both elemental analysis and low-resolution mass spectral data ($M^+ = 370$), which established the molecular formula as $C_{25}H_{38}O_2$ [11].

Also in 1972 Shibata and coworkers reported the isolation of two new hopanoids and an unknown sesterterpene (tentatively named L-A) from the lichens *L. retigera* [12] and *L. subretigeria* [13], collected in the eastern Himalayas (Bhutan). The physical properties of L-A closely resembled those of retigeranic acid, and it was identified as such by mixed melting point and comparisons of IR spectra and TLC elution characteristics. Shibata confirmed the presence of the acrylate moiety by IR spectroscopy (1,662 and 1,608 cm^{-1}) and the molecular formula by high-resolution mass spectroscopy (M^+ calcd. for $C_{25}H_{38}O_2$: 370.287; found 370.286). The analytical data obtained did not correspond to any known sesterterpene and did not provide any information as to the skeletal structure of the unknown natural product. A crystal structure was obtained by the heavy atom method from a sample of the *p*-bromoanilide derivative ($C_{31}H_{42}NOBr$, mp 276–278 °C) crystallized from acetone. The X-ray analysis determined the absolute configuration as that of retigeranic acid (**1**), which all chemists interested in its synthesis assumed to be the natural product. It became evident more than 10 years later that (**1**) was only a minor isomer in a mixture of (**1**) and (**2**); it was this compound, obviously, whose *p*-bromoanilide crystallized out of the mixture.

Upon the completion of the first total synthesis of *rac*-retigeranic acid by Corey in 1985 [14], the group compared a natural sample (3.5 mg) obtained from the Shibata laboratory with the newly synthesized compound. Thin-layer chromatography on silica gel indicated identical mobility characteristics ($R_f = 0.43$ in 4:1 hexanes/ethyl acetate); however, upon derivatization as the methyl ester and analysis by HPLC, the natural sample was found to be an isomeric mixture. The two-component mixture of methyl esters (components A and B) was separated and analyzed by IR, mass spectroscopy, and NMR, and it was determined that component B and Corey's synthetic sample were identical. [Note: Component "B" in this context was the minor constituent, i.e., retigeranic acid A (**1**), while component A had noticeable variations in NMR signals leading to the assumption of a stereoisomeric mixture.] In subsequent private communications with Corey, and later Paquette (1986) [15], Shibata supported the findings that natural retigeranic acid existed as a mixture of components A and B and provided further evidence via crystal structure that the two were stereoisomers at the isopropyl stereogenic center and hereby designated as retigeranic acid A (minor constituent, α-isopropyl (**1**)) and retigeranic acid B (β-isopropyl (**2**)).

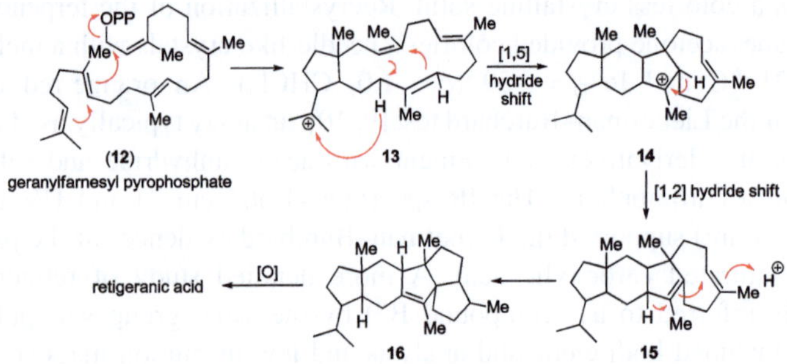

Fig. 10.2 Suggested biosynthesis of retigeranic acid skeleton

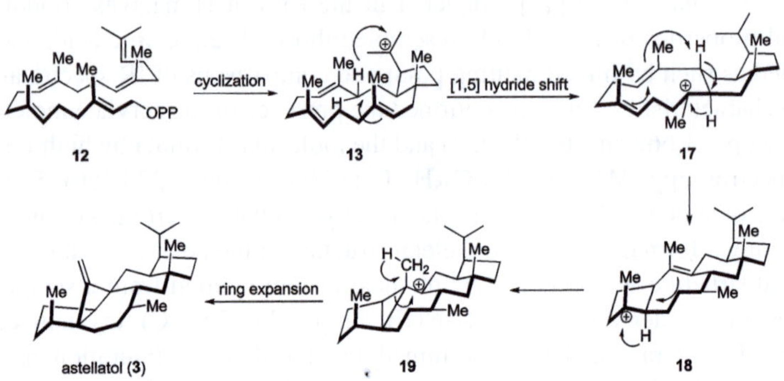

Fig. 10.3 Proposed biosynthetic pathway for astellatol

10.3 Biosynthesis

The seminal publication [12] on the isolation of retigeranic acid proposed a possible biosynthetic pathway by means of cyclization from geranyl farnesyl pyrophosphate (**12**) (Fig. 10.2) to tertiary carbocationic cyclopentane **13**. A series of [1, 5] and [1, 2] hydride shifts were proposed to establish the skeletal core of the sesterterpene, whose subsequent oxidation would yield retigeranic acid.

In 1994 Simpson [16] of the University of Bristol reported biosynthetic labeling studies of the sesterterpenoid astellatol (**3**), a secondary metabolite of *Aspergillus variecolor*, from which a more comprehensive analogy to the biogenetic pathway of retigeranic acid can be drawn. The pathway shown in Fig. 10.3 is consistent with Simpson's ^{13}C-labeling experiments in which astellatol (**3**) was enriched by feeding [1,2-^{13}C]-acetate to cultures of *A. variecolor*.

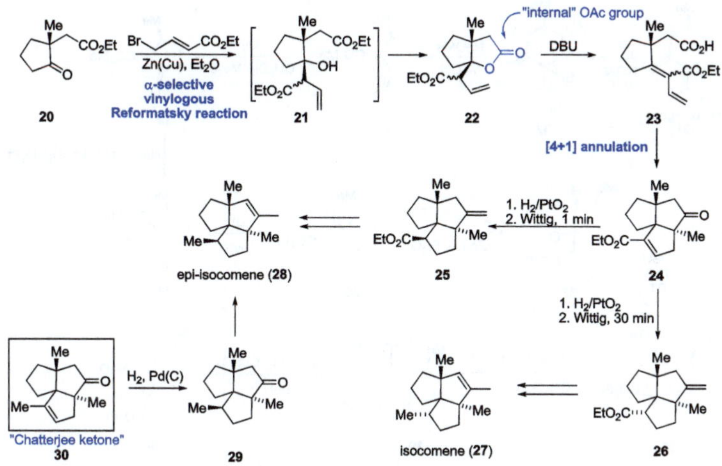

Scheme 10.1 Hudlicky's synthesis of isocomene, and *epi*-isocomene. Validation of "Chatterjee's ketone" fraud

10.4 Approaches to Total Synthesis

10.4.1 Hudlicky

The first disclosure of an approach to retigeranic acid was made in May 1982 at the ACS Regional Meeting in Dayton, Ohio, where Hudlicky reported the synthesis of the "right half" of retigeranic acid, namely, triquinane ester **42** (Scheme 10.2) [17]. The design followed the strategy toward triquinanes employed by the Hudlicky group at the time based on the [4+1] annulation of dienic diazoketones developed in the late 1970s and early 1980s [18]. At the same time as the approach to (**1**) was initiated, the Hudlicky group was pursuing the synthesis of isocomene and its epimer in an effort to validate the Chatterjee synthesis of isocomene. For this purpose, the carboxylate-substituted dienes were used, as both isomers of isocomene at the secondary methyl group stereogenic center would become available by hydrogenation, epimerization, and reduction (Scheme 10.1). The key strategy for the synthesis of carboxylate-functionalized dienes was based on the vinylogous Reformatsky reaction [19], which led to the lactones such as **22**. Base-catalyzed elimination of the internal acetate then furnished the mixtures of *E*/*Z* dienic acids **23**, which were transformed to the corresponding isomers of vinylcyclopropanes. Pyrolysis of the diastereomeric mixture provided the key triquinane **24**. Hydrogenation and a controlled Wittig reaction furnished ester **25**, with the β-configuration of the carboxylate, whereas isomerization gave the α-configured ester **26**. The isomers were transformed to epiisocomene and isocomene, respectively [20]. This accomplishment provided an indirect proof that the

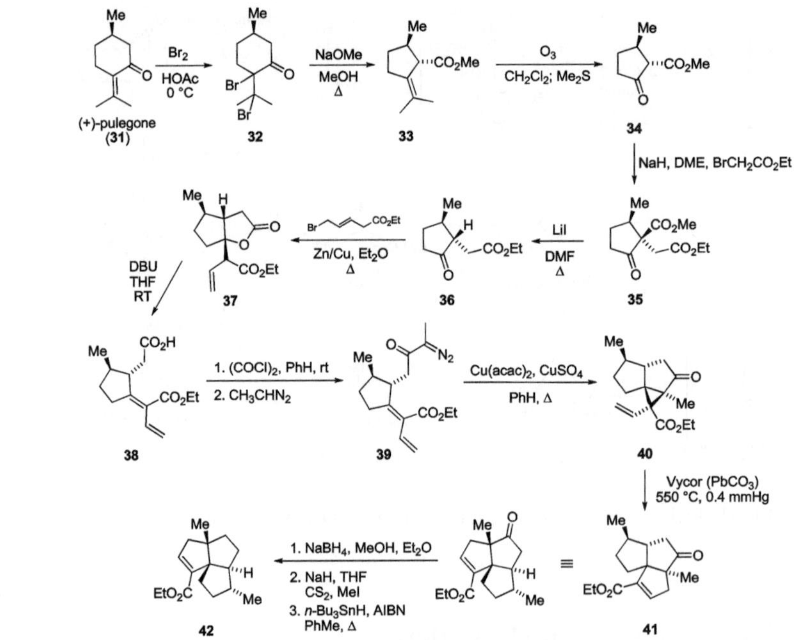

Scheme 10.2 Hudlicky's synthesis of the triquinane portion of retigeranic acid

Chatterjee synthesis of isocomene constituted fraud because of the low probability of hydrogenation occurring from the concave surface of ester **24**. The final direct proof was obtained by the synthesis of Chatterjee ketone **30**, its hydrogenation to **29**, and further conversion to epiisocomene [21].

The synthesis of triquinane acids, initiated by the preparation of isocomenic acid [22], thus provided a general method for control of the stereochemistry of secondary methyl groups in these terpenes. The [4+1] annulation based on the dienes of type **23** then laid the groundwork for the first-generation design and a model study for the approach to retigeranic acid [23].

The synthesis of the right half of **1** is shown in Scheme 10.2. The required chirality of the target was envisioned to originate from pulegone, which was transformed to keto ester **34** by the series of transformations shown. The key step was the Favorskii ring contraction of pulegone as reported by Wolinsky [24]. The original rearrangement leading to pulegonic acid was also reported by Wallach in 1918 [25]. The methyl ester of pulegonic acid, **33**, was converted to the keto ester **34** and, eventually, to the key starting material, keto ester **36**. The vinylogous Reformatsky reaction followed by base-catalyzed elimination provided the dienic acid **38**, which was converted to diazoketones **39**. The cyclopropanation gave a mixture of exo- and endo-vinylcyclopropanes **40** that furnished, upon flash vacuum pyrolysis, a single isomer of triquinane ester **41**. Finally, reduction of the ketone provided the right half of retigeranic acid, namely, acrylate **42**.

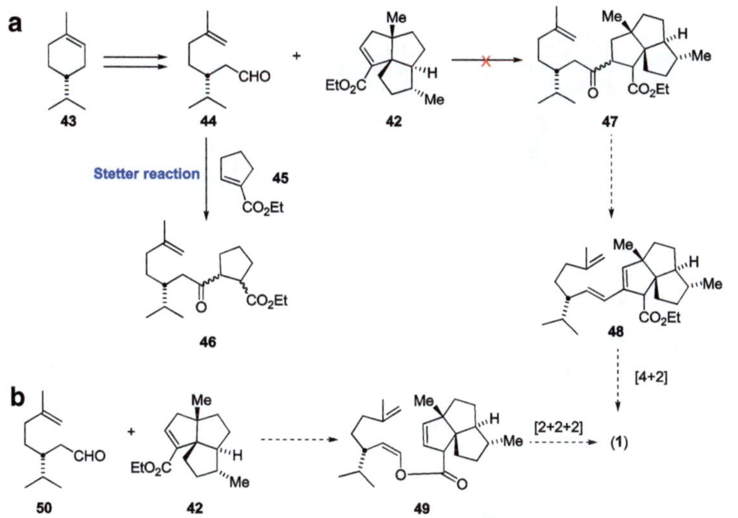

Scheme 10.3 Initial convergent designs for retigeranic acid by Hudlicky

With the completed model study, we considered two different approaches to a convergent synthesis of (**1**), neither of which proved successful. These are shown in an abbreviated fashion in Scheme 10.3. Both utilized the "left-half" moiety derived from menthene or limonene and containing the correct absolute configuration of the isopropyl group. The Stetter reaction strategy (umpolung) worked on a model system to provide keto ester **46** but was not successful with acrylate **42**, presumably because the olefin unit was out of alignment with the carboxylate (as was later affirmed by Paquette) and hence was unreceptive to a Michael addition. The strategy that would employ triene **48** was later implemented by Wender but led to a non-stereoselective cycloaddition. For these reasons, a new strategy was adopted involving the convergent [2+3] annulation, as described in Sect. 10.5.4.

10.4.2 Fallis

Along with several other research groups in the 1980s, the Fallis group added to the already plentiful methods for the construction of the bicyclo[4.3.0]nonane (hydrindane) skeleton with the preparation of *trans*-hydrindanone **60** in 1985 and alluded to its usefulness as an intermediate for the synthesis of retigeranic acid (Scheme 10.4) [26]. Indeed, during the proofing stages of Fallis' aforementioned manuscript, Corey et al. published the first total synthesis of (±)-retigeranic acid by means of the key intermediate hydrindanone **60** [14]. Fallis' study of the stereochemical control available in the intramolecular Diels–Alder (IMDA) (**55**/**58** → **56**/**59**) and Michael addition–aldol (**62** → **63**) sequence provided a direct comparison of

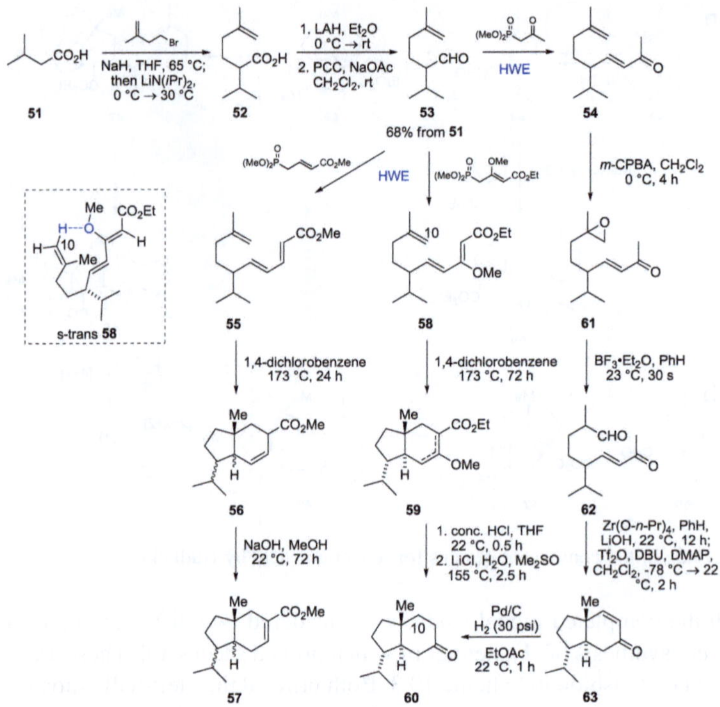

Scheme 10.4 Fallis's synthesis of hydrindanone in the approach to retigeranic acid

the utility of this strategy with other methods of synthesis of more highly substituted precursors reported elsewhere [27, 28]. Aldehyde **53** was prepared by alkylation of the mixed sodium/lithium salt of 3-methylbutanoic acid **51** with 2-methyl-4-bromobutene and a subsequent reduction–oxidation sequence. The Horner–Wadsworth–Emmons (HWE) protocol afforded the precursors for the intramolecular Diels–Alder reaction, namely, **55** and **58**; also prepared by the Wittig reaction was the precursor for the Michael–aldol, epoxide **61**. Triene ester **55** was cyclized in refluxing *p*-dichlorobenzene to provide a mixture of all four stereoisomers of **56** in high yield. The isomerization of **56** to α,β-unsaturated ester **57** with NaOH produced a 5:3 ratio of trans/cis ring systems, as identified by NMR analysis of the mixture; however, the cyclization of the methoxy-substituted triene **58** under the same conditions (72 h vs. 24 h), followed by hydrolysis and decarboxylation gave hydrindanone **60** (34 % from **53**). NMR analysis indicated only the single trans isomer. Fallis provided a tentative explanation for the observed selectivity based on hydrogen bonding of the C-10 vinyl proton to the s-trans-oriented methoxy oxygen in **58**, thus providing a transition state with a fixed orientation.

The IMDA reaction of **58** proved to be unreliable and was reported to suffer from decreased yields (15–20 %) when a fresh bottle of *p*-dichlorobenzene was used. It was hypothesized that a minor contaminant in commercial *p*-dichlorobenzene provided some catalytic activity. The use of pyridine in the reaction or the use of tri-*n*-butylamine as solvent did not lead to an increase in yield. A Lewis acid screen or the addition of a base to the solvent did not restore the previously observed yield or provide further insight. When 1,2,4-trichlorobenzene (bp 214 °C, 38 h) was employed as an alternative, the desired cycloadduct **59** was obtained in 54 % yield. Although the yield was good, the better alternative to acquire hydrindanone **60** on preparative scale was through enone **54**. The methylene in **54** was selectively oxidized to epoxide **61** (92 %) and then isomerized with Lewis acid to give aldehyde **62** (75 %). Following previous reports of Stork and coworkers [29, 30], the conjugate addition was effected using zirconium *n*-propoxide and LiOH in a one-pot sequence in which the intermediate ketols (epimeric at ring junction and isopropyl carbon) were dehydrated with trifluoroacetic anhydride (TFAA), 1,8-diazabicylo[5.4.0]undec-7-ene (DBU), and (dimethylamino)pyridine (DMAP) to yield **63**. Hydrogenation of **63** afforded hydrindanone **60** in 97 % yield (40 % from **53**). Fallis thus demonstrated the feasibility of constructing the left-hand portion of retigeranic acid (**1**) by intramolecular cycloadditions to form hydrindanone **60**. To our knowledge, no further communication toward the synthesis of (**1**) has been reported from the Fallis laboratories.

10.4.3 Fraser-Reid

In the mid-1980s Fraser-Reid and coworkers at Duke University demonstrated the tandem radical cyclization of 3-vinyl pyranosides [31] and the utility of the substituted pyranoside ring as a stereochemical influence in carbocycle annulation, as demonstrated in their synthesis of actinobolin [32]. Fraser-Reid's approach toward retigeranic acid assumed the coupling of a triquinane portion **65** or **66** with an intermediate such as **64** [33], similar to a compound that Paquette [15] successfully used in the total synthesis of (**1**) (Scheme 10.5). The preparation of key intermediate **70** began with Mander's methoxycarbonylation [34] protocol of (*R*)-(–)-carvone (**68**) using methyl cyanoformate [35]. The hydrogenation of **69** provided saturated α-keto ester **70** as an 85:15 mixture of epimers with one with the equatorial methyl group representing the major constituent. The isopropyl group in **70** effectively shielded the enolization at the α-position of the methyl ester and allowed for the regiocontrolled bromination as represented in stereoisomers **71** and **72**. Access to the isomer **71**, with axially oriented bromine (supported by IR and NMR), was provided by fractional crystallization from hot hexanes. Traditional Favorskii conditions were explored with intermediate **71** using sodium methoxide in methanol at 0 °C. The Favorskii product was not obtained, and instead, the substitution to yield a mixture of α-methoxy ketones **75** and **76** resulted in 80% combined yield. As an indirect method was required for the desired ring contraction, the protocol of Bordwell and Almy was employed in which a secondary amine-promoted Favorskii rearrangement enabled enamine formation [36]. Bordwell's halocyclohexanone substrate was activated by a phenyl group, while

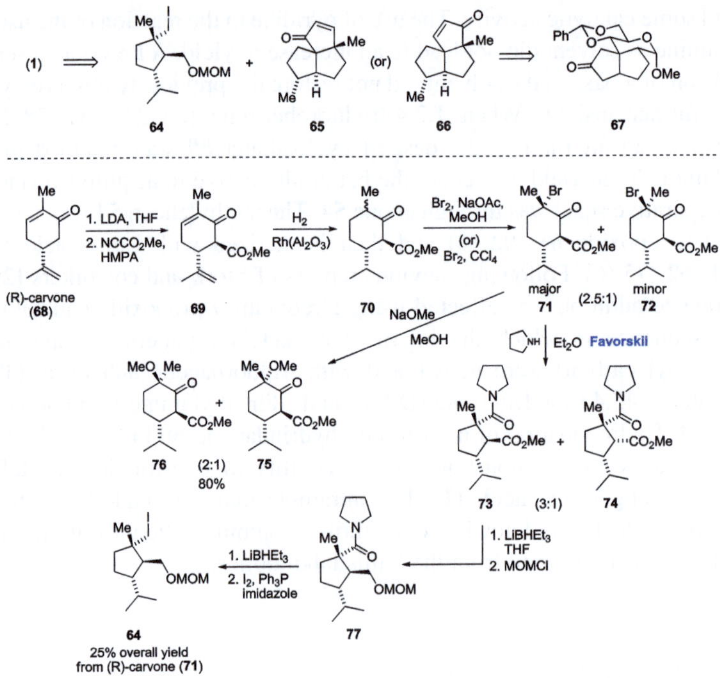

Scheme 10.5 Fraser-Reid's approach to retigeranic acid from carvone

71 was activated toward enamine formation by a more enolizable methyl ester. Treatment of **71** with pyrrolidine in Et_2O at room temperature for 5 min yielded a 3:1 mixture of Favorskii products **73** and **74** and provided the platform for subsequent selective reduction with lithium triethylborohydride (2.2 equiv.) and protection of the resulting primary alcohol as the methoxy methyl ether (MOM) derivative to afford **77** in high yield. The reduction of amide **77** with excess $LiBHEt_3$ and iodination via Garegg conditions provided primary iodide **64** [37]. To our knowledge no further studies toward retigeranic acid were reported.

10.4.4 Trauner

Trauner and coworkers released a report on a unified approach to isopropyl containing *trans*-hydrindane sesterterpenes via a common intermediate [38]. Their effort provided compounds that lend themselves to the substitution patterns embedded in retigeranic acids A (**1**) and B (**2**) as well as oxygenated precursors providing access to other sesterterpenes such as nitidasin (**6**), astellatol (**3**), and YW3699 (**89**).

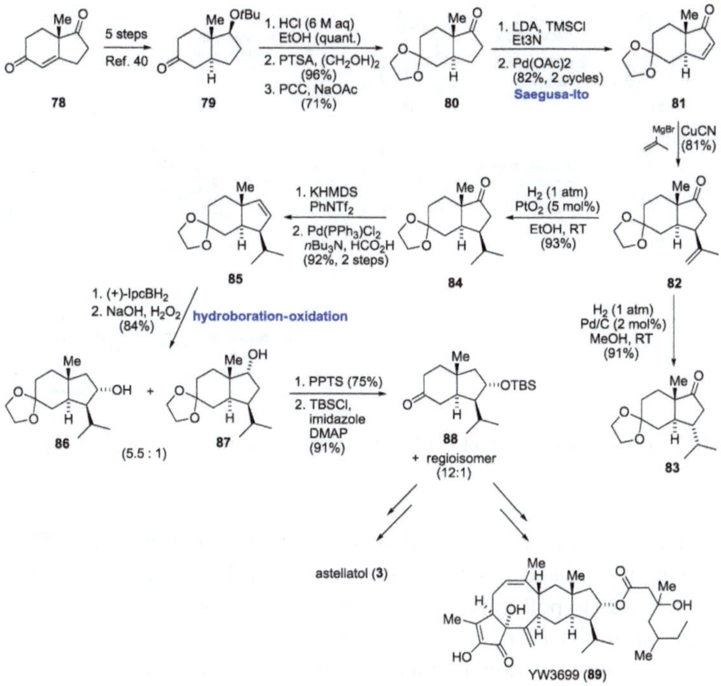

Scheme 10.6 Trauner's synthesis of hydrindanone intermediates for sesterterpenes

After the known intermediate **79** (contaminated with ca. 6 % *cis* isomer) [39] was prepared from Hajos–Parrish ketone [40] **78**, the *tert*-butyl ether was cleaved (quant.) and the ketone protected as the acetal (96 %). The secondary alcohol was oxidized by pyridinium chlorochromate (PCC) to provide ketone **80** in good yield (71 %) and after fractional crystallization afforded material absent of any *cis*-hydrindane (Scheme 10.6). [*NOTE: All compounds shown in Schemes 10.6 and 10.7 are shown in the ent-configuration, as published*]. The oxidation of protected hydrindane **80** under Saegusa–Ito conditions [41, 42] gave enone **81** (82 %), confirmed by X-ray crystallography.

As a straightforward introduction of an isopropyl group to enone **81** gave poor results, a tactical change to a conjugate addition–reduction procedure was used (i.e., **81** → **83**, **81** → **84**). Treatment of enone **81** with vinyl magnesium bromide in the presence of cuprous cyanide resulted in an unexpected stereochemical outcome. The attack of the vinyl nucleophile occurred from the *re*-face (*syn* to the angular methyl group) to afford **82** (81 %) in good yield. The result was validated by X-ray crystallography and substantiated by reports published by Molander [42] and by Bull and Loedolff [43] in studies toward variecolin and a derivative of estrone,

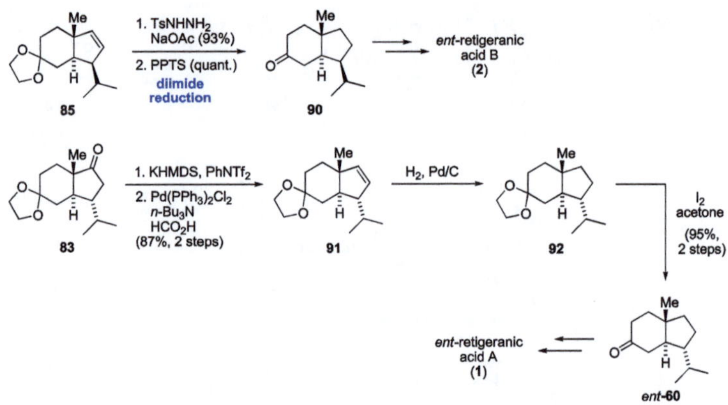

Scheme 10.7 Trauner's synthesis of *ent*-hydrindanone as intermediate for *ent*-retigeranic acid

respectively. Intermediate **82** proved versatile in that epimers **84** and **83** could be accessed in high yield and as single diastereomers by the variation of hydrogenation conditions. It was hypothesized that inversion of the stereocenter to provide **83** occurred by an initial Pd-mediated isomerization [44] and subsequent reduction. The isomerization of the double bond was circumvented by the use of a different hydrogenation catalyst, namely, PtO_2, and yielded hydrindane **84**. Intermediate **84** was further developed to provide building block **88** by formation of the enol triflate of **84** followed by reduction under Pd-catalyzed conditions to give **85** (92 %). After extensive experimentation, the regio- and stereoselective hydroboration of **85** was achieved in a 5.5:1 ratio (**86**:**87**, 84 % combined) using a matched chiral hydroboration reagent, (+)-monoisopinocampheylborane [(+) $IpcBH_2$] [45], and oxidative workup. Treatment of isomeric mixture **86/87** with pyridinium *p*-toluenesulfonate (PPTS) cleaved the acetal to liberate the carbonyl (75 %), and the secondary alcohol was silylated with TBSCl to yield building block **88** (91 %) that could be used in synthetic endeavors toward YW3699 (**89**) and astellatol (**3**).

It was envisioned that hydrindanone **83** and cyclopentene **85** could be used as intermediates in the synthesis of *ent*-retigeranic acid A (**1**) and *ent*-retigeranic acid B (**2**), respectively. To prepare the building block **90**, cyclopentene **85** was reduced with diimide (93 %) in order to prevent isomerization and subsequently deprotected with PPTS to yield hydrindanone **90** (quant.), which could provide access to *ent*-retigeranic acid B (**2**) (Scheme 10.7). Hydrindanone **83** was reduced via an enol triflate and then subjected to Pd-catalyzed reduction to provide cyclopentene **91** (87 % from **83**). Upon hydrogenation of **91** with Pd/C and cleavage of the acetal with iodine, protected hydrindanone **92** (95 % from **91**) was obtained. The deprotection of **92** provided *ent*-**60**, whose enantiomer was used in previous syntheses of retigeranic acid A (**1**) by Corey [14] and Hudlicky [46, 47].

10.5 Total Syntheses

Despite the numerous model studies published and discussed above, as well as various PhD theses addressing synthetic approaches [48], there have been only four total syntheses of retigeranic acid reported to date. The first, by Corey, is a linear synthesis; the following three are convergent. The design of all three convergent approaches suffered from erroneous assumptions regarding the electronic nature of α,β-unsaturated functionalities on the triquinane unit (For detailed discussion of these issues, see reference [7f]). No new approaches to the total synthesis of retigeranic acid have been reported in recent years.

10.5.1 Corey

The first total synthesis of (±)-retigeranic acid was reported by Corey and coworkers [14] in 1985 and was achieved in 27 steps from hydrindenone **93**. Enone **93** was accessible in two steps from 2,6-dimethyl-5-heptenal [49] and was subjected to a reduction–inversion sequence to provide allylic alcohol **94** (89 % from **93**), providing the grounds for the stereospecific hydrogenation previously reported by the Corey group (Scheme 10.8) [27]. The allylic alcohol **94** was treated with rhodium (norbornadiene) (DIPHOS-4) tetrafluoroborate (1.8 mol%) under a hydrogen atmosphere (950 psi) to deliver the *trans*-bicyclo[4.3.0]nonane **95**, which was oxidized to hydrindanone **60** (93 %, two steps). Ketone **60** was reacted with vinyl magnesium bromide to yield a 3:1 mixture of carbinols **96** (97 %) that was directly dehydrated via distillation from potassium bisulfate (150 °C, 40 Torr) to form diene **97** (82 %). Upon heating **97** in the presence of methyl 3-formyl-*cis*-crotonate, cycloadduct **98** (61 %) was obtained from a mixture of products and structurally confirmed by X-ray analysis of a crystalline bromolactone derivative (not shown). The selective *Z* olefination of aldehyde **98** with the appropriate orthoester ylide provided orthoester **99** (80 %), which was reduced to primary alcohol **100** (quant.) and subjected to Grieco's dehydration protocol [50] to form exocyclic olefin **101** (75 %, two steps). Hydrolysis of orthoester **101** (quant.) and activation of the resulting carboxylic acid as the acid chloride **102** provided cyclobutanone **103** (80 %, two steps) via [2+2] cycloaddition of the transient ketene species. Cyclobutanone **103** underwent a ring expansion via the initial addition of lithiated acetaldehyde dimethylthio acetal to form **104** (73 %), and subsequent cuprous triflate promoted rearrangement [51] to yield **105**. Oxidation of sulfur with periodate and desulfurization with amalgam afforded cyclopentanone **106** (65 %, three steps).

The selective reduction of the D-ring olefin in **106** using a partially poisoned catalyst (Pd/C, 0.25 % pyridine) provided intermediate **107** (83 %), which was epimerized at –78 °C with sodium methoxide (HOAc quench at –78 °C, 89 %) (Scheme 10.9). Deoxygenation by means of tosyl hydrazone **108** and subsequent treatment with catechol borane and tetrabutylammonium acetate gave pentacyclic

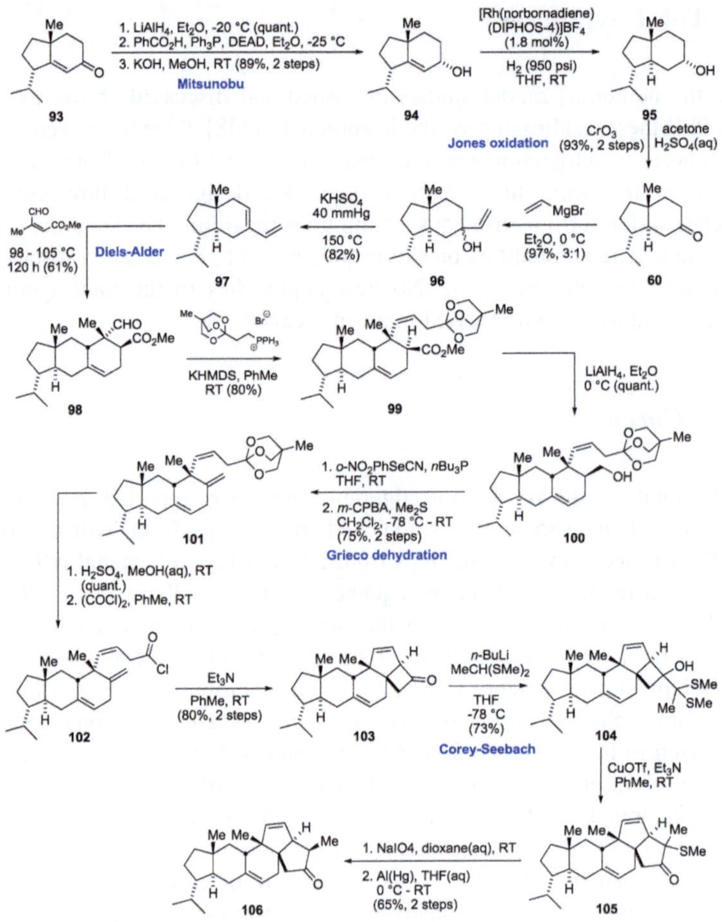

Scheme 10.8 Corey's synthesis of advanced intermediate for retigeranic acid

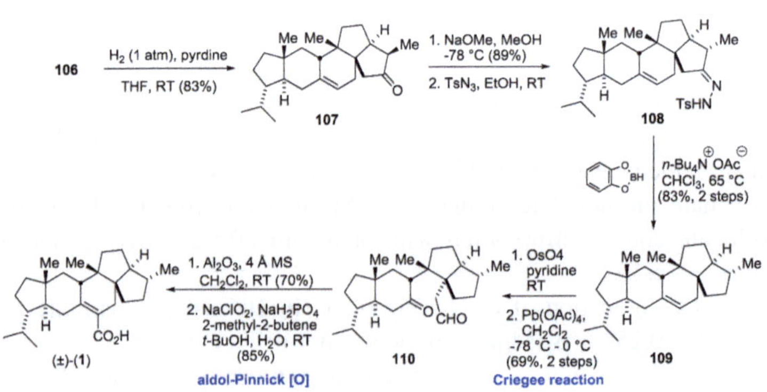

Scheme 10.9 Completion of total synthesis of retigeranic acid by Corey

olefin **109** (83 %, two steps) [52]. The oxidative cleavage of the double bond in **109** was accomplished by dihydroxylation and subsequent cleavage with lead tetraacetate to yield **110** (69 %, two steps), whose aldol condensation (70 %) catalyzed by neutral alumina established the skeletal core of the target. Finally, Pinnick oxidation of the unsaturated aldehyde completed the synthesis of (±)-**1** (85 %).

10.5.2 Paquette

At the time of Paquette's synthesis, it was known that natural retigeranic acid consisted of two stereoisomeric compounds differing in the stereochemistry of the isopropyl group in ring A [retigeranic acids A (**1**) and B (**2**)] [14]. Paquette et al. presented a convergent, enantioselective synthesis of retigeranic acid A (**1**) that also provided access to stereoisomers **129** (Scheme 10.12) and also to **137** (Scheme 10.13) [15, 53]. The synthetic scheme was designed to generate an optically pure "southern" angular triquinane **116** (Scheme 10.10) by a methodology previously demonstrated by the synthesis of silphiperfol-3-ene and 5-oxosilphiperfol-6-ene [54–56]. The "northern" cyclopentane intermediate **123** (Scheme 10.11) was prepared through a coupling reaction.

The synthesis began with ring contraction of (*R*)-(+)-pulegone [57], followed by enolate addition and ozonolysis to yield the desired β-keto ester **111** (Scheme 10.10). Aldol condensation of **111** and decarboxylation provided a mixture of separable epimers (66:34) containing **112** as the major component. Studies indicated that enrichment of **112** by base-induced equilibration of the mixture of isomers was impractical. The conjugate addition of acetaldehyde acetal to **112** provided ketal **113** whose deprotection, aldol addition to form the tricyclic intermediate (72 % from **112**, not shown), and subsequent acylation gave thiocarbonate **114** (73 %) all occurred in reasonable yields. The thermal elimination of the thiocarbonate in **114** and Wolff–Kishner reduction gave angular triquinane **115** (72 % from **114**), which was oxidized in the allylic position to yield the southern fragment **116** (64 %).

In order to establish the correct absolute stereochemistry in cyclopentanoid **123** (Scheme 10.11), a chirality transfer strategy was employed with aldehyde **117**, obtained from (*S*)-(–)-limonene (Scheme 10.11). A modified procedure for the conversion of (*S*)-(–)-limonene to cyclopentene **117** (58 % from limonene) was used [58], and aldehyde **117** was reduced with diisobutylaluminium hydride (DIBAL) (quant.) and alkylated to provide tributylstannane ether **118**. This compound underwent a Still–Wittig rearrangement upon treatment with *n*-butyl lithium (*n*-BuLi) to yield **119** (75 %, two steps) [59]. The extent to which the chirality transfer was successful was deemed quantitative on the basis of conversion of alcohol **119** to its (+)-*O*-methylmandelic acid ester and subsequent analysis of optical purity. The ozonolysis (70 %) of **119**, protection of the free alcohol as the silyl ether (85 %), and reduction of the ketone with DIBAL (quant.) gave alcohol **120**. Elimination of the alcohol in **120** with phosphorus oxychloride–pyridine

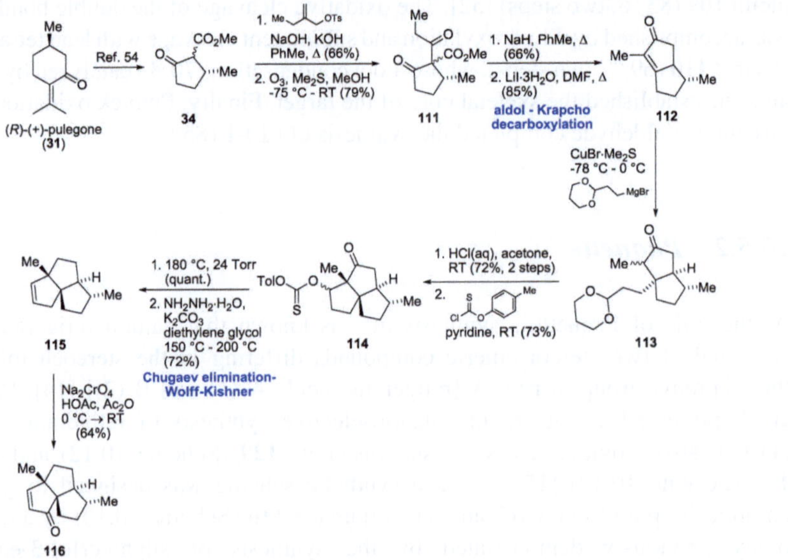

Scheme 10.10 Paquette's synthesis of triquinane portion of retigeranic acid

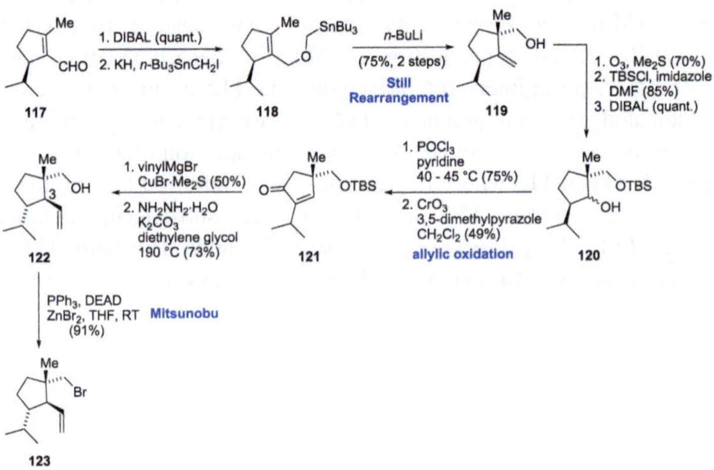

Scheme 10.11 Paquette's synthesis of "left half" precursor for retigeranic acid

(75 %) followed by allylic oxidation with chromium trioxide-3,5-dimethylpyrazole complex [60] yielded enone **121** (49 %). The previously established quaternary center now played a pivotal role in the outcome of stereochemical events at C-3 in terms of the stereoselectivity of the divinyl cuprate addition reaction (50 %, 77:23 β:α) [61] to give **122**. After Wolff–Kishner reduction and concurrent desilylation

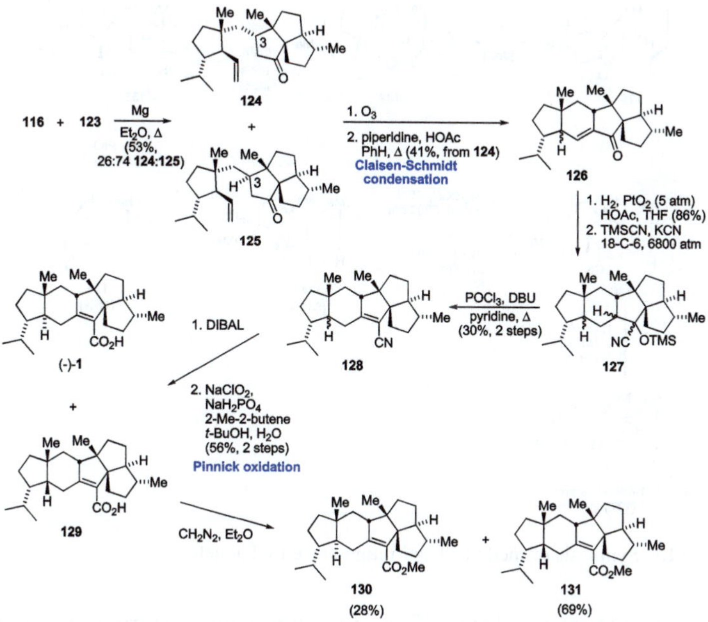

Scheme 10.12 Completion of synthesis of retigeranic acid and its isomers by Paquette

(73 %), a Mitsunobu reaction [62] with added zinc bromide [63] provided the northern fragment **123** (91 %) in good yield.

The Grignard reagent derived from **123** was used to connect the northern and southern fragments, but not without problems. Ultimately, the Grignard compound was formed from **123** with activated magnesium turnings in conjunction with a suitable initiator (ethylene dibromide). The solution was introduced by the slow addition to enone **116** to yield exclusively 1,4-products as a mixture of C-3 epimers **124/125** (26:74, 53 %) (Scheme 10.12). The isolation of the individual epimers proved difficult at this stage, and the mixture of **124/125** was subjected to ozonolysis and separated at the stage of the intermittent aldehydes (not shown). The transformation of coupled product **124** to the retigeranic skeleton of **126** via ozonolysis and aldol cyclization occurred in moderate overall yield (41 %) and allowed for the alkene reduction over platinum oxide (PtO_2, 86 %) and homologation by 1,2-addition with cyanide anion (6,800 atm) to give **127** as the precursor to the carboxylic acid. The elimination of the cyanohydrin **127** with $POCl_3$/DBU/pyridine gave nitrile **128** (30 %, two steps) [64]. A final redox manipulation of **128** with DIBAL/sodium chlorite provided retigeranic acids (–)-(**1**) and **129** that were subsequently derivatized as methyl esters **130** (28 %) and **131** (69 %), respectively. The direct comparison of methyl ester **131** with natural methyl retigeranate B by NMR spectroscopy indicated that these were different compounds, further supporting Shibata's empirical data that retigeranic acid B (**2**) was epimeric only at the isopropyl carbon.

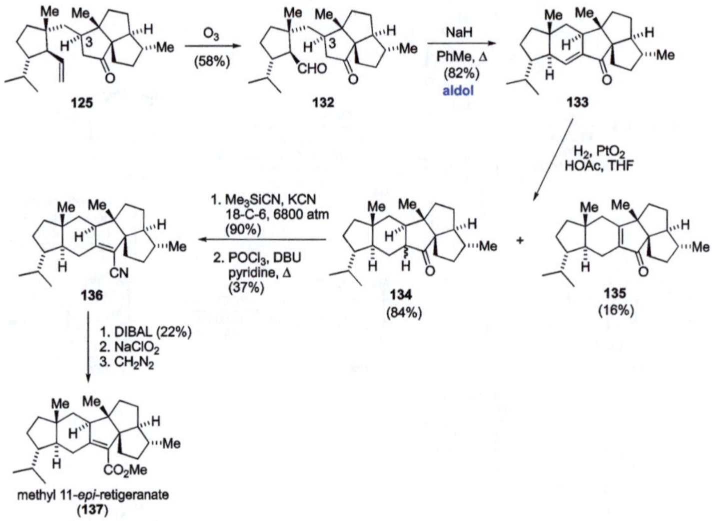

Scheme 10.13 Synthesis of methyl 11-*epi*-retigeranate by Paquette

With access to intermediate **125** (as a mixture with **124**) from the coupling of **116** and **123**, ozonolysis was performed to provide aldehyde **132** (now separable from its C-3 epimer, 57 %). Aldol cyclization yielded enone **133** in moderate yield (47 % from **125**) (Scheme 10.13). The reduction of enone **134** was explored using a variety of reagents (diimide, CuH, Li/NH_3(l), $RhCl_3 \cdot 3H_2O$) before suitable conditions (Adams catalyst, H_2) provided a usable mixture of epimers **134** (84 %) and isomerized enone **135** (16 %). A sequence analogous to the one yielding (–)-(**1**)/**129** was used to provide epimeric retigerate (**137**), namely, cyanation/elimination of the cyanohydrin, reduction and reoxidation, and methylation.

The synthesis reported by Paquette was plagued by several steps of very low stereoselectivity and is, at 26 steps, the longest of the four approaches. However, the analysis of the various stereoisomers related to retigeranic acid helped to finally end the confusion regarding the mixture of stereoisomers in the original isolate of retigeranic acid.

10.5.3 Wender

Although Wender's synthesis was completed prior to Hudlicky's, it was published later [65]. It utilized the technique of *meta*-photocycloaddition of arenes that Wender developed into a general method of synthesis for both angular and linear triquinanes. The overall strategy was centered around an intramolecular Diels–Alder cycloaddition for construction of the internal six-membered ring of the target.

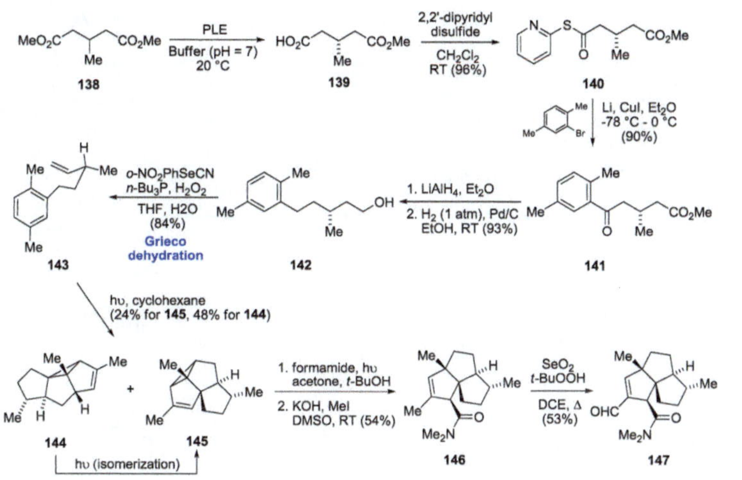

Scheme 10.14 Wender's synthesis of triquinane part of retigeranic acid by meta-photocycloadditon

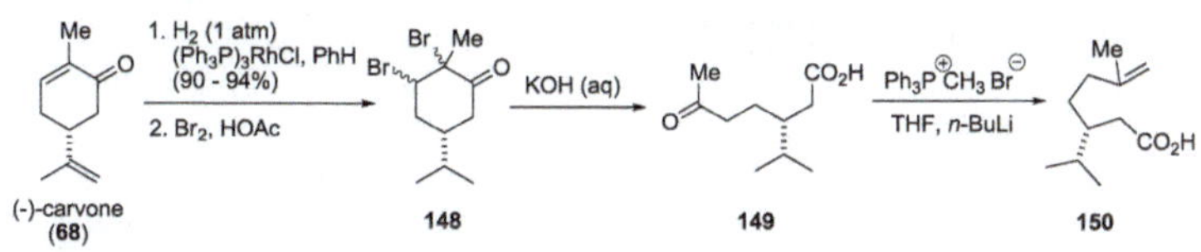

Scheme 10.15 Synthesis of "left half" precursor for retigeranic acid by Wender

Wender's synthesis began with the hydrolysis of the methyl ester of 3-methylglutaric acid **138** using pig liver esterase (PLE) [66] and fractional crystallization of its cinchonidine salt to provide enantiopure half-ester **139** (recrystallized to >99 % e.e.) (Scheme 10.14). The carboxylic acid **139** was converted to thioester **140** (96 %) [67] and subsequently reacted with the cuprate of bromo-*p*-xylene to give ketone **141** (90 %). Reduction of **141** and deoxygenation of the resulting benzyl alcohol gave intermediate **142** (93 %), which was subjected to Grieco's conditions [50] to yield enantiopure alkene **143** (84 %). Photolysis of arene **143** furnished a mixture of linear and angular triquinane vinylcyclopropanes **144** and **145** (2:1 mixture, 72 % combined yield), respectively. The unfavorable preference for the linear cycloadduct **144** was resolved by photo-equilibration via the vinylcyclopropane to the angularly fused isomer **145**. This method enabled the preparation of multigram quantities of **145** in eight synthetic steps (including photoisomerization, **144** → **145**). The addition of the carbon-centered free radical of formamide [68, 69] to vinylcyclopropane **145** followed by dimethylation with methyl iodide (MeI) gave triquinane **146** in moderate yield. An allylic oxidation converted **146** to aldehyde **147** (53 %) and completed the triquinane portion of the synthesis.

The carboxylic acid **150** (Scheme 10.15) was prepared from commercial (–)-carvone (**68**) via a chemoselective reduction of the terminal olefin using tris

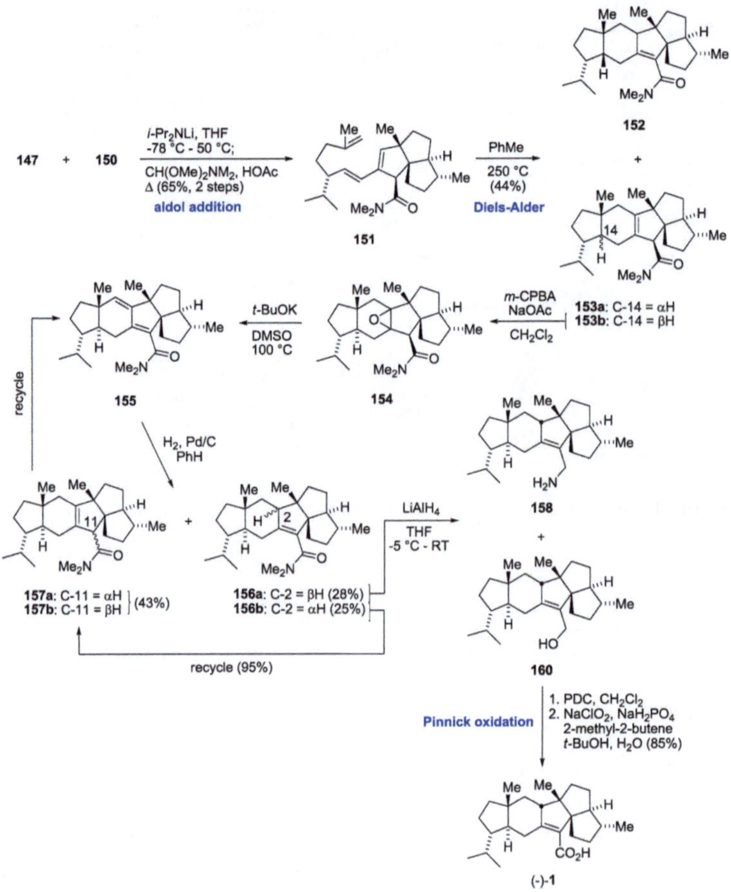

Scheme 10.16 Completion of retigeranic acid synthesis by Wender

(triphenylphosphine)chlororhodium (I) [70, 71] as catalyst under an atmosphere of hydrogen (Scheme 10.15). The bromination of dihydrocarvone to give **148** and subsequent Grob-like fragmentation gave acyclic keto acid **149**, which was treated with methyl phosphonium ylide to provide alkene **150** as the acting diene fragment for the convergent route.

The Diels–Alder adduct was prepared by the addition of the dianion of carboxylic acid **150** with aldehyde **147** to give a mixture of hydroxy acids that upon decarboxylative dehydration [72] yielded the single alkene isomer **151** (65 %, two steps) (Scheme 10.16). The thermolysis of triene **151** gave a mixture of cycloadducts **152**, **153a**, and **153b** (64 % overall, 1:8.6:3 ratio). The isomerization of the C-2 hydrogen in **152** failed, and so an indirect, two-step conversion of **153a** to diene **155** was utilized via epoxide **154** and subsequent treatment with potassium *t*-butoxide (*t*-BuOK) in DMSO at 100 °C. The hydrogenation of **155** yielded a mixture of C-2 epimers **156a/b** (1:1.1) and C-11 epimers **157a/b** in high yield

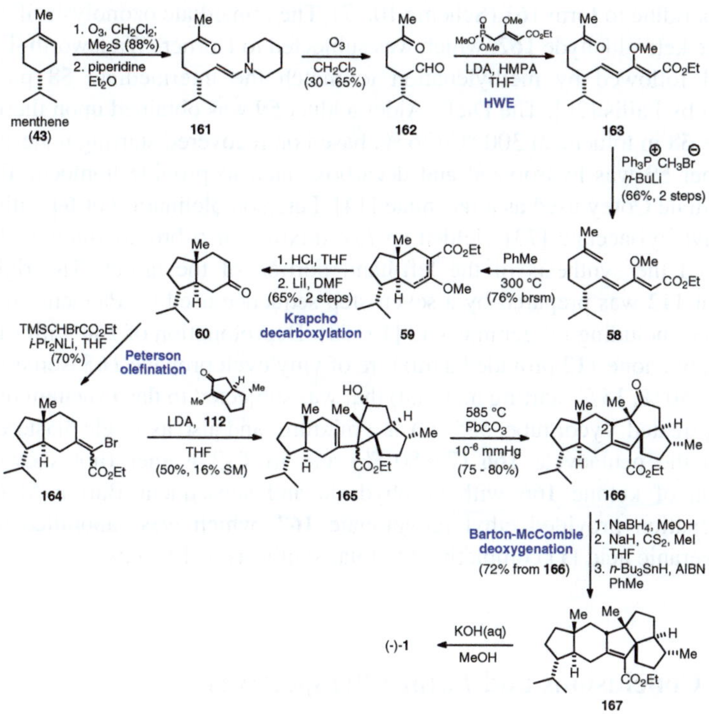

Scheme 10.17 Hudlicky's convergent synthesis of retigeranic acid

(95 %). As all attempts at improving the selectivity of this process failed, a recycling sequence was established to convert the undesired isomer **156b** to **157a,b** and then to diene **155** (as given for **153a,b** → **155**). The pentacyclic amide **156a** was subjected to a redox process to complete the synthesis and provide (–)-retigeranic acid (**1**) by oxidation of **160**.

10.5.4 Hudlicky

After unsuccessful attempts at connecting the triquinane unit with the left-hand moiety by the Stetter reaction, the entire strategy had to be changed. In the mid-1980s a new [2+3] cyclopentene annulation protocol was developed by the Hudlicky group. It was more efficient than the previously employed [4+1] annulation method and allowed for a convergent approach to the target. The known hydrindanone **60** was chosen as the left-hand component and a precursor for the synthesis of α-bromocrotonate **164** (Scheme 10.17), required for the construction of the key vinylcyclopropane **165**. The synthesis of hydrindanone **60** began by oxidation of menthene (**43**) via ozonolysis and subsequent enamine formation by reaction with piperidine to form **161** (Scheme 10.17). The immediate ozonolysis of enamine **161** gave keto aldehyde **162**, which was subjected to Horner–Wadsworth–Emmons protocol followed by methylenation to match the intermediate **58** previously reported by Fallis [26]. The Diels–Alder adduct **59** was obtained upon thermolysis of triene **58** in toluene at 300 °C (76 %, based on recovered starting material). The enol ether **59** was hydrolyzed and decarboxylated to provide homochiral **60**, an intermediate Corey used as a racemate [14]. Peterson olefination of **60** with bromo (trimethylsilyl)acetate [73] yielded an *E*/*Z* mixture of α-bromocrotonate **164** and completed the synthesis of the left-hand portion of the target. The right-hand fragment **112** was prepared by a seven-step sequence used by Paquette in several syntheses, including retigeranic acid [15, 53]. Deprotonation of **164** with LDA and addition of enone **112** provided a mixture of vinylcyclopropane **165** isomers (endo/exo, 1:1; 50 %, 15 % starting material) that was subjected to flash vacuum pyrolysis ($PbCO_3$-treated Vycor tube, 585 °C) as a mixture (and also as single diastereomers) to yield the pentacycle **166** (75–80 %) and its C-2 isomer (not shown). The reduction of ketone **166** with borohydride and subsequent Barton–McCombie deoxygenation provided ethyl retigeranate **167**, which was saponified to yield (–)-retigeranic acid (**1**) completing the total synthesis in 15 steps.

10.6 Conclusions and Future Perspectives

Retigeranic acid attracted a lot of attention of organic chemists, most likely because of the "esthetic" nature of its structure. It is a bit surprising that only four total syntheses emerged from the interest displayed by the rather large community of chemists who pursued triquinane-containing terpenoids in the 1970s and 1980s. Its synthesis is still a challenge. Perhaps this chapter will stimulate further interest and new approaches to this very interesting and esthetically pleasing molecule.

Acknowledgments The authors are grateful to the following agencies for financial support of this work: Natural Sciences and Engineering Research Council of Canada (NSERC) (Idea to Innovation and Discovery Grants), Canada Research Chair Program, Canada Foundation for Innovation (CFI), TDC Research, Inc., TDC Research Foundation, and Brock University. In addition, TH is grateful to all of the students who participated in the triquinane program at Illinois Institute of Technology (1978–1982) and Virginia Tech (1982–1992). Their names appear in the cited references.

References

1. Hog DT, Webster R, Trauner D (2012) Nat Prod Rep 29:752
2. Zalkow LH, Harris RN, Van Derveer D, Bertrand JA (1977) J Chem Soc Chem Commun 456
3. Oppolzer W, Bättig K, Hudlicky T (1979) Helv Chim Acta 62:1493
4. Chatterjee S (1979) J Chem Soc Chem Commun 620

5. For discussion of the Chatterjee synthesis and the proof of fabrication see: Hudlicky T, Reed JW (2007) The way of synthesis (Chapter 3.7.1). VCH, pp 363–380
6. For reviews on the methodology of cyclopentanoid synthesis see: (a) Hudlicky T, Price JD (1989) Chem Rev 89:1467; (b) Hudlicky T, Becker DA, Fan RL, Kozhushkov S (1996) Houben–Wyel, vol. E 17c, p 2538; (c) Wong HNC, Hon MY, Tse CW, Yip YCh, Tanko J, Hudlicky T (1989) Chem Rev 89:165; (d) Hanson JR (1986) Nat Prod Rep 3:123
7. For reviews of triquinane syntheses see: (a) Paquette LA, Doherty AM (1987) Synthesis and reactions. Polyquinane chemistry, vol. 26. Springer-Verlag, Berlin; (b) Paquette LA (1984) Recent synthetic developments in polyquinane chemistry. Topics in current chemistry, vol. 119. Springer-Verlag, Berlin; (c) Hudlicky T, Natchus MG, Sinai-Zingde G, Ranu BC, Rigby HL, Andre T (1986) Progress in terpene chemistry. In: Joulain D (ed) Proceedings of the conference on terpene chemistry, Grasse (France); Editions Frontières: Gif-sur-Yvette, France, pp 221–236; (d) Hudlicky T, Rulin F, Lovelace TC, Reed JW (1989) In: Atta-ur-Rahman (ed) Studies in natural products chemistry, vol. 3, Part B. Elsevier Science, pp 3–72; (e) Hudlicky T, Reed JW (2007) The way of synthesis (Chapter 3.7). VCH, pp 361–429; (f) Hudlicky T, Reed JW (2007) The way of synthesis (chapter 3.7.3). VCH, pp 392–420
8. Rap PS, Sarma KG, Seshadri TR (1965) Curr Sci 34:9
9. Liebermann NC (1885) Ber Dtsch Chem Ges 18:1803
10. Burke RW, Diamondstone BI, Velapoldi RA, Menis O (1974) Clin Chem 20:794
11. Aiyar VN, Rao PS, Seshadri TR (1972) Curr Sci 41:813
12. Kaneda M, Takahashi R, Iitaka Y, Shibata S (1972) Tetrahedron Lett 13:4609
13. Takahashi R, Chiang H-C, Aimi N, Tanaka O, Shibata S (1972) Phytochemistry 11:2039
14. Corey EJ, Desai MC, Engler TA (1985) J Am Chem Soc 107:4339
15. Paquette LA, Wright J, Drtina GJ, Roberts RA (1987) J Org Chem 52:2960
16. Simpson TJ (1994) J Chem Soc Perkin Trans 1 3055
17. Short RP, Hudlicky T (1982) Abstr Pap Am Chem Soc 183:12
18. (a) Hudlicky T, Sheth JP, Gee V, Barnvos D (1979) Tetrahedron Lett 20:4889; (b) Hudlicky T, Koszyk FJ, Kutchan TM, Sheth JP (1980) J Org Chem 45:5020
19. (a) Hudlicky T, Rice L, Boston C, Finklea H, Frazier J, Suder B (1984) J Org Chem 49:1845; (b) Hudlicky T, Natchus MG, Kwart LD, Colwell B (1985) J Org Chem 50:4300
20. Higgs L, Kavka M, Ranu BC, Hudlicky T (1984) Tetrahedron Lett 25:2447
21. (a) Kwart LD, Tiedje M, Frazier JO, Hudlicky T (1986) Synth Commun 16:393; (b) Hudlicky T, Kwart LD, Tiedje MH, Ranu BC, Short RP, Frazier JO, Rigby HL (1986) Synthesis 716
22. Short RP, Ranu BC, Revol JM, Hudlicky T (1983) J Org Chem 48:4453
23. Hudlicky T, Short RP (1982) J Org Chem 47:1522
24. Wolinsky J, Chan D (1965) J Org Chem 30:41
25. Wallach O (1918) Annalen 414:233
26. Attah-Poku SK, Chau F, Yadav VK, Fallis AG (1985) J Org Chem 50:3418
27. Corey EJ, Engler TA (1984) Tetrahedron Lett 25:149
28. Evans DA, Morrissey MM (1984) J Am Chem Soc 106:3866
29. Stork G, Shiner CS, Winkler JD (1982) J Am Chem Soc 104:310
30. Stork G, Winkler JD, Shiner CS (1982) J Am Chem Soc 104:3767
31. Tsang R, Fraser-Reid B (1986) J Am Chem Soc 108:2116
32. Rahman MA, Fraser-Reid B (1985) J Am Chem Soc 107:5576
33. Llera JM, Fraser-Reid B (1989) J Org Chem 54:5544
34. Mander LN, Sethi SP (1983) Tetrahedron Lett 24:5425
35. Childs ME, Weber WP (1976) J Org Chem 41:3486
36. Bordwell FG, Almy J (1973) J Org Chem 38:571
37. Garegg PJ, Samuelsson B (1980) J Chem Soc Perkin Trans 1 2866
38. Hog DT, Mayer P, Trauner D (2012) J Org Chem 77:5838
39. Micheli RA, Hojos ZG, Cohen N, Parrish DR, Portland LA, Sciamanna W, Scott MA, Wehrli PA (1975) J Org Chem 40:675
40. Jajos ZG, Parrish DR (1985) Org Synth 63:26

41. Ito Y, Hirao T, Saegusa T (1978) J Org Chem 43:1011
42. Molander GA, Quirmbach MS, Silva LF, Spencer KC, Balsells J (2001) Org Lett 3:2257
43. Bull JR, Loedolff MC (1996) J Chem Soc Perkin Trans 1 1269
44. Smit C, Fraaije MW, Minnaard AJ (2008) J Org Chem 73:9482 and references cited therein
45. Brown HC, Jadhav PK, Mandal AK (1982) J Org Chem 47:5074
46. Hudlicky T, Radesca-Kwart L, Li L-Q, Bryant T (1988) Tetrahedron Lett 29:3283
47. Hudlicky T, Fleming A, Radesca L (1989) J Am Chem Soc 111:6691
48. (a) Dedopoulou D (1991) Studies directed toward the total synthesis of (±)-Retigeranic acid A. Ph.D. Thesis, University of North Carolina, Chapel Hill, NC; (b) Radesca LA (1989) Vinylcyclopropanation of enones in [2+3] cyclopentene annulation. Application to the total synthesis of (–)-retigeranic acid. Ph.D. Thesis, Virginia Polytechnic Institute and State University, Blacksburg, VA; (c) Singh, SK (1988) Synthetic studies of the arene-olefin meta photocycloaddition: total synthesis of the silphiperfolenes and (–)retigeranic acid. Ph.D. Thesis, Stanford University, Stanford, CA; (d) Hobbs SH (1986) Approaches toward the total synthesis of retigeranic acid. Ph.D. Thesis, Duke University, Durham, NC; (e) Polt RL (1986) Part I. Beta-lithioenamines in synthesis. Novel variations of the aldol and Claisen condensations. Part II. Metalation under conditions of poor solvation. The X-ray structure of an internally solvated vinyllithium. Part III. Trans-hydrindanones via zirconium catalyzed intramolecular Michael reaction. Synthesis of a key intermediate to retigeranic acid. Ph.D. Thesis, Columbia University, New York, NY; (f) Roberts RA (1983) Studies directed toward a convergent approach to the synthesis of (–)-retigeranic acid. Ph.D. Thesis, Ohio State University, Columbus, OH
49. Snider BB, Rodini DJ, Van Straten J (1980) J Am Chem Soc 102:5872
50. Grieco PA, Gilman S, Nishizawa M (1976) J Org Chem 41:1485
51. Cohen T, Kuhn D, Falck JR (1975) J Am Chem Soc 97:4749
52. Kabalka GW, Chandler JH (1979) Synth Commun 9:275
53. Wright J, Drtina GJ, Roberts RA, Paquette LA (1988) J Am Chem Soc 110:5806
54. Paquette LA, Roberts RA, Drtina GJ (1984) J Am Chem Soc 106:6690
55. Wender PA, Singh SK (1985) Tetrahedron Lett 26:5987
56. Curran DP, Kuo SC (1986) J Am Chem Soc 108:1106
57. Marx JN, Norman LR (1975) J Org Chem 40:1602
58. Newhall WF (1958) J Org Chem 23:1274
59. Still WC, Mitra A (1978) J Am Chem Soc 100:1927
60. Salmond WG, Barta MA, Havens JL (1978) J Org Chem 43:2057
61. Gadwood RC, Lett RM (1982) J Org Chem 47:2268
62. Mitsunobu O (1981) Synthesis 1
63. Ho PT, Davies N (1984) J Org Chem 49:3027
64. Sugihara Y, Wakabayashi S, Saito N, Murata I (1986) J Am Chem Soc 108:2773
65. Singh SK, Wender PA (1990) Tetrahedron Lett 31:2517
66. Lam LKP, Hui RAHF, Jones JB (1986) J Org Chem 51:2047
67. Mukaiyama T, Matsueda R, Suzuki M (1970) Tetrahedron Lett 11:1901
68. Elad D, Rokach J (1964) J Org Chem 29:1855
69. Elad D, Rokach J (1965) J Org Chem 30:3361
70. Ireland RE, Bey P (1973) Org Synth 53:63
71. Brown M, Piszkiewicz LW (1967) J Org Chem 32:2013
72. Hara S, Taguchi H, Yamamoto H, Nozaki H (1975) Tetrahedron Lett 16:1545
73. Zapata A, Ferrer GF (1986) Synth Commun 16:1611

Chapter 11
Total Synthesis of the Lycopodium Alkaloid Complanadine A

Richmond Sarpong and Daniel F. Fischer

11.1 Introduction

Mankind's fascination with symmetry can be traced back to early civilizations in art, sculpture, architecture, music, and many other forms. Certainly, the bilateral symmetry of the human form and other life forms supports the idea that symmetry confers significant evolutionary advantages. It is therefore not surprising that in the world of complex molecules, many examples of symmetrical molecules can be found, which undoubtedly also provide an evolutionary advantage to the producing organism. Yet, as one more critically analyzes the structures of natural products that possess some element of symmetry, it becomes apparent that pseudosymmetry may, in fact, be more common and a key to an as of yet underappreciated subtlety of evolution. It was in the wake of these musings that we first started to ponder the pseudosymmetric, complex *Lycopodium* alkaloid complanadine A (**1**, Fig. 11.1).

As is evident from its structure, complanadine A is comprised of two molecules of the natural product lycodine (**2**), which are joined at the 2 and 3′ positions of the two monomers, respectively (see **1** for numbering). Although there are other examples of dimeric *Lycopodium* alkaloids, the pseudosymmetric variants are not very common. Therefore, a synthesis of complanadine A could offer opportunities to study the properties and intricacies of the reactivity of the pseudodimeric *Lycopodium* alkaloids.

11.2 Biosynthesis

As a whole, the *Lycopodium* alkaloids consist of over 270 members, and the number of natural products in this family continues to grow [1]. The natural products are divided into four structural classes, three of which are represented by the parent

R. Sarpong (✉) • D.F. Fischer
Department of Chemistry, University of California, Berkeley, CA 94720, USA

J.J. Li and E.J. Corey (eds.), *Total Synthesis of Natural Products*,
DOI 10.1007/978-3-642-34065-9_11,

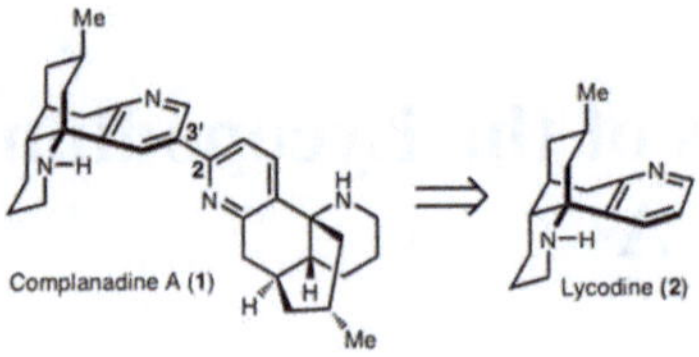

Fig. 11.1 Complanadine A and lycodine

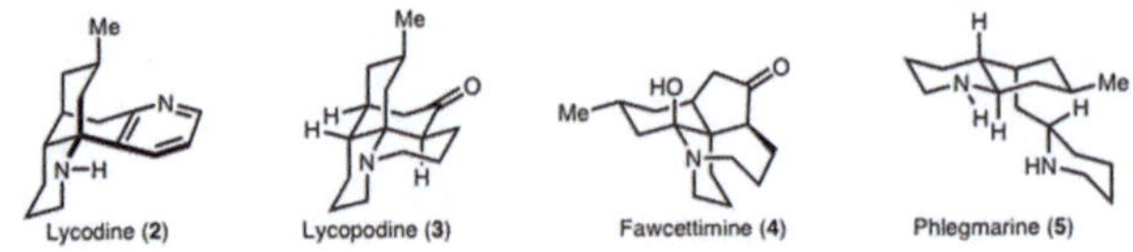

Fig. 11.2 Representative members of the *Lycopodium* structural classes

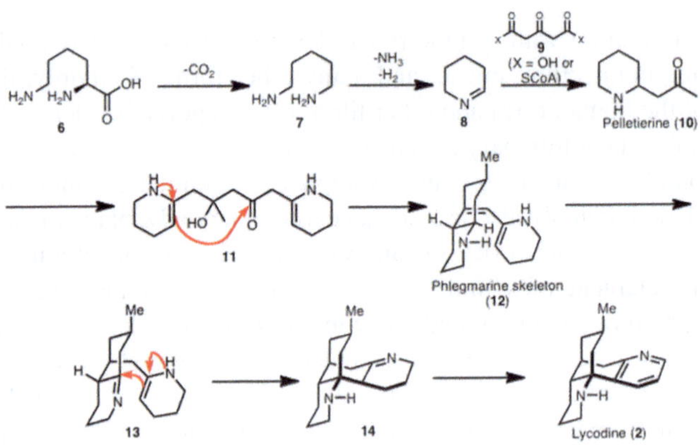

Scheme 11.1 Proposed biosynthesis of lycodine

natural products lycodine (**2**, Fig. 11.2), lycopodine (**3**), and fawcettimine (**4**). The fourth structural class is a conglomerate of unusual *Lycopodium* natural products, and so, this class is referred to as the miscellaneous class. Members of the miscellaneous class, like all the other *Lycopodium* alkaloids, likely arise from phlegmarine (**5**).

Complanadine A may be regarded as a member of the lycodine class. While the latter biosynthetic steps for the genesis of complanadine A are not well established, a proposal for the biosynthesis of lycodine, aided by extensive feeding studies carried out by Spenser and his coworkers, has been advanced as outlined in Scheme 11.1 [2]. Overall, biosynthetic elucidation of the *Lycopodium* alkaloids is challenging because cultivation and culturing of plant tissue from the producing club mosses is difficult [3]. The biosynthesis of the *Lycopodium* alkaloids is believed to begin with lysine (**6**), which undergoes decarboxylation to afford

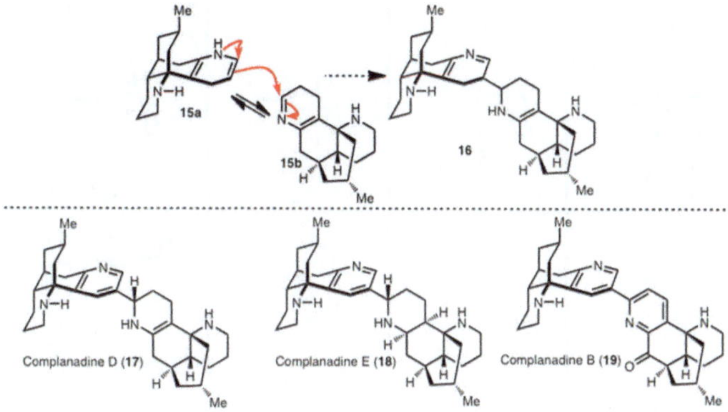

Scheme 11.2 Potential biosynthetic origin of pseudodimeric complanadines

diamine **7**. Oxidation of one of the amine groups of **7** followed by cyclization yields Δ^1-piperidine **8**, which is advanced to pelletierine (**10**) by the addition of a three-carbon unit that could emanate from acetone dicarboxylic acid or its corresponding bis-coenzyme A ester (see **9**). Pelletierine is a versatile and important intermediate that has been implicated in the biosynthesis of the *Lycopodium* alkaloids [4]. Dehydrogenation of pelletierine and attendant self-aldol reaction (or vice versa) leads to **11**, which undergoes further cyclization to yield the phlegmarine skeleton (**12**). As recently discussed by Hemscheidt [5], phlegmarine is believed to be a central branching point to the different structural classes of the *Lycopodium* alkaloids. In one scenario toward lycodine, imine **13**, which may be derived from **12**, undergoes a Mannich cyclization to yield cyclic imine **14**, which upon further oxidation gives lycodine.

Interestingly, the 2,2′ or 3,3′ conjoined dimers of lycodine have not been isolated to date. While complanadine A could arise from a dehydrogenative coupling of two lycodine molecules, there is no laboratory or biological precedent for this type of union. Instead, what appears to be more plausible and consistent with the precedent of Leete and Slattery [6] in their biosynthetic studies of natural products such as anatabine, α,β-dipyridyl [7], and nicotelline [8] is that complanadine A arises from **16**, which is in turn derived from a Mannich-type addition of enamine **15a** to imine **15b**. As is clearly evident, **15a** and **15b** are tautomers, and so it would seem that possibly, the secret to the biosynthesis of the pseudosymmetrical complanadine A is in fact rooted in symmetry, where **15b** is the likely biosynthetic progenitor. Circumstantial support for this assertion can also be gleaned from the structures of complanadines D (**17**) and E (**18**), which likely arise from dehydrogenation and disproportionation scenarios from **16**, respectively. The plausible biogenesis scenario presented here for complanadine A was first recognized and advanced by Professors Morita and Kobayashi in 2005 [9] (Scheme 11.2).

From a laboratory synthesis standpoint, it was difficult to envision a synthesis of complanadine A along the proposed biosynthetic lines, given the propensity with which compounds such as **15a** and **15b** undergo disproportionation or oxidation to afford **2** and **14** as recognized in the work of Schumann and coworkers [10].

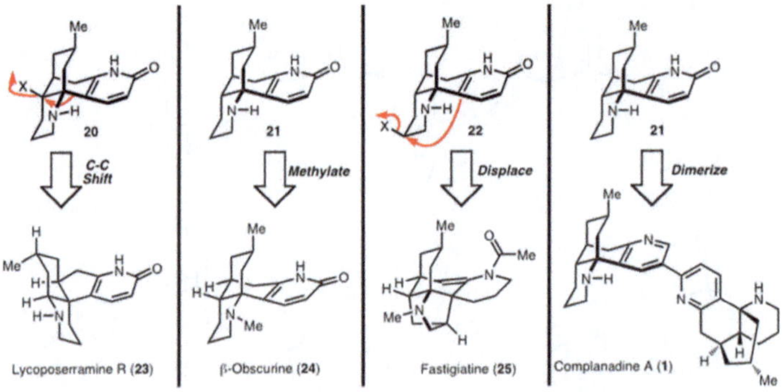

Fig. 11.3 Potential synthetic link of selected *Lycopodium* alkaloids to obscurines

While only a snapshot of the proposed biosynthetic picture for complanadine A has been presented above, we considered an alternative approach in our synthetic strategy analysis (as shown below in Fig. 11.3) that also put into focus the close connections between complanadine A, which would arise from a dimerization of the β-obscurine precursor **21**, and the *Lycopodium* alkaloids lycoposerramine R (**23**) and fastigiatine (**25**). The details of this strategic approach are the subject of this chapter.

11.3 Biological Activity

Complanadine A was isolated from the club moss *Lycopodium complanatum* (Lycopodiaceae) by the group of Professor Jun'ichi Kobayashi (Hokkaido University) [11]. The report of the isolation of this unique dimer of the well-known alkaloid lycodine (**2**) appeared in 2000 and was followed by subsequent reports of the isolation of the related compounds complanadine B (**19**) [9], D (**17**) [12], and E (**18**) [13]. The original report of its isolation stated that complanadine A was cytotoxic against murine leukemia L1210 cells with an IC_{50} of 5.6 μg/mL in an in vitro assay. However, like many *Lycopodium* alkaloids, the pseudodimeric alkaloids **1**, **17**, **18**, and **19** are more interesting from a bioactivity standpoint because of their recognized ability to slow and reverse the effects of cognitive decline. Many of the *Lycopodium* alkaloids, exemplified by huperzine A [14], are potent inhibitors of acetylcholinesterase (AChE). This activity leads to increased memory and learning in many murine models [15]. As a result, several AChE inhibitors, including Aricept® [16, 17], are currently marketed to combat, for example, cognitive decline resulting from Alzheimer's disease. As has emerged over the years, there is an acute need to identify natural products that not only slow the progression of neurodegenerative disease (which is only a palliative treatment) but also those that reverse the effects of cognitive decline by promoting the development of new neural networks, especially since a large number of

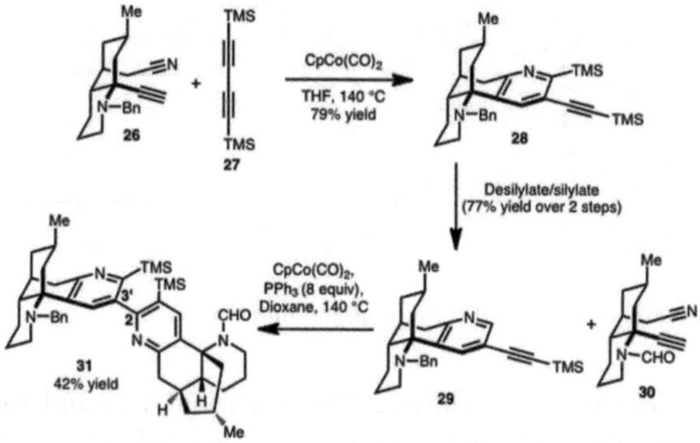

Scheme 11.3 Siegel approach to the pseudodimeric complanadine framework

neurodegenerative diseases are characterized by neurite atrophy [18]. The complanadines are very interesting in this regard because they enhance the expression of mRNA in glial cells that codes for neurotrophins (especially nerve growth factors)—small proteins that maintain the health of neurons and also promote neuronal growth [19]. Furthermore, a large majority of the currently investigated and administered neurotrophins are large proteins that do not easily cross the blood–brain barrier (BBB), whereas small molecules like the complanadines are expected to easily cross the BBB [20]. For these reasons, we viewed the synthesis of **1** as an opportunity to gain access to small molecule leads that may reverse the effects of neurodegeneration.

11.4 The Siegel Synthesis of Complanadine A

Contemporaneous with our completion of the total synthesis of complanadine A, as described herein, the group of Prof. D. Siegel at the University of Texas at Austin also completed a synthesis of **1**. The Siegel synthesis featured a very different approach from our own and was highly innovative in its creative use of sequential [2+2+2] cyclotrimerization reactions [21] to construct the two pyridine rings that are found in complanadine A (see Scheme 11.3). The details of the Siegel synthesis, which are only summarized here, are described in more detail in a communication that appeared at the same time as our report on the total synthesis of **1** [22]. As shown in Scheme 11.3, the key to synthesizing the 2,3′ conjoined lycodine units in the Siegel approach was to take advantage of the steric and electronic influences of the alkyne substituents in guiding the regioselectivity of the cyclotrimerization reaction. Thus, the initial [2+2+2] cyclotrimerization would be driven by the directing influences of a TMS group vs. an alkyne group in **27**, whereas in the second [2+2+2] cyclotrimerization, a pyridine group would be pitted against a TMS group in **29** to exert regiocontrol. In the event, excellent regioselectivity was observed in the cyclotrimerization involving **26** and **27** (25:1), whereas only modest

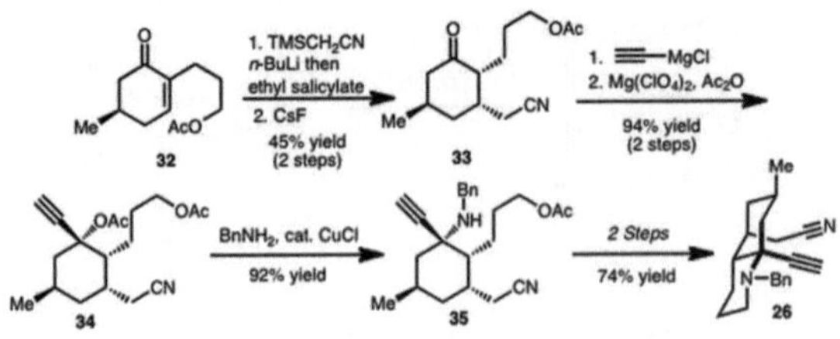

Scheme 11.4 Synthesis of the Siegel bicyclic piperidine derivative

regioselectivity was observed in the cyclotrimerization of **29** and **30** (3:1). Interestingly, cyclotrimerization to favor formation of **31** was only observed upon addition of 8 equiv of triphenylphosphine. Additionally, the use of a formyl-protecting group for the alkynyl nitrile partner (see **30**) as opposed to a benzyl-protecting group (as in **26**) was important in order for the cyclotrimerization to occur in the presence of the excess of triphenylphosphine. The influence of triphenylphosphine on the regioselectivity is not obvious and is the subject of further study in the Siegel group.

Bicyclic piperidine derivative **26** is in its own right a very interesting compound, which was prepared from **32** (available in four steps from pulegone) by the Siegel group as described in Scheme 11.4. Particularly noteworthy in the preparation of **26** was the utility of $TMSCH_2CN$ as an acetonitrile synthon for conjugate addition [23] and the implementation of conditions that likely generate an allenylidine intermediate [24] to effect a propargylic substitution reaction in converting **34** to **35** in accord with the precedent of Tomioka and Koga.

11.5 Strategy and Retrosynthesis

We recognized at the outset of our studies that the synthetic challenge in achieving a direct synthesis of complanadine A would be to identify ways to exploit two molecules of lycodine or a closely related precursor. With appropriately positioned functional handles (at C2 and C3′, respectively; see **36** and **37**, Scheme 11.5), the coupling of the pyridine moieties would yield the complanadine skeleton.

We were drawn to *N*-desmethyl α-obscurine (**38**), a *Lycopodium* alkaloid isolated in 1944 from *Lycopodium* obscurum by Manske and Marion [25], as a starting point for our studies. This natural product derivative differs only in the oxidation pattern of one of its nitrogen-containing rings (dihydropyridone vs. pyridine) from lycodine. A synthesis of **38** had been reported previously in racemic form by Schumann using a highly creative cascade sequence [10]. Therefore, our synthetic studies commenced by targeting **38**, following closely the precedent of Schumann.

The elegant procedure of Schumann to prepare **38** required two compounds, aminoketal **42** (Scheme 11.6a) and dihydropyridone **45** (Scheme 11.6b). As with many syntheses of *Lycopodium* alkaloids, the synthesis of aminoketal **42** began with

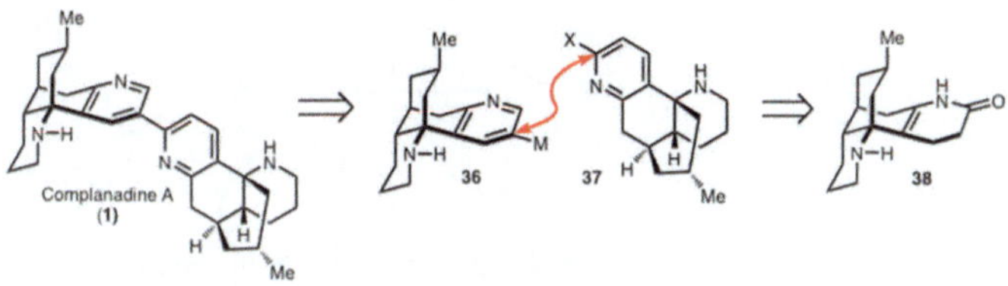

Scheme 11.5 Retrosynthesis of complanadine A

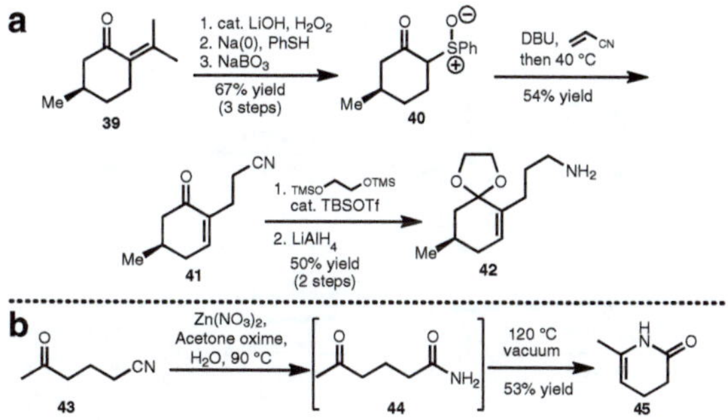

Scheme 11.6 Preparation of the components for tetracycle formation

pulegone, which upon Weitz-Scheffer epoxidation [26] and treatment with sodium phenylthiolate produced an intermediate phenyl sulfide (not shown) via an addition/retroaldol sequence with loss of acetone. Oxidation of the phenyl sulfide with sodium perbromate yielded sulfoxide **40** in 67 % yield over three steps. This procedure has been conducted on 25-g scale without event. Enone nitrile **41** was formed by conjugate addition of ketosulfoxide **40** into acrylonitrile followed by sulfoxide extrusion [27] upon warming to 40 °C. This procedure was partly inspired by the preparation of a closely related compound that was employed in Dake's synthesis of fawcettidine [28]. Ketal protection of the enone carbonyl group of **41** was best achieved by enlisting the procedure of Noyori [29] using the bis-trimethylsilyl ether of ethylene glycol. This step is plagued by incomplete conversions and remains one of the bottlenecks of the synthesis. Reduction of the nitrile functional group in **41** following installation of the ketal group gave aminoketal **42**. The synthesis of dihydropyridone **45** was much more straightforward and began with ketonitrile **43**, which, although commercially available, can be prepared on large scale according to the one-pot procedure of Lawesson from *t*-butyl acetoacetate [30]. Subjecting **43** to Zn(II)-catalyzed hydrolysis of the nitrile group followed by distillation gave dihydropyridone **45** directly. Presumably, **45** is formed via ketoamide **44**, which undergoes dehydration. As would become evident further on, ketoamide **44** was in fact our desired coupling partner; however, cyclodehydration to **45** occurs rapidly, and so we chose to employ **45** in our studies.

Access to **42** and **45** set the stage for a remarkable cascade event, first described by Schumann, where the lone stereocenter in **42** directs the installation of three

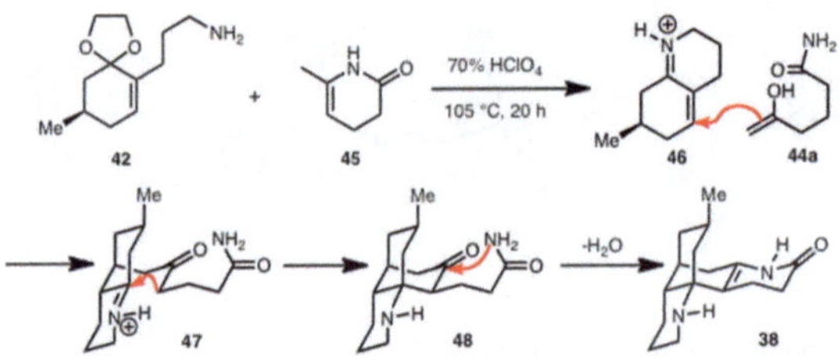

Scheme 11.7 Synthesis of tetracycle **38**

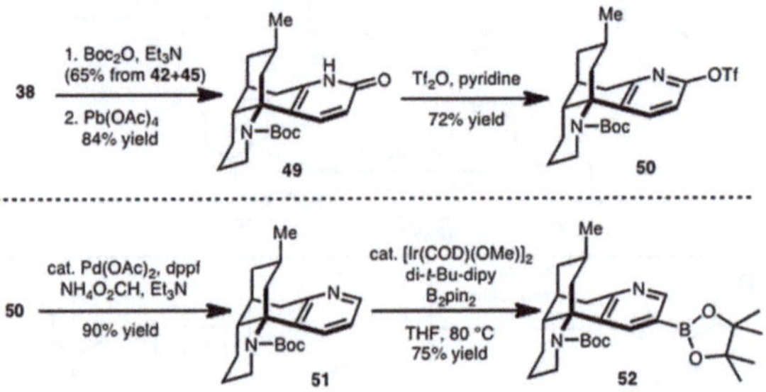

Scheme 11.8 Synthesis of the triflate and boronic ester coupling partners

additional stereocenters in the construction of *N*-desmethyl α-obscurine (**38**). In the event (Scheme 11.7), subjecting a mixture of aminoketal **42** and dihydropyridone **45** in dioxane with perchloric acid and heating at 105 °C for 20 h gave **38**. A mechanistic rationalization, as presented in Scheme 11.7, can be constructed from the studies of Schumann and our observations. Upon exposure to aqueous acid conditions, aminoketal **42** likely undergoes ketal hydrolysis and subsequent iminium ion formation to generate **46**. Conjugate addition of **44a**, the enol tautomer of **44**, at this stage would yield the Michael adduct **47**. Intramolecular Mannich cyclization would then lead to tricycle **48**, which upon cyclodehydration would give **38**. The sequence has been carried out on up to 10-g scale.

Access to **38** set the stage for the preparation of the two coupling partners that would be joined to give the complanadine A skeleton (Scheme 11.8). The preparation of the C(2)-functionalized partner began with the Boc protection of *N*-desmethyl-α-obscurine (65 % yield from **42** to **45**), which then required oxidation (dehydrogenation) to pyridone **49**. This dehydrogenation has been found to be difficult in related systems. For example, after investigating several direct oxidation conditions, Wu and Bai settled on a two-step procedure that involved chlorination with sulfuryl chloride and elimination (by heating to 120 °C) in the synthesis of the related huperzine B [31]. Our desire to effect the dehydrogenation using less forcing conditions led to a reinvestigation of oxidation conditions. After an extensive study, we found that this could be best achieved using lead (IV) tetraacetate (84 % yield), which gave pyridone **49** from **38**. Triflation of pyridone **49** using standard procedures provided pyridine triflate **50**, which would serve as the

C(2) coupling partner. Our plan to prepare the C(3) coupling partner, boronic ester **52**, relied on using triflate **50** as a starting point. Thus, removal of the triflate group using standard Pd(0)-mediated reduction conditions followed by a site-selective borylation of the pyridine ring gave C(3)-functionalized coupling partner **52**. The remarkable position-selective borylative C–H functionalization, which had been described by the groups of Hartwig and Miyaura [32] and also by Maleczka and Smith [33], is the topic of the next section.

11.6 Borylative C–H Functionalization

Borylative C–H functionalization is a kinetic and thermodynamically favorable process that has been used to great effect to convert sp^2 and sp^3 C–H bonds into useful functional groups. With the identified importance of the Suzuki-Miyaura cross coupling in modern C–C bond-forming events, borylative C–H functionalization has taken on even more added significance. Pioneering stoichiometric studies by Marder and by Hartwig appeared in 1993 and 1995, respectively, as discussed in a leading review published by these two investigators and their coworkers in 2010 [34]. An updated account on this general area has also recently appeared [35]. The success of these studies and the emergence of Rh(I)- and Ir(I)-catalyzed borylation reactions has brought these methodologies to a stage where practitioners of complex molecule synthesis are actively applying them in their syntheses. Especially exciting are applications that involve the site-selective borylation of heterocycles including indoles, pyrroles, and pyridines, often at positions that differ from the traditional sites for Friedel-Crafts-like reactivity. Three examples that illustrate the emerging applications of borylative C–H bond functionalization in complex molecule synthesis are summarized here.

11.6.1 Benzene Ring Functionalization: Hartwig Synthesis of Taiwaniaquinol B

In 2011, Hartwig and coworkers reported the total synthesis of taiwaniaquinol B (**55**, Scheme 11.9), a member of a family of diterpenoids that are derived from the abietane skeleton [36]. A key aspect of the Hartwig synthesis of taiwaniaquinol B was the use of the iridium-catalyzed borylation reaction to accomplish the C(5) functionalization of resorcinol derivative **53**. This regioselectivity for the overall bromination is complementary to that which would be obtained using a standard electrophilic aromatic substitution (EAS) reaction. In the transformation of **53** to **54**, a sterically controlled borylation was first accomplished, which was then followed by treatment of the boronic ester intermediate with cupric bromide to

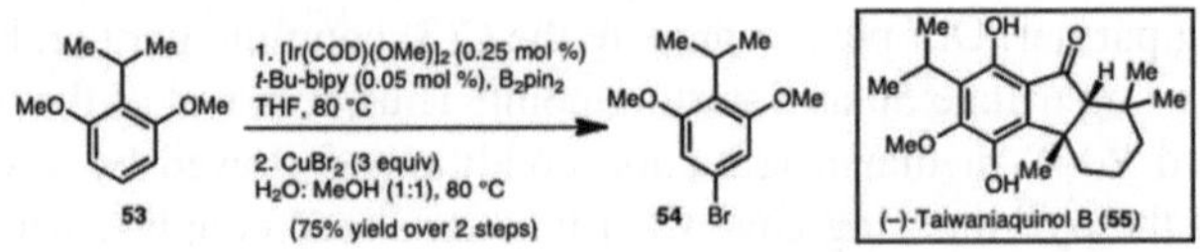

Scheme 11.9 Arene borylation in the synthesis of taiwaniaquinol B by Hartwig

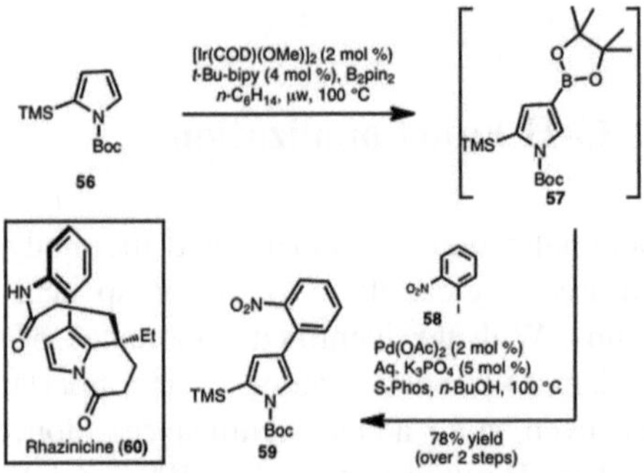

Scheme 11.10 Pyrrole borylation in the synthesis of rhazinicine by Gaunt

effect the overall conversion. Ultimately, brominated resorcinol derivative **54** was utilized in the preparation of the arene portion of taiwaniaquinol B.

11.6.2 Pyrrole Ring Functionalization: Gaunt Synthesis of Rhazinicine

In what was the first synthesis of the pyrrole alkaloid rhazinicine (**60**, Scheme 11.10), Gaunt and coworkers utilized the Ir-catalyzed borylation chemistry to great effect to functionalize a pyrrole nucleus [37]. In order to streamline the synthesis of **60**, silylated Boc pyrrole **56** was subjected to the Ir-catalyzed borylation conditions, which effected regioselective borylation to afford **57**. The Boc group likely directs borylation to the C(4) position (see **57**) consistent with related precedent from Maleczka and Smith using N-TIPS silylated pyrroles [38]. Boronic ester **57** was directly subjected to Suzuki-Miyaura cross coupling (without isolation) to iodonitroarene **58** to afford **59**. Access to **59** set the stage for the synthesis of rhazinicine. The Gaunt synthesis of rhazinicine was an early application of the borylation of heteroaromatics in complex molecule synthesis.

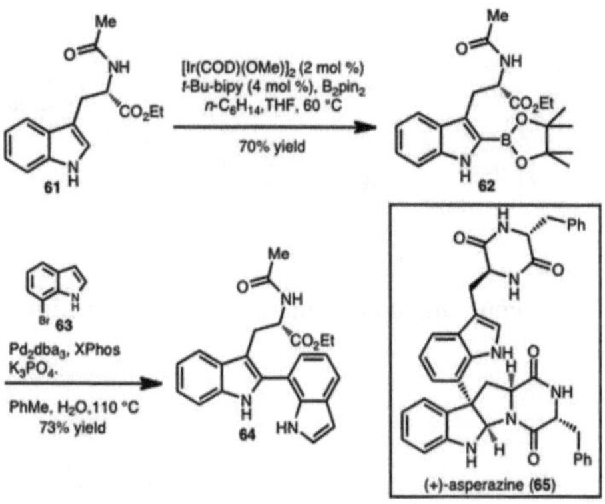

Scheme 11.11 Indole borylation toward a synthesis of asperazine by Movassaghi and Miller

11.6.3 Indole Ring Functionalization: Movassaghi Synthesis of the Asperazine Core

The Ir-catalyzed borylation of the indole nucleus is another important development that promises to find widespread use in complex molecule synthesis. Early reports include the functionalization of C(7) and also of C(2), reported by Malezcka and Smith and by Hartwig, respectively [39, 40]. In a report in 2011, Movassaghi, Miller, and coworkers demonstrated the borylation of tryptamine derivative **61** to afford **62** in 70 % yield [41]. This material was subjected to Suzuki-Miyaura cross coupling with 7-bromoindole (**63**) to set the stage for studying the oxidative rearrangement of **64**, which would eventually provide diketopiperazine indole alkaloids such as asperazine (Scheme 11.11).

11.7 Completion of the Complanadine A Synthesis

In our own studies toward complanadine A, with triflate **50** and boronic ester **52** in hand, the stage was set for the key regioselective coupling to yield the complete natural product framework. In the event, subjection of pyridine triflate **50** and boronic ester **52** to Suzuki cross-coupling conditions gave adduct **66** in 53 % isolated yield. Several aspects of this sequence are of note. First, over the course of our screening of precatalysts, we have found that $PdCl_2$(dppf) is uniquely effective for this coupling reaction. Second, adding 25 mol% of triethylsilane to the precatalyst (presumably to effect reduction to the active Pd(0) species) has emerged to be important in realizing high yields of the cross-coupling product. In the absence of this additive, a substantial amount of pyridone **49**, which likely arises from the hydrolysis of triflate **50**, is formed. This observation is consistent with one of the challenges that has been observed with the cross coupling of heteroaromatic

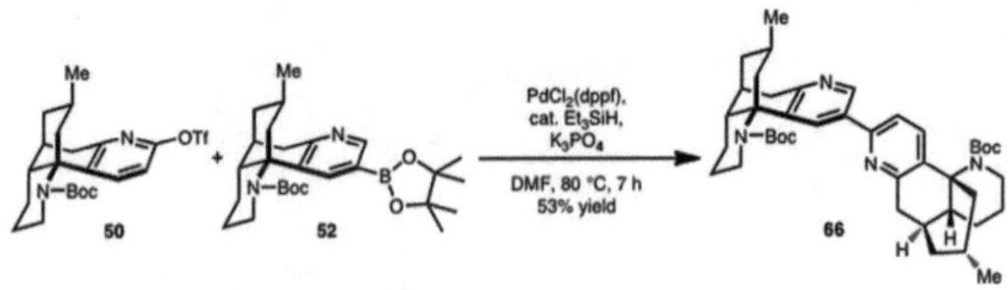

Scheme 11.12 Suzuki cross coupling to build the core of complanadine A

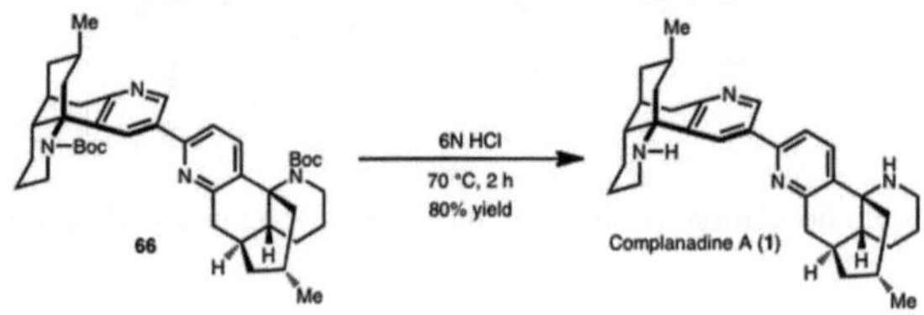

Scheme 11.13 Acid-mediated Boc removal to afford complanadine A

pseudohalides and halides. Furthermore, along with the desired product, small amounts of Boc lycodine, **51**, likely arising from the reduction of triflate **50** or proto-deborylation of **52** under the cross-coupling conditions, were also isolated (Scheme 11.12).

The Suzuki adduct **66** was subjected to 6 N HCl to afford complanadine A as its HCl salt in 80 % yield (Scheme 11.13). The ^{1}H and ^{13}C NMR spectral data of complanadine A were consistent with that which had been obtained by isolation (Kobayashi) and in Siegel's total synthesis.

11.8 Application of the Strategy to Lycopladines F and G

Perhaps one of the most important outcomes of our synthetic studies on complanadine A is our ability to access boronic ester **52**, which we believe should serve as a versatile intermediate en route to many related natural products. In a preliminary study, we have investigated the conversion of **52** to the natural products lycopladines F and G [42]. Strategically, we envisioned that a direct approach to these *Lycopodium* alkaloids would involve a cross coupling of an acid halide (e.g., **67**) and boronic ester **52**. However, despite our best efforts, this desired cross coupling has not been successful. As an alternative, Suzuki cross coupling with vinyl bromide **68**, followed by dihydroxylation using the Upjohn method [43] and periodate cleavage, affords Boc-protected lycopladine G (**70**) (Scheme 11.14).

Similarly, we anticipate that boronic ester **52** should prove highly useful for the synthesis of other lycodine-derived *Lycopodium* alkaloid pseudodimers including complanadines D and E (see Scheme 11.2) as well as unnatural analogues for biological studies. This is the direction of our ongoing research in this area.

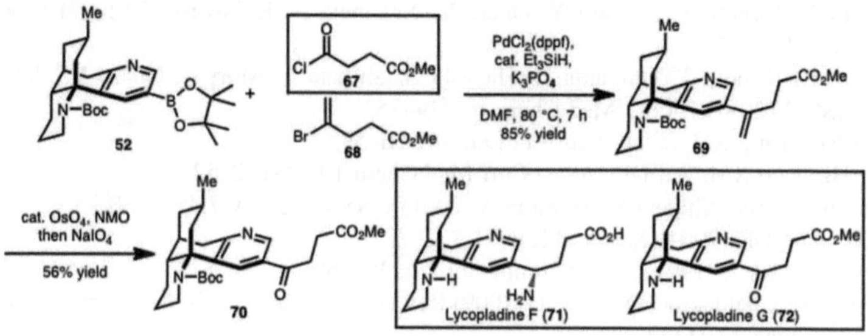

Scheme 11.14 Application of boronic ester **52** to the synthesis of lycopladines F and G

11.9 Conclusion

The total synthesis of complex natural products presents myriad opportunities to test the scope and limitations of new methods. This is clearly borne out in our synthetic studies of the *Lycopodium* alkaloid complanadine A, which highlighted a powerful method for site-selective C–H borylation developed by the Hartwig/Miyaura and Smith/Maleczka laboratories. The pseudosymmetry of complanadine A challenged us to develop a strategy for its synthesis that relied on a common monomer to maximize synthetic efficiency. An effective method to diverge the reactivity of the common intermediate at a late stage was important in realizing the total synthesis goal. Our studies provide one example of a pseudosymmetrical dimerization tactic where positioned functional groups determine the sites of reactivity. A more ambitious goal, and as yet unrealized in our hands, is to achieve the 2,3′ dimerization from lycodine in a single pot. The pace of innovation of modern C–H functionalization methods makes us optimistic that this "dream sequence" may be orchestrated in the not too distant future.

References

1. For a recent review, see: Hirasawa Y, Kobayashi J, Morita H (2009) Heterocycles 77:679–729
2. (a) Gupta RN, Castillo M, MacLean DB, Spenser ID (1970) Can J Chem 48:2911–2918; (b) Castillo M, Gupta RN, MacLean DB, Spenser ID (1970) Can J Chem 48:1893–1903; (c) Gupta RN, Castillo M, MacLean DB, Spenser ID, Wrobel JT (1968) J Am Chem Soc 90:1360–1361
3. Ma XQ, Gang DR (2008) Phytochemistry 69:2022–2028
4. Hemscheidt T, Spenser ID (1996) J Am Chem Soc 118:1799–1800
5. Hemscheidt T (2000) Top Curr Chem 209:175–206
6. Leete E, Slattery SA (1976) J Am Chem Soc 98:6326–6330
7. Späth E, Zajic E (1936) Chem Ber 69:2448–2452
8. Kuffner F, Kaiser E (1954) Monatsh Chem 85:896–905
9. Morita H, Ishiuchi K, Haganuma A, Hoshino T, Obara Y, Nakahata N, Kobayashi J (2005) Tetrahedron 61:1955–1960
10. Schumann D', Naumann A (1983) Liebigs Ann Chem 220–225
11. Kobayashi J, Hirasawa Y, Yoshida N, Morita H (2000) Tetrahedron Lett 41:9069–9073

12. Ishiuchi K, Kubota T, Mikami Y, Obara Y, Nakahata N, Kobayashi J (2007) Bioorg Med Chem 15:413–417
13. Ishiuchi K, Kubota T, Ishiyama H, Hayashi S, Shibata T, Mori K, Obara Y, Nakahata N, Kobayashi J (2011) Bioorg Med Chem 19:749–753
14. Liang YQ, Tang XC (2004) Neurosci Lett 361:56–59
15. Jiang HL, Luo XM, Bai DL (2003) Curr Med Chem 10:2231–2252
16. For a review, see: Shigeta M, Homma A (2001) CNS Drug Rev 7:353–368
17. Ma X, Gang DR (2004) Nat Prod Rep 21:752–772
18. Braak H, Braak E (1991) Acta Neuropathol 82:239–259
19. (a) Fiore M, Chaldakov GN, Aloe L (2009) Rev Neurosci 20:133–145; (b) Levi-Montalcini R (1987) Science 237:1154–1162
20. Gill SS, Patel NK, Hotton GR, O'Sullivan K, McCarter R, Bunnage M, Brooks DJ, Svendsen CN, Heywood P (2003) Nat Med 9:589–595
21. For pertinent reviews, see: (a) Varela JA, Saá C (2003) Chem Rev 103:3787–3801; (b) Louie J, Chopade PR (2006) Adv Synth Catal 348:2307–2327
22. Yuan C, Chang C-T, Axelrod A, Siegel D (2010) J Am Chem Soc 132:5924–5925
23. Tomioka K, Koga K (1984) Tetrahedron Lett 25:1599–1600
24. Imada Y, Yuasa M, Nakamura I, Murahashi S-I (1994) J Org Chem 59:2282–2284
25. Manske RHF, Marion L (1944) Can J Res 22:53–55
26. (a) Weitz E, Scheffer A (1921) Ber Dtsch Chem Ges 54:2327–2344; (b) For a recent review, see: Corma A, Iborra S, Mifsud M, Renz M (2005) ARKIVOC 9:124–132
27. (a) Caine D, Procter K, Cassell RA (1984) J Org Chem 49:2647–2648; (b) Or neat in the presence of $CaCO_3$: Mutti S, Daubié C, Decalogne F, Fournier R, Rossi P (1996) Tetrahedron Lett 37:3125–3128
28. Kozak JA, Dake GR (2008) Angew Chem Int Ed 47:4221–4223
29. Tsumoda T, Suzuki R, Noyori R (1980) Tetrahedron Lett 21:1357–1358
30. Näslund G, Senning A, Lawesson S-O (1962) Acta Chem Scand 16:1324–1328
31. Wu B, Bai D (1997) J Org Chem 62:5978–5981
32. Ishiyama T, Takagi J, Ishida K, Miyaura N, Anastasi NR, Hartwig F (2002) J Am Chem Soc 124:390–391
33. Cho J-Y, Tse MK, Holmes D, Maleczka RE Jr, Smith MR (2002) Science 295:305–308
34. Mkhalid IA, Barnard JH, Marder TB, Murphy JM, Hartwig JF (2010) Chem Rev 110:890–931
35. Hartwig JF (2012) Acc Chem Res 45:864–873
36. Liao X, Stanley LM, Hartwig JF (2011) J Am Chem Soc 133:2088–2091
37. Beck EM, Hatley R, Gaunt MJ (2008) Angew Chem Int Ed 47:3004–3007
38. Tse MK, Cho JY, Smith MR III (2001) Org Lett 3:2831–2833
39. (a) Paul S, Chotana GA, Holmes D, Reichle RC, Maleczka RE Jr, Smith MR III (2006) J Am Chem Soc 128:15552–15553; (b) Kallepalli VA, Shi F, Paul S, Onyozili EN, Maleczka RE Jr, Smith MR III (2009) J Org Chem 74:9199–9201
40. (a) Takagi J, Sato K, Hartwig JF, Ishiyama T, Miyaura N (2002) Tetrahedron Lett 43:5649–5651; (b) Ishiyama T, Takagi J, Nobuta Y, Miyaura N (2005) Org Synth 82:126–131
41. Kolundzic P, Noshi MN, Tjandra M, Movassaghi M, Miller SJ (2011) J Am Chem Soc 133:9104–9111
42. For the isolation of lycopladines F and G, see: Ishiuchi K, Kubota T, Hayashi S, Shibata T, Kobayashi J (2009) Tetrahedron Lett 50:4221–4224
43. (a) Schneider WP, McIntosh AV (Upjohn) (1956) US Patent 2,769,824, 1956; (b) VanRheenen V, Kelly RC, Cha DY (1976) Tetrahedron Lett 17:1973–1976

Index

A
Allyl alcohol route, 203
Asperazine core, Movassaghi synthesis of, 269
Aspergillus variecolor, 238
Astellatol, retigeranic acid, 238

B
Baylis–Hillman reaction, 47
Borylative C–H functionalization, complanadine A
  benzene ring, 267–268
  indole ring, 269
  pyrrole ring, 268
Broka phleichrome synthesis, 161
Brook rearrangement, strychnine
  C–C bond formation, 96
  copper-assisted, 95
  dolabelide D, 96, 97
  model system, 97
  polyfunctional building block, 96
  siloxacycle ring opening, 96
  six-step linear synthesis, 97, 98
  vinylsilanes, 95
Bryostatin 1
  architectural features, 104
  clinical applications, 105
  discovery, 103
  human cancer therapy, 105
  pharmacological activity, 105
  pharmacophore model, 105
  PKC isozymes, 104–106
  structures of, 103, 104
  synthesis of
    aldehyde allylstannations, 115
    A-ring aldehyde to hydroxyallylsilane conversion, 116
    A-ring fragments, 115
    C-ring fragments, 115
    Fuji's chiral phosphonate, 117
    intermolecular Prins pyran annulation, 115
    olefinic ester preparation, 115–116
    Yamaguchi macrolactonization, 117
  in vitro/in vivo antineoplastic activity, 104
Bryostatin 2 synthesis, 110–111
Bryostatin 3 synthesis, 112, 113
Bryostatin 7 synthesis
  AB-ring fragment construction, 109–110
  A-ring fragment synthesis
    β acetoxy aldehyde, 120
    cyclic ketal formation, 120–121
    diols, 120
    enantioselective double allylation, 119
    homoallylic alcohol formation, 120
    iridium-catalyzed transfer hydrogenation, 120
    neopentyl alcohol, 120
    ortho-cyclometalated iridium catalyst, 120
  asymmetric aldol addition methodology, 108
  biosynthetic pathway, 106–108
  C–C bond forming hydrogenation, 118, 119
  chiral auxiliaries, 126, 127
  C-ring fragment synthesis
    β-ketoaldehyde, Horner–Wadsworth–Emmons olefination, 122
    α-bromoketone, 122
    chromatographic purification, 122

J.J. Li and E.J. Corey (eds.), *Total Synthesis of Natural Products*,
DOI 10.1007/978-3-642-34065-9, © Springer-Verlag Berlin Heidelberg 2012

crude allylic alcohol, 122
1,3-enyne preparation, 122–123
glyoxal preparation, 121–122
hydrogen-mediated reductive coupling, 120, 123, 124
ketone-enol silane conversion, 122
Mitsunobu inversion, 122
fragment union
Keck–Yu silyl-Prins pyran annulation, 123, 124
Piers'stannylcupration method, 110
protecting groups, 126, 127
sulfone containing C-ring fragment, 110
synthetic efficiency, 127
transfer hydrogenation, 118, 119
triol elaboration, 125, 127
Julia–Lythgoe olefination, 108
Bryostatin 9 synthesis, 117, 118
Bryostatin 16 synthesis
B-ring pyran, 114, 115
C19 alcohol, Dess–Martin oxidation of, 115
C-ring pyran installation, 114
metal catalysis, 114
substrate-directed vinylogous aldol reaction, 114
Yamaguchi esterification, 113–114

C

Calphostin A,D
biomimetic approach, 165, 166
Broka syntheses, 162
Chiral lithium, 163
Coleman naphthalene, 163
Hauser synthesis, 162
Calycanthaceous alkaloids, 212, 213
C_{19}-diterpenoid alkaloids, 1
C_{20}-diterpenoid alkaloids, 1, 2
Cercospora kikuchii, 158
Cercosporin
atropisomerization, 166–167
biological activity, 159–161
bisacid, 178
calphostin A,D
biomimetic approach, 165, 166
Broka syntheses, 162
Chiral lithium, 163
Coleman naphthalene, 163
Hauser synthesis, 162
cancer therapy, 175
C7,C72′ and C2,C2′ substitution, 176, 177
C. kikuchii, 158
Cladosporium cladosporioides, 159
Cladosporium phlei, 159
description, 157
enantioselective synthesis, 168
Kita's method, 173
organocuprate-mediated epoxide, 173
phleichrome synthesis, 161–164
Wilkinson's catalyst, 175
Cladosporium cladosporioides, 159
Cladosporium phlei, 159
Class B perylenequinones, 157, 158
C-nor-D-homosteroids
nakiterpiosin
biological functions, 34
carbonylative Stille coupling, 30
convergent synthesis approach, 30
coupling component synthesis, 32–33
halogen atoms, biogenesis of, 31
photo-Nazarov cyclization reaction, 30, 31
structural revision, 31
synthetic strategy, 31
transannular Diels–Alder reactions, 29
6,6,5,6 steroidal skeleton synthesis
biomimetic approach, 27–28
miscellaneous approach, 28–31
ring-by-ring approach, 28, 29
Coleman synthesis, 164
Complanadine A
acid-mediated Boc removal, 270
application to lycopladines F and G, 270, 271
bicyclic piperidine derivative, 264
biological activity, 262–263
biosynthetic elucidation, 260, 261
borylative C–H functionalization
benzene ring, 267–268
indole ring, 269
pyrrole ring, 268
cyclotrimerization reactions, 263, 264
dehydrogenative coupling, 261
pelletierine intermediate, 261
phlegmarine, 261
pseudosymmetry, 259, 271
retrosynthesis
aminoketal, 264, 265
dihydropyridone, 265
triflate and boronic ester coupling, 266
Siegel synthesis, 263–264
structural class, 259, 260
Suzuki cross-coupling condition, 269, 270

D

(+)-11,11′-Dideoxyverticillin A
biosynthesis
dimeric epipolythiodiketopiperazine alkaloids, 214–215
PLP cofactor, 216

C3-C3′ dimeric linkage, 216
enantioselective hexahydropyrroloindole formation, 218
Hendrickson's oxidative dimerization, 217
meso-linkage, 218
Overman's asymetric alkylation, 217
C–H abstraction method, 220
classification, 211–213
diketopiperazine, 218
gliotoxin, 220, 221
hyalodendrin, 221
pharmacology, 213–214
retrosynthetic analysis
dithiol intermediate, 229
EDC peptide coupling, 223, 224
monomeric model systems, 225
N-acyliminium ion, 228
octacyclic dimer, 224
push–pull arrangement, 225
radical-clock hydantoin, 226–227
radical-mediated rebound process, 227
tetraols, 227
WIN 64821, 223
Woodward–Prévost method, 228
Trown method, 218–219
Diels–Alder reaction
(±)-*O*-methylkinamycin C synthesis, 51, 54
retigeranic acid, 241–242
strychnine
dehydrotubifoline synthesis, 81
indoles intramolecular cycloaddition, 75–77
structural challenges, 71–73
tryptamine derived Zincke aldehydes, 77–79
Zincke aldehydes, 73, 74
Diene route, phomactin A, 202–203
Dimeric hexahydropyrroloindole alkaloids, 212

F
Favorskii products, retigeranic acid, 243, 244

H
Heck reaction
norfluorocurarine synthesis, 81–82
strychnine
β-H elimination selectivity, 89
dehydrodesacetylretuline synthesis, 88
D-ring formation
valparicine synthesis, 88
Hedgehog antagonists, 25, 26
Hendrickson's oxidative dimerization, 217
Hetisine alkaloids
aldehyde, α-arylation of, 4
azomethine dipolar cycloaddition, 9, 10
biomimetic strategies, 9
diterpene biosynthesis of, 3, 4
endocyclic azomethine ylide, 9, 10
Hofmann–Loffler–Freytag reaction, 7
Hoveyda method, 19
imide synthesis , intramolecular alkylation, 6
intramolecular Diels–Alder reaction, 18, 19
intramolecular disconnections, 8
Lewis acid-catalyzed acetal-ene reaction, 4
nominine, total synthesis of, 4–5, 20, 21
4-oxidoisoquinolinium betaine, intramolecular dipolar cycloaddition, 15–18
3-oxidopyridinium betaine
dipolar cycloaddition of, 11
ene-nitrile substrate preparation, 11, 12
nitroalkene oxidopyridinium betaine preparation, 12
vinyl sulfone cycloaddition substrate, 11, 12, 14
pentacyclic intermediate synthesis, 7
pharmacological investigations, 2–3
pyrrolidine core access, 7
retro-cycloaddition key elements, 8
retrosynthetic strategy, 8–9
structures, 1, 2
synthetic challenge, 7
vinyl sulfone cycloaddition substrate, 7
HF6-31G*/B3LYP6-31G* calculations, 192, 193
Hofmann–Loffler–Freytag reaction, 7
Homologation C5a, phomactin A
enone synthesis, 200
experimental realization, 200–202
intramolecular delivery, 198
MeLi addition, 197, 198
NMR observation, 198–199
pseudo-axial position, 200
Huperzine A, 132–133
Hypocrellin
atropisomerization, 166–167
biological activity
Broka synthesis, 161
chemical structures, 160
photodynamic therapy, 159
PKC inhibitors, 160
vitiligo, 159
bisacid, 178
calphostin A,D
biomimetic approach, 165, 166
Broka syntheses, 162

Chiral lithium, 163
Coleman naphthalene, 163
Hauser synthesis, 162
cancer therapy, 175
C7,C72′ and C2,C2′ substitution, 176, 177
description, 157
diketone aldol transition, 171
enantioselective synthesis, 168
Kita's method, 173
Kozlowski retrosynthesis, 169
organocuprate-mediated epoxide, 173
palladium-mediated decarboxylation, 170
phleichrome synthesis, 161–164
purple speck disease, 158
Shiraia bambusicola, 159
Hypocrellin bambusae, 159

I
Isocomene, Chatterjee synthesis of, 239

K
Kinamycins
biological effects, 43
biomimetic approach, 46
cyclopentadienone formation, 45, 46
cytotoxic effects, 44, 45
dehydrorabelomycin formation, 45, 46
diazofluorene functional group, 43, 44
DNA-cleaving pharmacophore, 45
D-ring dearomatization, 46
isoprekinamycin, 44
(–)-kinamycin C synthesis
aryl aldehyde preparation, 54, 55
arylstannane synthesis, 48, 49
asymmetric nucleophilic epoxidation, 47, 48
Baylis–Hillman reaction, 47
carbene-catalyzed benzoin reaction, 57
diazofluorene antitumor antibiotics, 50
enantioselective, 54
enone, Porco's synthesis, 47, 48
epoxide-opening conditions, 49
α-iodoenone synthesis, 55, 56
ketoaldehyde, intramolecular benzoin reaction, 54
modified Ullmann reaction, 56
samarium-mediated deoxygenation mechanism, 57, 58
silyl hydrazone formation-oxidation sequence, 50
tetracyclic ketone, 47, 50
transformations of, 58
vinyl bromide fromation, 47
(–)-kinamycin F synthesis
β-trimethylsilylmethyl-α,β-unsaturated ketone, 61
diazofluorene development, 59, 60
Heck-type pathway, 60
juglone derivatives, 62
palladium-mediated cyclization mechanism, 60, 61
TASF(Et), 59, 62, 63
trans 1,2-diol, 63, 64
two-step annulation procedure, 59–60
(–)-kinamycin J synthesis, 54, 55, 58
vs. lomaiviticins, 40–41
macrofunctional groups, 39, 40
(±)-*O*-methylkinamycin C synthesis
cyclohexanone function, diacetate, 53–54
drawbacks, 53
enoxysilane formation, 53
α-hydroxyketone, 53
indenone synthesis, 51, 52
Ishikawa's retrosynthetic analysis, 51
mixed borate ester intermediate, 54
substrate-directed dihydroxylation reaction, 52
tetraol formation, 53
vicinal diol function, 54
ortho-quinone methide intermediates, 44, 45
prekinamycin analogs, hydrogenation of, 45
quinone function, 43
structural elucidation, 41–43
vinyl radicals, 44

L
Lomaiviticins
biological effects, 43
cytotoxic effects, 44, 45
description, 43
vs. kinamycins, 40–41
structures of, 40, 41
L-Selectride®, 195–196
Lycopodium alkaloids
biosynthesis
^{14}C-labeling experiments, 133
lysine incorporation, 133
Mannich reaction, 134
N-C13 bond, 135
one-pot transformation, 136
pelletierine precursor, 133, 134
phlegmarine, 135
Polonovski-Potier rearrangement, 136
complanadine A (*see* Complanadine A)
fawcettimine class, 132

huperzine A, 132–133
lycodine class, 131–132
lycopodine class, 131
miscellaneous class, 132
serratezomine A (*see* Serratezomine A)
serratinine
Inubushi synthesis, racemic, 137–138
synthetic approach, 138–139

M

Merlic naphthalene synthesis, 165, 166

N

Nakiterpiosin
biological functions, 34
carbonylative Stille coupling, 30
convergent synthesis approach, 30
coupling component synthesis
carbonylative conditions, 33
electrophilic, 32
nucleophilic, 32–33
halogen atoms, biogenesis of, 31
photo-Nazarov cyclization reaction, 30
Smo protein, 25
structural revision, 31
synthetic strategy, 31
transannular Diels–Alder reactions, 29
Nominine, total synthesis of, 4–5, 20, 21. *See also* Hetisine alkaloids
Norfluorocurarine synthesis
bicyclization reaction, limitations of, 82, 83
dehydrohalogenation, 82
dehydrotubifoline intermediacy, 81
exocyclic C19-C20 alkene formation, 81
Heck reaction, vinyl halides, 81–82
history, 80–81
macrocyclic ketone, 80, 81
Ni-mediated reductive cyclization, 81
oxidative approach, 81
photoisomerization, 81
vinylsilanes, 82–83
Wittig reaction, 81

O

Organocuprate-mediated epoxide, 173

P

Perylenequinone analogs. *See* Cercosporin; Hypocrellin
Phomactin A
allyl alcohol route, 203
biosynthesis, 185
C8a and C8b at AB-ring junction, 196–197
C3 and C3a in B-ring
drawback, 195
ketone reduction, 196
L-Selectride®, 195–196
diene route, 202–203
endoperoxide ring opening, 195
homologation C5a
enone synthesis, 200
experimental realization, 200–202
intramolecular delivery, 198
MeLi addition, 197, 198
NMR observation, 198–199
pseudo-axial position, 200
medicinal chemistry, 185
oxa-annulation and D-ring atropisomerism
α-alkylations with iodide 9, 189
A-ring transformation, 190
Diels–Alder cycloaddition, 190, 194
diketo-enal 8 synthesis, 192
HF6-31G*/B3LYP6-31G* calculations, 192, 193
oxa-[3+3] annulation, 192
Suzuki–Miyaura cross coupling, 189, 191
vinylogous ester 12, 188
Phoma sp., 183
retrosynthetic analysis, 186
head-to-head *vs.* head-to-tail, 187
regiochemical issues, 188
Spartanx model, 187
structure, 184
synthetic challenges, 185–186
types, 184
vinyl epoxide route
epoxidation, 205
Lewis acid, 204
oxidation states, 206
S_N2' addition pathway, 204, 205
Photodynamic therapy (PDT), 159
Pictet–Spengler reactions, 69
Platelet-activating factor (PAF), 185
Protein kinase C (PKC) isozymes, 104–106
Purple speck disease, 158

R

Racemic serratinine, Inubushi synthesis, 137–138
Retigeranic acid
biosynthesis, 238–239
Corey synthesis, 247–249
description, 235
Fallis approach

IMDA reaction, 241–242
Lewis acid, 243
Michael–aldol, 241–242
Fraser–Reid approach, 243–244
Hudlicky approach
Barton–McCombie deoxygenation, 256
α-bromocrotonate, 255
cyclopentene annulation protocols, 255
Horner–Wadsworth–Emmons protocol, 256
Stetter reaction strategy, 241
triquinane ester, 239, 240
vinylogous Reformatsky reaction, 239, 240
isolation and structure
Lieberman–Burchard test, 237
terpenoid acid B, 236, 237
X-ray analysis, 237
Paquette's synthesis
cyclopentanoid, 249, 251
Grignard compound, 251
ozonolysis, 252
Wolff–Kishner reduction, 249–251
Trauner approach
enone, 245
hydrindanone, 246
pyridinium *p*-toluenesulfonate, 246
sesterterpenes, 244
Wender's synthesis, 252
carboxylic acid, 253–254
Grob-like fragmentation, 254
pig liver esterase, 253
Rhazinicine, Gaunt synthesis of, 268

S

Scott's oxidative dimerization of tryptamines, 217
Serratezomine A
pharmacology, 132
retrosynthesis
allyl group installation, 140
β-stannylenamine synthesis, 141
Gabriel amine synthesis, 141
piperidine and lactone ring formation, 140
schematic representation, 140
structure of, 140
synthesis, 151, 152
α-alcohol strategy, 149–151
β-alcohol reactions, 147
Brown crotylation, 141
β-stannylenamine formation and coupling reaction, 142
carboxylic acid, 141
Claisen rearrangement, 145
diastereoselective allylation, 146
ester reduction, 146
features, 151
Michael reaction, 143
oxidative deprotection, 143
Pcmodel, 145
spirocyclic stereocenter, late stage epimerization, 147, 148
terpene alcohol, 141–142
tricyclic system formation, 144
Sesterterpenes, 235, 244. *See also* Retigeranic acid
Shiraia bambusicola, 159
Staphylococcus aureus, 213, 214
Strychnine
Brook rearrangement
C–C bond formation, 96
copper-assisted, 95
dolabelide D, 96, 97
model system, 97
polyfunctional building block, 96
siloxacycle ring opening, 96
six-step linear synthesis, 97, 98
vinylsilanes, 95
C15–C20 bond formation, 91–93
complex propargylic alcohol, 94
C7 quaternary stereogenic center, 71
Diels–Alder reaction, 71–73
D-ring formation strategy
β-hydride elimination, 87–89
cycloreversion reaction, 91, 92
dehydrodesacetylretuline synthesis, 87, 88
electrophilic iminium species, 91
Heck reaction, 87–89
nickel-mediated reductive cyclization, 90
Sakurai allenylation, 90
valparicine synthesis, 87, 88
E-hydroxyethylidene chain construction, 72
Heck approach, 72, 73
historical perspective
aromatic ring oxidative cleavage, 69
elemental composition, 68
Pictet–Spengler reactions, 69
Wieland–Gumlich aldehyde, 69–70
indoles intramolecular cycloaddition
Boger's cascade cycloaddition chemistry, 75–76
challenges, 75, 77
deprotonation, 75
Lewis acid-catalyzed bicyclization reaction, 76
stoichiometric generation, 75
triplet photosensitization, 75

Zincke aldehydes, 77
N-alkylation, potential reagents for, 94
norfluorocurarine synthesis
bicyclization reaction, limitations of, 82, 83
dehydrohalogenation, 82
dehydrotubifoline intermediacy, 81
exocyclic C19–C20 alkene formation, 81
Heck reaction, vinyl halides, 81–82
history, 80–81
macrocyclic ketone, 80, 81
Ni-mediated reductive cyclization, 81
oxidative approach, 81
photoisomerization, 81
vinylsilanes, 82–83
Wittig reaction, 81
protecting groups and limitations, 84–86
protodesilylation, 95
structural elucidation, 67–68
trans-hydrometallation of alkynes, 94
vinyl iodide synthesis, 94
vinylmetal species, conjugate addition of, 93
Zincke aldehydes
α,β-unsaturated aldehyde, 74–75
Diels–Alder reaction, 73, 74
indole monoterpene alkaloids, 73
Wieland–Gumlich aldehyde, 74
Suzuki–Miyaura cross coupling, 189, 191

T

Taiwaniaquinol B, Hartwig synthesis of, 267–268
Terpios hoshinota, 26
Trown method, 218
Tryptamine-derived Zincke aldehydes
aminium catalysis, 78
base-mediated cycloaddition reaction, 79, 80
bicyclization reactions, 79
counter-cation dependence, 79
^{1}H-NMR spectra, 77–78
Pictet–Spengler-type reactions, 78
thermal rearrangement, 77, 78

V

Vinyl epoxide route
epoxidation, 205
Lewis acid, 204
oxidation states, 206
S_N2' addition pathway, 203, 204

W

Wilkinson's catalyst, 137, 147, 174, 175
Woodward–Prévost method, 228

Z

Zincke aldehydes. *See also* Tryptamine-derived Zincke aldehydes
α,β-unsaturated aldehyde, 74–75
Diels–Alder reaction, 73, 74
indole monoterpene alkaloids, 73
Wieland–Gumlich aldehyde, 74

MIX
Papier aus verantwortungsvollen Quellen
Paper from responsible sources
FSC® C105338
FSC
www.fsc.org

If you have any concerns about our products,
you can contact us on
ProductSafety@springernature.com

In case Publisher is established outside the EU,
the EU authorized representative is:
Springer Nature Customer Service Center GmbH
Europaplatz 3, 69115 Heidelberg, Germany

Printed by Libri Plureos GmbH
in Hamburg, Germany